Foundations of Engineering Mechanics

Series editors

More information about this series at http://www.springer.com/series/3582

Boris F. Shorr

Thermal Integrity in Mechanics and Engineering

Springer

Boris F. Shorr
Central Institute of Aviation Motors (CIAM)
Moscow
Russia

ISSN 1612-1384 ISSN 1860-6237 (electronic)
Foundations of Engineering Mechanics
ISBN 978-3-662-46967-5 ISBN 978-3-662-46968-2 (eBook)
DOI 10.1007/978-3-662-46968-2

Library of Congress Control Number: 2015936688

Springer Heidelberg New York Dordrecht London

Printed on acid-free paper

Springer-Verlag GmbH Berlin Heidelberg is part of Springer Science+Business Media
(www.springer.com)

To memory of my teachers:
Sergey D. Ponomarev, Jacob B. Fridman,
Yury N. Rabotnov, and Isaac A. Birger

Preface

The design and analysis of high-temperature behavior of engineering structures play an important role for aviation and transport gas turbine engines, stationary steam and gas turbine power plants, including gas-transfer stations, and also for aerospace, atomic and chemical engineering, and other branches of industry.

The pioneering work by J. Duhamel in the middle of the nineteenth century inspired investigation of temperature stresses within the framework of solid mechanics. Interest in engineering applications of this phenomenon manifested only in the early twentieth century with the advent of steam and then gas turbines (A. Stodola, R. Bailey, F. Norton, N. Lebedev, and others). Toward the second half of the last century, the theory of thermal integrity (thermal strength) was established as a branch of engineering science studying high-temperature material behavior. It became especially apparent in connection with the invention of aviation gas turbine engines.

In the West, related problems were embodied in publications by A. Freudenthal, R. Larson and J. Miller, L. Coffin, N. Hoff, W. Prager, B. Gatewood, B. Boley and J. Weiner, R. Hill, and other English-speaking authors.

In the USSR, analysis of material strength at high temperatures in application to aviation gas turbines was initiated by V. Uvarov. Beginning from the middle of the twentieth century, this topic was intensively developed at the Moscow Central Institute of Aviation Motors (CIAM) by I. Birger and S. Serensen, and also by their followers B. Shorr, R. Sizova, R. Dul'nev, I. Dem'janushko, Yu. Temis, and some others. This research was carried out in close collaboration with a number of Russian and Ukrainian colleagues, including Yu. Rabotnov, Ya. Fridman, R. Shnejderovich, N. Malinin, and G. Pisarenko. Main results on the subject were delivered at major scientific conferences and widely published.

A relatively complete exposition of the aforementioned problems until the mid-1970s has been presented in the monograph "Thermal Strength of Machine Parts" edited by I. Birger and B. Shorr. Although 12,000 copies were published, the book gained rare status soon after publication in 1975 in USSR.

However, Russian publications in the area of thermal strength remained unknown to researchers from other countries, except for a small number of contributions translated into English.

During the second half of the twentieth century, both in the West and in Russia, numerous papers and books reporting the novel theoretical and experimental results on the strength of materials and structures under elevated temperatures have been published. The range of applications of these developments has been expanded since non-uniform temperature stresses arise not only in mechanical systems, but also in civil structures, road coatings, and even in the Earth's core.

Thermal integrity is a multidisciplinary field combining joint efforts of mechanical engineers, material scientists, and applied mathematicians approaching the problem from their specific viewpoint. This monograph draws on the research of a broad scientific community including the author's contribution.

The scope of thermal strength analysis was considerably extended due to making use of modern computers along with the implementation of FEM codes. However, the author believes that there is a sort of disparity between the power of advanced software, and lack of easy-to-follow books on the theoretical and experimental aspects of thermal integrity. The author endeavors to ensure the rigor of the underlying assumptions along with sufficient simplicity of presentation making the book compelling to a wide audience.

The book is targeted at engineers, university lecturers, postgraduates, and final-year undergraduate students involved in computational modeling and experimental and theoretical analysis of high-temperature behavior of engineering structures. The book is also intended for attention of researchers developing the thermal strength theory as a branch of continuum mechanics.

The author expresses deep gratitude to Prof. V. Babitsky for support of issuing of the present monograph, Dr. V. Arulchandran and Dr. L. Prikazchikova for their contribution to editing of the English text, Dr. Ch. Baumann and the staff of Springer-Verlag for assistance in publication. The author is grateful to Prof. Yu. Kaplunov, Prof. Yu. Temis, Prof. E. Golubovskiy, and colleagues from CIAM, and also from Keele and Brunel Universities for fruitful discussions.

Moscow Boris F. Shorr

Contents

About the Author

Prof. Boris F. Shorr D.Sc. is a renowned expert in the field of thermal strength analysis. He has played a key role in the dynamical and stress analysis of the numerous aviation and stationary gas turbines of different generations. For many years Prof. Boris Shorr presented the advanced courses in solid and applied mechanics at Moscow Institute of Physics and Engineering, Moscow Institute of Physics and Technology, and some other leading technological universities. Since long ago and to this time he has been affiliated with Central Institute of Aviation Engines in Moscow. He is a winner of Zhukovsky Award, Honourable Scientist of Russian Federation and a member of Russian National Committee of Theoretical and Applied Mechanics, an author of several professional books in Russian and English.

Abbreviations and Basic Symbols

Abbreviations

APS	Accumulated plastic strain
CH	Axis of current anisotropic hardening
FEM	Finite element method
FH	Axis of former anisotropic hardening
HCF	High cyclic fatigue
LCF	Low cyclic fatigue
RAPS	Relative accumulated plastic strain
SAH	Stabilized anisotropic hardening
SARH	Stabilized anisotropic-ray hardening
SSS	Stress–strain state

Basic Symbols

$\sigma_{ij},\ i,j = 1,2,3$	Stress tensor and its components
s_{ij}	Deviator of stress tensor
$\boldsymbol{\sigma}_i$ or $\{\sigma\}$	Stress vector
$\lvert\sigma_i\rvert = \sigma_i$	Magnitude of stress vector equal to stress intensity
σ_{ij}^*	Active stresses
r_{ij}	Residual micro-stresses
$r_{\lim}$	Limiting value of micro-stress unsteady part
β_p	Parameter of hardening stabilization in plasticity
$\tau_{\max}$	Maximum tangential stress
$\varepsilon_{ij},\ i,j = 1,2,3$	Strain tensor and its components
e_{ij}	Deviator of strain tensor
$\boldsymbol{\varepsilon}_i$ or $\{\varepsilon\}$	Strain vector
$\lvert\varepsilon_i\rvert = \varepsilon_i$	Magnitude of strain vector equals to strain intensity
ε_{ij}^e	Elastic strains
ε_{ij}^p	Plastic strains

ε_{ij}^{c}	Creep strains
ε_{st}^{c}	Steady creep strains
ε_{*}^{c}	Stabilized part of unsteady creep
ε_{*0}^{c}	Transmitting part of unsteady creep
$\varepsilon_{\lim}$	Limiting value of unsteady creep strain
β_c	Parameter of hardening stabilization in creep
v_i^{st}	Intensity of the steady creep rate
$\gamma_{\max}$	Maximum shear
$\Delta\sigma, \Delta\varepsilon$	Stress and strain increments or ranges
$d\boldsymbol{\sigma}_i$	Increment of the stress vector
$\lvert d\boldsymbol{\sigma}_i\rvert = (d\sigma)_i$	Magnitude of the vector $d\boldsymbol{\sigma}_i$ equal to value of stress increments intensity
$d\sigma_i$	Increment of stress intensity
$d\boldsymbol{\varepsilon}_i$	Increment of the strain vector
$\lvert d\boldsymbol{\varepsilon}_i\rvert = (d\varepsilon)_i$	Magnitude of the vector $d\boldsymbol{\varepsilon}_i$ equal to value of strain increments intensity
$d\varepsilon_i$	Increment of the strain vector
E, E_p	Moduli of elasticity and plasticity
G, E_t	Shear and tangential moduli
K	Bulk (voluminal) modulus
μ	Poisson's ratio
$[D]$	Elasticity matrix
E_p''	Modulus of unsteady plasticity
E^*	Modulus of isotropic strain hardening
E_{p0}	Minimum (limiting) value of plasticity modulus
t	Time—physical or formal quantity
T	Temperature
α_T, β_T	Coefficients of linear and volume thermal expansion
ε_T	Thermal expansion (dilatation)
σ_B	Ultimate stress in tension
σ_y	Yield point in tension
N	Number of cycles or half-cycles
D	Damage
f_p or f^p	Parameter referred to plastic yielding
f_c or f^c	Parameter referred to creep
f_d	Parameter referred to damage or failure
$\hat{v}_i$	Projection of any vector $\mathbf{v}_i$ onto the axis CH
ζ	Factor of restitution (recovery) in creep

Chapter 1
Introduction

1.1 General

Analysis of high-temperature behavior of engineering structures, or in a wider scope, solids (deformable bodies), is a multidisciplinary problem. It is of particular importance for numerous industrial applications but is also of great academic interest as a branch of continuum mechanics.

A rise in temperature of a working medium (steam, gas, and so forth) of various machine parts and other units, leading to the growth of structure temperature even in the presence of cooling, is a mainstream feature of modern machinery. It is critical for aviation, shipping, and vehicular gas turbine engines, and also for stationary steam and gas turbine power plants including gas transfer stations. A similar trend is typical for aerospace technology, and atomic and chemical engineering, as well as for many other branches of industry. Non-uniform temperature fields have a strong effect on the durability of buildings, bridges, tunnels, and other engineering structures, as well as on geological processes in the Earth core. Even small temperature deformations may result in inadmissible surface distortion of large telescope lenses.

The term "thermal integrity" covers a whole group of problems connected with the maintenance of the working capacity of machine parts and structures depending on their thermal state. In particular, these include the following:

- Evaluation of temperature stresses due to material expansion.
- Analysis of material strength and deformation under elevated and lowered temperatures, including new properties specific to high-temperature ranges.
- An experimental justification of theoretical approaches to the calculation of structure deformability and integrity while taking into consideration variable loads and temperature fields.
- An establishment of permissible margins for structure strength and durability.
- The development of authentic mathematical models and reliable numerical methods utilizing modern computing technologies.

B.F. Shorr, *Thermal Integrity in Mechanics and Engineering*,
Foundations of Engineering Mechanics, DOI 10.1007/978-3-662-46968-2_1

The main objective of the monograph is to present a clear notion of the features of thermal integrity necessary for practical application to design, testing, and operation of machine parts and mechanical structures. Another objective is to study the effects induced in solids by non-uniform temperature fields. Some theoretical issues can be of interest to specialists studying thermal integrity within the framework of solid mechanics.

The monograph consists of 12 chapters and Appendix.

The present chapter (Introduction) states the basic scientific and methodological elements of thermal integrity from an engineering standpoint: a phenomenological approach to the process analyzing, a role of thermal expansion, effect of temperature on the material behavior, operating and "formal" time, the tensor theory of stresses and strains, their vector representation, etc.

Chapter 2 is devoted to thermoelasticity. The difference between "static" and "dynamic" elastic moduli and the reversibility of elastic deformations are demonstrated, including a simultaneous variation of temperatures and stresses. The equations in the theory of thermoelasticity for a transient non-uniform temperature field are presented along with effective numerical methods. The energy principles and the variational equations in non-isothermal elasticity used in FEM packages are considered.

In Chaps. 3–5, the main results are presented for numerous tests on elastoplastic material behavior under various loading and heating conditions when creep can be neglected. Theoretical insight into analysis of experimental data and computational models for basic deformation regimes is given.

In Chap. 3, plastic deformations under combined stress for proportional (simple) loading and constant or monotonously increasing temperatures are studied while taking into consideration a non-isothermal yielding surface. Unlike an isothermal loading, the conditions and rate of plastic strain evolution depend on a relationship of rates of the temperature and stress intensity. It is shown that a role of temperature stresses induced by a non-uniform heating distinctly weakens when plastic strains considerably exceed the thermal expansion.

The cyclic thermal plasticity dealing with multiple loading and heating cycles (to 10^4–10^5) is discussed in Chap. 4. Typical results of cyclic tests under isothermal and non-isothermal conditions are presented along with the major factors affecting thermal fatigue. It is shown that a great extent of the long-term working capacity of machines is ensured by cycle stabilization. Methods for evaluating non-isothermal cyclic durability (low-cyclic fatigue—LCF) based on cyclic deformation charts, and damage summation for complex cycles, are illustrated by calculations and verified by experiments.

General approaches to calculations involving machine parts and structures on sign-varying elastoplastic deformation are described in Chap. 5.

Examining application ranges of the known methodologies, such as theories of plastic deformation, plastic flow, and strain hardening, using experimental results, shows that these theories describe material behavior accurately enough only in the case of proportional unidirectional loading and provided that stresses and temperatures increase simultaneously.

The advanced version of the stabilized anisotropic hardening model (SAH model) developed over the last several decades provides a better estimation for the direction and rate of plastic deformation under sign-variable proportional loading.

A new and fairly simple approach based on the stabilized anisotropic-ray hardening (SARH) model is proposed for calculating elastoplastic deformations under non-proportional, non-isothermal loading. The model initially begins with the idea of a rotating axis of anisotropy hardening. In this case, all constituents of stress and strain tensors and their increments are projected onto this axis. Calculations demonstrate a sufficient agreement of the results predicted by the model and experiments. A step-by-step procedure of numerical solution of a thermal plasticity problem using the SARH model is also described.

Damage accumulation and material failure are briefly discussed within the context of sign-varying elastoplastic regimes.

Chapters 6–8 deal with the most general and reliable creep and durability experiments for both steady and unsteady loading and heating and also with calculations of creep hardening incorporating the effect of material heredity and adaptability.

In Chap. 6, creep associated with steady or monotonically varying mechanical and thermal disturbances is considered for statically determinate and indeterminate systems, and also for structures with prescribed kinematic variations of creep velocities under uni- and multi-axial stressed states. Examples of stress redistribution and relaxation induced by non-isothermal loading of constrained structures, and formation of residual stresses due to unloading and cooling, are demonstrated.

The possibility of predicting the durability of systems designed to work over a long time is discussed.

Numerical methods for cyclic creep analyzing, in particular, for thermocyclic creep, are considered in Chap. 7. Straightforward calculation of a creep steady-state cycle, alternative to the widespread step-by-step procedure, is proposed and illustrated by examples. Methods for evaluating thermocyclic durability, taking into account high-temperature exposure and the simultaneous effect of creep and plastic deformations, are described.

In Chap. 8, the most validated experimental data for creep and durability of structural materials subjected to unsteady loading and heating are presented. Various methods for unsteady creep calculation are discussed, and a new approach also starting from the model of stabilized anisotropic-ray hardening is formulated. It is somewhat similar to the plasticity SARH model, which considers the loading history and material adaptability and reveals increase in rate of creep development after repeated loadings, spatial variation of the creep rate evolution at rotation of the stress tensor, and many other features of the unsteady creep. A step-by-step procedure of numerical solution of a creep problem using the SARH model is also described.

Chapter 9 contains the concepts of strength and durability margins (safety factors) that provide the working capacity of machines under inevitable deviations of structure parameters from their expected values due to technical or other reasons. The key role of combined deterministic- and probabilistic-statistical approaches for establishing safety factors and also for developing related design principles is stated. Ideas concerned with designing non-isothermal "full-strength" structures and inserting "a weak link" into the system are discussed. Methods for calculating structure-bearing ability along with local strength and durability margins are formulated.

The problems of elastic and elastoplastic stability loss are covered in Chap. 10. The loss of stability and buckling under gradually increasing load and temperatures, as well as under temperature stresses, are studied. The effect of the latter is compared with that of external loading. Finally, the loss of stability induced by creep and the non-uniform heating of rods and plates, and nonlinear analysis of the jump of an elastic system into a new equilibrium state caused by the variation of the temperature field, are presented.

Chapter 11 focuses on the thermal strength of anisotropic bodies with reference to single-crystal nickel alloys, which are widely used as a material for the blades of high-temperature gas turbines. The influence of the material anisotropy on elastic and damage characteristics is alluded to. The role of the "Schmid factor" describing the relative level of tangential stresses in different slip systems and also the use of maximum tangential stress as failure criterion are discussed.

The experimentally justified procedures for calculating deformation and damage of single crystals at combined stress are proposed starting from the concept of "equivalent" orientation under uni- and multi-axial loadings.

Chapter 12 is oriented to the audience interested in gaining a better understanding of several nontrivial problems in solid mechanics specific for elevated temperatures. They are presented in a relatively simple form and do not require knowledge of advanced mathematics.

Section 12.1 is devoted to non-isothermal dynamics. It is shown that elevated temperatures may change resonance frequencies not only due to lower elasticity moduli but also due to temperature expansion, which sometimes affects the contact interactions between structure members. In some cases, especially for thin-walled structures, temperature stresses influence the frequency spectrum.

Section 12.2 shows a possibility for the directive change of the shape of a body via successive cyclic heating of its parts, having different temperature-dependent elastoplastic properties. A similar procedure of thermocyclic forming is used in some industrial technologies.

Section 12.3 deals with the "coupled" theory of thermoelasticity, which is based on the laws of thermodynamics and takes into consideration the interaction of thermal and elastic fields. The influence of heat transfer to the environment on the elastic strain scale is illustrated for the case of a rod extension. In addition, the physical reason for elastic vibration attenuation in absence of external dissipative forces (so-called thermoelastic energy dissipation) is clarified.

Section 12.4 gives a short exposition of a wave theory of heat conduction, which incorporates a finite speed of heat and temperature propagation in contrast to the classical theory of heat conduction. As an example, temperature and stress fields in a semi-infinite body induced by a thermal shock are analyzed.

Appendix gives examples of analytical solutions of thermal strength problems for some structure elements (rods of a different kind, plates, and trusses), which can be useful for demonstrating some sides of uneven heated structure behavior and testing numerical programs.

1.2 Experimental Foundations of Thermal Integrity

1.2.1 Phenomenological Approach

1.2.1.1 Phenomenological Description of Material Behavior

Mechanical properties of materials can be investigated at different scales (on a molecular or atomic level, and on the level of crystals, grains, and so forth). Local damage phenomena developing into global cracks and causing further destruction are studied at the same scales.

Since we consider large region of material in comparison with the size of its particles, individual features of "micro-particles" behavior are neutralized and averaged, and "micro-discontinuities" between particles cease to influence the global material behavior. This effect allows the material to be considered as a *continuum deformable solid medium*, usually briefly named a *solid*. The mechanical properties of a solid can be studied experimentally and described theoretically without concerning their physical origin.

This approach, usually referred to as "*phenomenological,*" underlies the most theoretical and applied works on solid mechanics and enables to obtain results suitable for practical use in machine design.

A solid, having identical mechanical properties at every point and identical value of a property in all directions, is named *homogeneous and isotropic*. Since material properties depend on temperature, the isotropic solid under non-uniform temperature field becomes nonhomogeneous. This book is primarily concerned with the investigation of behavior of such solids.

Real metal alloys used in machine parts have crystalline structure and include a large number of grains (sometimes of rather large sizes). In spite of this, experience shows that in most cases reliable calculation results can be achieved, considering these materials within the framework of continuum mechanics.

However, there are a number of *anisotropic* materials (composites, metal alloys of special structure, etc.) where distinction of properties in different directions plays an important role and should be considered in calculations (see Chap. 11).

1.2.1.2 Phenomenological Dependencies

Primary data on material properties are obtained by processing the particular test results, and the intermediate values necessary for further calculation have to be found by an interpolation procedure. For this purpose, approximate analytical formulas are widely used.

In some cases, a form of phenomenological dependencies is prompted by theoretical reasons, and their approximate correlation with real material behavior must be provided by experimental coefficients.

A considerable number of the coefficients usually provide good approximate results of each specific test. However, the more coefficients are involved in the formula (the number may sometimes exceed 20), the more difficult to reveal the underlying dependencies of these coefficients on the governing parameters, so one has to average their values. Moreover, the use of formulas with a large number of averaged coefficients makes calculation results unreliable.

It is necessary to be careful while applying the phenomenological dependencies to property extrapolation out of ranges within which these dependencies have been experimentally established, especially, into the domain of high temperatures.

Approximate phenomenological dependencies should finally be verified during operation experience of the machine parts designed according to these dependencies.

1.2.2 Basic Experimental Data

1.2.2.1 Test Aims and Methods

Information on specimens' geometry, construction of testing devices, and test methods, used for obtaining various thermal strength characteristics, can be found in specialized engineering literature (see, e.g., Touloukian et al. 1975; Troshchenko et al. 2000). The requirements for testing equipment can differ depending on the purpose of experiment.

To provide the integrity of a designed machine part, material properties of the specimens should approximate the properties in real operating conditions as close as possible. These include material composition and structure, conditions of mechanical and thermochemical processing, possible non-uniformity in separate parts of a detail, inherent anisotropy, and so on. Values of loads and temperatures during testing should correspond to operational ones, including the most severe. An example of similar testing is presented in Wilshire and Evans (1994).

For laboratory investigation of general material characteristics, it is important to ensure stress and strain uniformity and material isotropy within sample zones where the measurements are taken. Stability and control of test parameters should also be provided, including the rates of loading and heating.

The experiment duration can be restricted by the time needed to obtain temporal dependences between stresses, strains, and temperatures within some range of their measurement, or continue up to full rupture of a specimen.

Sometimes, a machine part can entirely be an object of the trail. For example, a turbine disk can be subjected to spin up to rotation frequency at which it will collapse at the set temperature distribution over the disk.

In case of low temperatures, it is required to use appropriate test methods, see, e.g., Pisarenko et al. (1969).

1.2.2.2 Basic Experimental Data for Temperature-Dependent Material Properties

There are several basic experimental sources of data on temperature-dependent material properties, used in thermal integrity theory, namely:

- Curves of sample relative elongation $\varepsilon_T = \Delta l_T / l_0$ versus temperature change $\varepsilon_T = f(\Delta T)$ serving for evaluation of linear and volume thermal expansion coefficients.
- Diagrams of elastoplastic deformation of the material under tension obtained at various constant temperatures as $\sigma = f(\varepsilon, T)$. Tests are usually carried out at a gradual increase in strains with a rate $(1-5) \times 10^{-2}\,\%/s$ before specimen rupture. Sometimes the tension tests are executed at gradual rate of stress or at other kinds of loading. These tests define a *short-term strength* of the material.
- Diagrams of material behavior under tension–compression or torsion cyclic loading for a considerable deformation range. The tests cease with destruction due to LCF.
- Diagrams of material creep at various constant values of stress and temperature, ending by destruction from exhaustion of *long-term durability*.
- Creep diagrams under alternating stresses and various constant temperatures. These tests define resistance of the material to *cyclic creep*.
- Experimental data for the samples under loading, affected by alternating temperatures, defining resistance to *thermal fatigue*.

For a more complete study of material properties at elevated temperatures, a number of other tests also are carried out.

Experimental foundations of material mechanical properties under various conditions of loading and heating are considered in the subsequent chapters in more detail. Since a material behavior at elevated temperatures essentially depends on temperature and operating time, these factors can be considered as "governing" parameters or "key factors" of thermal integrity.

1.2.3 Role of Temperature

1.2.3.1 Current Temperature of a Material

All physical–mechanical properties of a material, which correspond to construction thermal integrity, depend on the current body temperature.

As temperature increases, all strength properties of metals usually decrease, while plasticity characteristics grow. As temperature decreases, the properties alter in opposite direction. Under non-uniform heating, a body acquires a nonhomogeneous feature.

At moderate temperatures, material properties change slightly, which enable those to be averaged over the body volume V. Hence the Young's modulus can be written as $E_m = \{\int_V E[T(x,y,z)]dV\}/V$ or in the form $E_m = E(T_m)$, where $T_m = [\int_V T(x,y,z)dV]/V$ is the average body temperature.

If there is insufficient information on temperature field, imposed on a body, a rough approximation for the temperature T_m can be taken as the average of the minimum and maximum temperatures $T_m \approx 0.5(T_{\min} + T_{\max})$.

While studying failure of a structure, it is necessary to estimate the strength characteristics for temperatures at which the utilized criteria of material destruction reach their critical values.

Analyzing the majority of thermal integrity problems, the temperature field of a structure is supposed to be known and independent of stresses and strains arising in the body, although there is a relation between these parameters due to the thermodynamic laws. This approach is applicable in most engineering problems and is justified by the so-called coupled theory of thermoelasticity (see Chap. 12). It also proves to be true by numerous experiments.

1.2.3.2 Thermal Expansion of a Body

Thermal Strains

The majority of bodies increase in their volume at heating, i.e., obtain a *thermal* (or *temperature*) *expansion* (*dilatation*). At a small rise in temperature dT, a volume V becomes

$$V + \frac{\partial V}{\partial T}(T)dT = V[1 + \beta_T(T)]. \tag{1.1a}$$

The material characteristic

$$\beta_T(T) = \frac{1}{V}\frac{\partial V}{\partial T}, \tag{1.1b}$$

named a coefficient of *volume thermal expansion* describes relative change of volume at heating.

The effect of thermal expansion is usually estimated experimentally according to measurements of change of a specimen's linear dimension L at heating. In an isotropic body, the linear dimensions in all directions increase equally and are defined by a coefficient of *linear thermal expansion* $\alpha_T(T)$.

Since $V \approx L^3$, then $V + dV \approx (L + dL)^3 \approx L^3 + 3L^2 dL$ and

$$\alpha_T(T) = \frac{1}{L}\frac{\partial L}{\partial T} = \frac{1}{3}\beta_T(T). \tag{1.2}$$

The coefficients $\alpha_T(T)$ and $\beta_T(T)$, as the other material properties, are the functions of current body temperature. With increase in temperature, their values usually grow.

Temperature rise dT leads to increase of *thermal expansion*

$$d\varepsilon_T = \alpha_T(T)dT. \tag{1.3}$$

At finite change of the temperature from T_1 to $T_2 = T_1 + \Delta T$, a current increment of thermal strain is defined as

$$\Delta\varepsilon_T = \int_{T_1}^{T_2} \alpha_T(T)dT = \alpha_{T,m}(T_1, T_2)\Delta T, \tag{1.4a}$$

where $\alpha_{T,m}(T_0, T)$ is an average coefficient of thermal expansion in a considered temperature range.

Thermal strain $\varepsilon_T = \varepsilon_{T0} + \Delta\varepsilon_T$ is usually referred to body's dimensions at initial temperature T_0 when there is no initial strain ($\varepsilon_{T0} = 0$). Then

$$\varepsilon_T = \alpha_{T,m}(T_0, T)\Delta T, \tag{1.4b}$$

where $\Delta T = T - T_0$.

Coefficients α_T for structural materials are of order $\sim 10^{-5}$ 1/°C, so for temperature variation ΔT within several hundreds of degrees centigrade a magnitude of thermal strain is $\varepsilon_T \approx 10^{-3}$–$10^{-2}$. This allows estimating a possible temperature change in a body's initial dimensions.

During the tests, thermal strains are measured at a gradual temperature increase, so for each temperature range from T_1 to T_2, the average value can be found as

$$\alpha_{T,m}(T_1, T_2) = \frac{\varepsilon_T(T_2) - \varepsilon_T(T_1)}{T_2 - T_1} \tag{1.5}$$

and a current value of the coefficient $\alpha_T(T)$ used in the step-by-step calculation methods can be defined due to Eq. (1.3) as a derivative

$$\alpha_T(T) = \frac{d\varepsilon_T(T)}{dT}. \tag{1.6}$$

The reference books do not usually provide a primary experimental curve $\varepsilon_T(T)$, but average values $\alpha_{T,m}(T_1, T_2)$, calculated by Eq. (1.5) for several temperature intervals T_1, T_2. Assigning the values $\alpha_{T,m}(T_1, T_2)$ to the mean temperature of each interval, an experimental curve $\alpha_T(T)$ can be restored by interpolation.

A free thermal expansion is usually a reversible process when after heating in ΔT and subsequent cooling at the same value the total thermal strain is equal to zero. However, this can happen only if the material structure has not changed during heating.

Ceramic materials (and some special metal alloys) have insignificant thermal expansion.

Thermal Stresses

Thermal stresses arise when a free thermal expansion of any part of a body is interfered by its other parts or other bodies. In some cases (in the absence of external loads), the stresses can be measured experimentally, but mostly they have to be evaluated by calculation.

Thermal stress calculations are based on the following:

- According to the "uncoupled" theory of thermal strength, a temperature field of a body is considered to be known from the experimental data or a result of a heat transfer calculation.
- Deformation of any character (elastic, elastoplastic, viscous) caused by action of external loads and thermal expansion are linearly summarized, i.e., satisfy a superposition principle.

Thermal stress calculation methods, depending on mechanical properties of the material, configuration of a system, along with conditions of loading and other factors, are published in numerous works and will be partially considered in the subsequent chapters.

If boundary conditions of a body restrict displacements, thermal stresses under heating arise in most cases. However, in seldom situations, when the temperature of various parts of such body changes in opposite directions and the thermal expansion of heated area and compression of cooled area are identical, thermal stresses might not occur.

Rods, strings, and other one-dimensional systems, in the absence of external restrictions, have the possibility to be freely expanded, and thermal stresses in them do not arise. In more complex systems, for example in circular plates with free boundaries, thermal stresses appear at any non-uniform heating along the radius (see Appendix).

Typical methods of thermal stress reduction are considered by Gatewood (1957), Manson (1966), and others, and will be partially discussed further.

Thermal expansion is volumetric, i.e., it occurs in all directions. Clearly, thermal stress does not arise only in case of free expansion of a body in all directions. Otherwise, a so-called edge effect within the boundary layer can occur with the system of the self-equilibrated stresses, which quickly fade out of the edge.

1.2.4 Role of Time

1.2.4.1 The Influence of Time on Material Properties

Various Scales of Temporal Processes

All physical processes, including mechanical ones, proceed in time; however, time intervals within which the processes occur differ considerably.

Waves of elastic and plastic deformations in constructional materials propagate at a speed of several thousand meters per second. Therefore, stresses and strains (including plastic ones) arise in a body practically simultaneously with the application of external loads and occurrence of a temperature field.

On the other hand, it usually takes hours, and sometimes years, to observe effects of material creep. Even duration of the so-called short-term creep is measured in minutes, and more rarely in seconds.

In the majority of cases, above-mentioned distinctions in time allow to consider a total deformation process as a combination of alternating stress and elastoplastic strain without material creep and long-term creep evolution under constant loading conditions. However, it is worth noting that the accumulation of nonelastic deformations of one type influences sometimes the behavior of a material at deformation of another one.

Situations when creep develops simultaneously with increase of plastic deformations are discussed in Chap. 8.

Operating Time and Service Life

At moderate temperatures (specified for a given material), an actual duration of operation regimes does not usually exert influence on deformation and destruction conditions. As temperature rises, time period for an object being subjected to stress (*operating time*) becomes more and more important, influencing creep and decrease of the long-term material strength.

As time passes, the structural changes might occur in the material (the so-called *aging*), which sometimes takes place not only under stresses but also after the loads have been removed.

In some cases, dynamic processes might also influence thermal integrity (see Chap. 12).

Rapid changes in temperature can cause a thermal shock. In this case, it is required to consider simultaneous changes in temperature and stress–strain state (SSS) of a body, taking into account inertial forces.

The material properties might also be functions of the speed of loading.

During long storage of machines in nonworking state (in so-called the preserved condition), some parts can lose the ability to operate normally. Thus, an appropriate *service life* parameter should be specified for these parts.

1.2.4.2 Time as a Formal Quantity

When analyzing the processes of gradually changing loads and temperatures, including transitions from one stationary regime to another, the speed of the process usually allows the inertial forces to be neglected. This allows processes to be considered as quasi-steady. At moderate temperatures, if creep does not occur or creep effect can be neglected, time may be considered as some formal quantity fixing only the sequence of loads and temperatures applied to the body. This aspect is referred to both processing of test data and calculations.

As a process does not depend on time, this parameter can be replaced by any other one, convenient for the analysis, for example, it could be changed in the rotor rpm, etc.

1.2.5 Full-Scale Tests of Machine Parts

Along with the experimental investigations of material mechanical properties at the elevated temperatures using specimens, full-scale tests of machine parts or their models are sometimes conducted. These tests ought to imitate loading and heating conditions, similar to those, expected during real-time operation of a machine.

Testing of thermal cyclic fatigue of the gas turbine blades (Dul'nev 1979) can serve as an example. A block of 5–7 blades or their models are exposed to cyclic action of a gas stream. Tensile load is periodically applied to the central blade, imitating real-life effect of the centrifugal forces at cyclic change of rotation frequency of the turbine wheel. Thermal stresses in the blade are caused by means of cooling air, blown through the blade.

At the elevated temperatures, the full-scale fatigue tests of machine parts are carried out (see, e.g., Balashov and Kozlov 1981). For this purpose, the part or its model is subjected to resonant oscillations, caused by the electrodynamic vibrator. Since it is located in the heating furnace, the temperature can also be cyclically changed.

The full-scale tests can be used not only for a direct estimation of thermal strength of the tested objects, but also for verification of the applied methods for calculation of destruction time and number of cycles. Thus, it is necessary to take into consideration a possible significant scatter of test results.

1.3 Theoretical Foundations of Thermal Integrity

The general approach to theoretical study of solids under various internal and external conditions (including elevated temperatures and non-uniform heating) is based on the theory of continuous media (see Prager 1937; Sedov 1973, and others). The concept of a continuum does not exclude the possibility for accounting that some "micropores" can gradually be accumulated in a deformed body decreasing its strength (Kachanov 1958; Volkov and Korotkikh 2008, and others). Local damage, formed in some finite area of a body, breaks the continuity in this area.

Theoretical research of thermal integrity has developed in two directions: (a) the calculation of SSS of a body under various operating regimes, and (b) the evaluation of a body's failure conditions.

1.3.1 Basic Concepts

The basic concepts of stress and strain calculations, taking into account a thermal expansion of a solid, were specified by Duhamel (1837) and were further developed by many researchers (see, e.g., Lebedev 1934; Boley and Weiner 1960; Lemaitre and Chaboche 1990, 2004, and others).

1.3.1.1 Linear Summation of Strains

All theories of thermal strength start with a principle of linear summation of small strains. According to this principle, a total strain ε is equal to the sum of a "mechanical" strain ε_σ caused by the action of stresses, and the thermal expansion ε_T arising due to temperature change

$$\varepsilon = \varepsilon_\sigma + \varepsilon_T. \tag{1.7}$$

The mechanical part of the strain is usually decomposed as the sum of elastic ε_e and inelastic ε_{ne} parts

$$\varepsilon_\sigma = \varepsilon_e + \varepsilon_{ne}, \tag{1.8}$$

though in some theories (Il'ushin 1963; Valanis 1971), they are unified (in Chap. 5, it will be shown that the latter approach demonstrates reliable results only in case of a sign-constant isothermal loading).

At elevated temperatures, the inelastic deformation appears as a sum of the "instant" plastic strain ε_p arising almost simultaneously with the increase of applied stress, and the creep strain ε_c growing over time even under constant stress

$$\varepsilon_{ne} = \varepsilon_p + \varepsilon_c. \tag{1.9}$$

Equations (1.7)–(1.9) concern strain components for all cases of loading (torsion, bending, and so forth) and combined SSS.

Under specific operating conditions, total strains can include additional components, for example, radiating swelling of a material.

In non-stationary problems, the specified equations can also include derivatives $(\dot{\varepsilon}, \dot{\varepsilon}_p, \ldots)$ or finite increments $(\Delta\varepsilon, \Delta\varepsilon_p, \ldots)$ of the corresponding strain components.

1.3.1.2 Universal Equations of Solid Mechanics

The equations of motion connecting stress components with volume (bulk) and surface forces are identical for all solids. The same concerns the strain–displacement (geometrical) relation providing body continuity under deformation.

A material density in the equations of motion does not change significantly under temperature rise. Therefore, the specified universal equations of solid mechanics do not depend on mechanical properties of a material for static problems and may include the material density as a constant parameter for dynamic problems.

For example, for extension of a rod of a constant cross-section area A, the equation of static equilibrium of elementary volume $A\,dx$ looks like

$$A\frac{d\sigma_x}{dx} + f_x = 0,$$

and the strain–displacement relation is

$$\frac{du_x}{dx} = \varepsilon_x,$$

where f_x is a distributed load acting in the x direction, and u_x is a longitudinal displacement of a cross-section at a point x.

If the rod is fixed at the point $x = 0$ and is affected by a tensile force F_l at the boundary $x = l$, then the boundary conditions are

$$u_x(0) = 0 \quad \text{and} \quad A\sigma_x(l) = F_l.$$

In dynamics, a force f_x includes, besides static component $f_{x,st}$, an inertial part

$$f_x = f_{x,st} - \rho A \frac{\partial^2 u_x}{\partial t^2}.$$

The given equations are valid for any material of the rod.

The universal equations of motion and geometrical relations for 3D problems are presented below.

1.3.1.3 Constitutive Equations of Material Behavior

Unlike the universal equations of solid mechanics, relations between stresses and strains in a body depend significantly on the individual properties of materials, temperature, and operating time. The same concerns relations between stress and strain rates and increments.

In view of a complexity and variety of the mechanical processes affecting the behavior of materials at high temperatures, general theory of thermal strength can hardly be proposed. Therefore, theoretical studies of deformations of various kinds are usually carried out separately, though an interference of such processes can also be considered.

A current rate of the mechanical part of strain $\dot{\varepsilon}_\sigma$ may be presented as a functional of current stress σ and temperature T, their rates $\dot{\sigma}, \dot{T}$, and some number n of the "structural" parameters q_r reflecting influence of the previous process on material behavior at the present time.

At a stable material state, the structural parameters can depend on nonelastic strains of various characters, and for unstable conditions they also depend on time.

In case of 1D stress state, it yields

$$\dot{\varepsilon}_\sigma = f(\sigma, T, \dot{\sigma}, \dot{T}, q_1, q_2, \ldots, q_n). \tag{1.10}$$

Various forms of Eq. (1.10), usually named "constitutive equation," have been provided in specialized literature with respect to different problems (Andrade 1910; Bailey 1935; Noll 1958; Rabotnov 1966, and others).

A current rate of the elastic strain is defined by current rates of stress and temperature

$$\dot{\varepsilon}_e = f_e(\dot{\sigma}, \dot{T}). \tag{1.11}$$

The phenomenological correlations that can be applied in practice are obtained by analysis of experimental data, presented in a generalized form. To ensure that the generalized characteristics are not too complicated, it is necessary to neglect some minor distinctions, found out during experiments, and at the same time it is important not to lose sight of the most important liaisons between parameters.

In some cases, the character of correlations can be taken according to theoretical reasons, or it can be offered intuitively. However, quantitative values are always established for tested materials depending on specific sequence of loading and heating.

1.3.1.4 Calculation Methods

In the "pre-computer" epoch, great attention in solid mechanics was paid to the development of calculation methods. Analytical solutions for various linear problems of thermoelasticity were published by many researchers. A review of the earliest works on the subject is provided by Lebedev (1937) and Melan and Parkus (1953), many examples are in Timoshenko and Goodier (1970), and a summary of recent contributions can be found in Zhou (2008).

For problems referred to materials with nonlinear properties, pure analytical approaches were used only for the simplest models. For investigation concerning more complicated material behavior, methods of approximate numerical analysis were utilized (Il'ushin 1943; Birger 1957; Shorr 1958; Boley and Weiner 1960; Birger and Dem'yanushko 1968; Temis 1994, and many others).

The most widely adopted numerical techniques, including the finite element method, used in commercial software codes (see, e.g., Zienkiewicz and Taylor 2000) are briefly discussed in the subsequent chapters.

Authors of many theoretical works were interested in calculation of thermal stresses and strains, concentrating their attention on computing procedures without concerning an estimation of damage danger.

An impact of digital computers in the area of solid mechanics can be hardly overestimated. The main concern now is an accurate modeling of material behavior using numerical techniques, rather than effectiveness of calculations. A quality of a numerical method is sometimes estimated by comparison of the calculated results and some well-known "benchmark" solution. Such estimation is justified if the benchmark results have been checked experimentally.

There are a few publications providing the primary experimental data, which can be used as a benchmark for demonstrating reliability of numerical methods, see for example Zubchaninov et al. (2004).

One of the simplest ways for verification of numerical techniques, used for analysis of the SSS of a body, is measurement of residual strains after different operating time periods. Strain measurement during the operating conditions, usually demanding simultaneous temperature gauging, is more complicated.

From practical point of view, the main criterion of the numerical method, used for designing machine parts, is the correct estimation of time period or number of cycles before destruction.

1.3.2 Combined Stress–Strain State

1.3.2.1 Index Notations

Many equations in solid mechanics written explicitly differ only in indices of the components. These expressions become more concise, using index notation and summation convention, adopted for vector and tensor (see, e.g., Timoshenko and Goodier 1970).

Denoting the three coordinates through the variables $x_i (i = 1, 2, 3)$, a vector **f** with components f_1,f_2,f_3 may be represented as f_i (e.g., the displacement vector u_i). The second-rank tensor, e.g., the stress tensor with components

$$\sigma_{11}, \sigma_{12}, \sigma_{13}, \sigma_{21}, \sigma_{22}, \sigma_{23}, \sigma_{31}, \sigma_{32}, \sigma_{33},$$

is denoted by σ_{ij}, where $i = 1, 2, 3$ and $j = 1, 2, 3$. The strain tensor can be written as ε_{ij}, where the components at $i \neq j$ are associated with shear deformation γ_{ij} in the form $\varepsilon_{ij} = 0.5\gamma_{ij}$.

For conciseness, the following rules in index notation are used:

- If the same index occurs twice, summation over the values 1, 2, and 3 is assumed. Repeated index is called dummy index and can be replaced by any other Latin letter. Using this convention, the mean stress is represented in the form

$$\sigma = \frac{1}{3}(\sigma_{11} + \sigma_{22} + \sigma_{33}) = \frac{1}{3}\sigma_{ii},$$

the change of elementary volume is written as

$$\theta = \varepsilon_{11} + \varepsilon_{22} + \varepsilon_{33} = \varepsilon_{ii},$$

the work of internal forces affecting unit volume becomes

$$W = 0.5(\sigma_{11}\varepsilon_{11} + \sigma_{12}\varepsilon_{12} + \cdots + \sigma_{32}\varepsilon_{32} + \sigma_{33}\varepsilon_{33}) = 0.5\sigma_{ij}\varepsilon_{ij}.$$

- Partial derivative of the vector f_i with respect to x_j is denoted by

$$\frac{\partial f_i}{\partial x_j} = f_{i,j}.$$

Similarly, the second derivatives are denoted as follows:

$$\frac{\partial^2 f_i}{\partial x_j^2} = f_{i,jj}, \quad \frac{\partial^2 f_i}{\partial x_j \partial x_k} = f_{i,jk}.$$

- Kronecker symbol δ_{ij} is defined by expression

$$\delta_{ij} = \begin{cases} 1 & \text{at} \quad i = j, \\ 0 & \text{at} \quad i \neq j. \end{cases}$$

1.3.2.2 Equations of Motion and Equilibrium

Summing up all forces affecting an element of the body, including the volume forces F_i, we obtain three equations of motion that can be written in the form as

$$\sigma_{ij,j} + F_i = \rho \ddot{u}_i, \tag{1.12a}$$

where ρ is a material density, and the dots denote time derivatives. For static problems, Eq. (1.12a) yields the equations of equilibrium

$$\sigma_{ij,j} + F_i = 0. \tag{1.12b}$$

At a free surface of the body, the internal stresses have to balance the external load with components p_i. If $\mathbf{n}$ is the unit outer normal to the surface area of the body, its components n_i are equal to the direction cosines of the normal. Equation

$$p_i = \sigma_{ij} n_j \tag{1.13}$$

corresponds to static boundary conditions at the body's surface.

1.3.2.3 Strain/Displacement (Geometric) Relations

In a solid body, the displacement vector with components u_i is a continuous function of coordinates x_i that enable to define all of the components of the small strain tensor using the Cauchy equation

$$\varepsilon_{ij} = 0.5(u_{i,j} + u_{j,i}). \tag{1.14a}$$

For the coinciding indices $i = j$, Eq. (1.14a) yields

$$\varepsilon_{11} = \frac{\partial u_1}{\partial x_1}, \text{ etc,}$$

and for different indices,

$$\varepsilon_{12} = \frac{1}{2}\left(\frac{\partial u_1}{\partial x_2} + \frac{\partial u_2}{\partial x_1}\right) = \frac{1}{2}\gamma_{12}, \text{ etc.} \tag{1.14b}$$

At the part of the body surface S where displacements u_i^S are prescribed (in particular, $u_i^S = 0$ if a surface point is fixed), kinematical boundary conditions are

$$u_i = u_i^S. \tag{1.15}$$

The equations of equilibrium and strain–displacement relations do not depend on physical and mechanical properties of the body and heating conditions.

1.3.3 Theories of Stresses and Strains

General theories of stresses and strains are stated in many scientific publications and textbooks. Brief outline of these theories, concerning the problems of thermal integrity, is provided below.

1.3.3.1 Principal Stresses and Invariants of the Stress Tensor

Multi-axial Stress State

Stress tensor σ_{ij}, strain tensor ε_{ij}, and displacement vector u_i may be represented by their components in an arbitrary Cartesian coordinate system x_i. In different coordinate systems, all components will be changed in accordance with the formulas of coordinate transformation.

A magnitude of the vector

$$|u_i| = \sqrt{u_x^2 + u_y^2 + u_z^2} \tag{1.16}$$

does not depend on coordinate system, i.e., is *invariant.*

Consider an elementary tetrahedron limited by three faces normal to the three coordinate axes x_i and an inclined surface dA' with an outward unit normal vector **n**. Equilibrium conditions for this tetrahedron in this case lead to the following equations written in explicit form adopted in engineering texts

$$\begin{aligned} \sigma'_x &= \sigma_x n_x + \tau_{xy} n_y + \tau_{xz} n_z, \\ \sigma'_y &= \tau_{yx} n_x + \sigma_y n_y + \tau_{yz} n_z, \\ \sigma'_z &= \tau_{zx} n_x + \tau_{zy} n_y + \sigma_z n_z, \end{aligned} \tag{1.17}$$

where $\sigma'_x, \sigma'_y, \sigma'_z$ are projections of the total stress of the surface dA' onto the axes x_i.

Projecting the above-mentioned stresses onto the outward normal of the plane dA', we can find the normal stress on this plane

$$\begin{aligned}\sigma'_n &= \sigma'_x n_x + \sigma'_y n_y + \sigma'_z n_z \\ &= \sigma_x n_x^2 + \sigma_y n_y^2 + \sigma_z n_z^2 + 2\tau_{xy} n_x n_y + 2\tau_{yz} n_y n_z + 2\tau_{zx} n_z n_x.\end{aligned} \tag{1.18}$$

Projections of the normal stress σ'_n onto the axes x_i are $\sigma'_{ni} = \sigma'_n n_i$. Subtracting those from the total stress projections onto these axes, given by Eq. (1.17), we find the tangential stress projections

$$\tau'_{ni} = \sigma'_i - \sigma'_{ni},$$

or in explicit form

$$\begin{aligned}\tau'_{nx} &= (\sigma_x - \sigma'_n) n_x + \tau_{xy} n_y + \tau_{xz} n_z, \\ \tau'_{ny} &= \tau_{yx} n_x + (\sigma_y - \sigma'_n) n_y + \tau_{yz} n_z, \\ \tau'_{nz} &= \tau_{zx} n_x + \tau_{zy} n_y + (\sigma_z - \sigma'_n) n_z.\end{aligned} \tag{1.19}$$

At a certain direction of the normal n_i, all three projections (1.19), and hence the tangential stress itself τ'_n, vanish, and the normal stress becomes *principal*.

The principal stresses are usually ordered algebraically and denoted by subindices as $\sigma_1 \geq \sigma_2 \geq \sigma_3$, i.e., σ_1 is the algebraically maximal principal stress, and σ_3 is the algebraically minimal one.

Homogeneous system of Eq. (1.19) with the unknown value of principal stress σ'_n has a nontrivial solution if the determinant is zero, which leads to a cubic equation

$$(\sigma'_n)^3 - I_1(\sigma'_n)^2 + I_2\sigma'_n - I_3 = 0, \tag{1.20}$$

where the coefficients

$$\begin{aligned}I_1 &= \sigma_x + \sigma_y + \sigma_z, \\ I_2 &= \sigma_x\sigma_y + \sigma_y\sigma_z + \sigma_z\sigma_x - \tau_{xy}^2 - \tau_{yz}^2 - \tau_{zx}^2, \\ I_3 &= \sigma_x\sigma_y\sigma_z + 2\tau_{xy}\tau_{yz}\tau_{zx} - \sigma_x\tau_{yz}^2 - \sigma_y\tau_{zx}^2 - \sigma_z\tau_{xy}^2\end{aligned} \tag{1.21}$$

do not depend on the coordinate system, i.e., are *invariants of the stress tensor.*

It may be shown from analysis of Eq. (1.20) that the roots defining the values of the three principal stresses σ'_n are always real. In an unstressed body, all principal stresses are equal to zero.

Obviously, the values of the principal stresses are independent of coordinate systems, i.e., invariants. Expressions derived from Eq. (1.21) are also invariants, e.g., the average stress $\sigma = I_1/3$, parameters of the type $\hat{\sigma} = k\sqrt{I_2}$, having dimensions of stress (for any value of factor k), and various combinations of invariants, which are used in the theories of thermal plasticity and creep. Below all such characteristics are called "invariants."

Equation (1.19) accounting to the relation for direction cosines in any coordinate system

$$n_x^2 + n_y^2 + n_z^2 = 1, \tag{1.22}$$

allows the direction of all three principal axes of stress tensor (principal directions of stresses) to be found. These directions are of particular interest in analysis of the stresses for anisotropic bodies.

In the case of uniform tension or compression, all normal stresses are identical $\sigma_x = \sigma_y = \sigma_z = \sigma_0$, i.e., they become principal; thus, $I_1 = 3\sigma_0$ and all tangential stresses vanish. The system of Eq. (1.19) degenerates, and it is not possible to use it.

Note that there are only two stress states (uniform tension or compression) when there are no tangential stresses, whereas the states with only tangential stresses are not possible. The common term "pure shear" applies only to certain body faces, where at $\tau = \tau_0$ the normal stresses are zero, while they exist in other faces where $\sigma_1 = -\sigma_3 = \tau_0$.

Plane Stress State

In case of plane stress state, we assume $\sigma_z = 0$, $\tau_{xz} = \tau_{yz} = 0$. The other principal stresses are defined as

$$\sigma_{I,II} = \frac{1}{2}\left(\sigma_x + \sigma_y \pm \sqrt{(\sigma_x - \sigma_y)^2 + 4\tau_{xy}^2}\right). \tag{1.23}$$

The signs (+) or (−) correspond to algebraically larger and smaller stresses σ_I and σ_{II}, respectively.

The plane stress state always takes place at the free unloaded body surface where the corresponding normal and tangential stresses are absent.

Magnitudes of the normal and shear stresses on the plane with the normal inclined to the principal axis I at an angle ϕ are equal to

$$\begin{aligned} \sigma_\phi &= \sigma_I \cos^2\phi + \sigma_{II} \sin^2\phi, \\ \tau_\phi &= 0.5(\sigma_I - \sigma_{II})\sin^2\phi. \end{aligned} \tag{1.24}$$

It follows from Eq. (1.24) that normal stresses on any area are within the limits $\sigma_{II} \le \sigma_\phi \le \sigma_I$, i.e., the principal stresses are extreme.

Maximum Tangential Stress

Maximum tangential stress refers to the plane equally inclined to the principal axis at $\phi = \pi/4$ and is given by

$$\tau_{\phi,\max} = 0.5(\sigma_I - \sigma_{II}) \tag{1.25}$$

The analysis of equations related to the multi-axial stress shows that the given above relations remain valid for each pair of the principal planes. Along with definition $\sigma_1 \geq \sigma_2 \geq \sigma_3$, the magnitude of the maximum tangential stress for the whole element for any combined stress is equal to

$$\tau_{\max} = 0.5|\sigma_1 - \sigma_3| \tag{1.26}$$

Signs of the normal stresses reflect the different loading directions of an element (under tension or compression); the sign of the tangential stress $\tau_{\max}$ plays a role only when comparing the effects of different stress states.

1.3.3.2 Principal Strains and Invariants of the Strain Tensor

The structure of the strain tensor ε_{ij} is identical to that of the stress tensor σ_{ij}. Therefore, all relations for stresses σ_{ij} presented in the previous section remain valid for strains ε_{ij} with appropriate substitution of symbols.

The first invariant of the strain tensor characterizes the relative change in volume

$$I_{\varepsilon I} = \varepsilon_x + \varepsilon_y + \varepsilon_z = \theta, \tag{1.27}$$

and the second one is defined similar to the second invariant for the stress tensor in Eq. (1.21) as

$$I_{\varepsilon 2} = \varepsilon_x\varepsilon_y + \varepsilon_y\varepsilon_z + \varepsilon_z\varepsilon_x - \varepsilon_{xy}^2 - \varepsilon_{yz}^2 - \varepsilon_{zx}^2. \tag{1.28}$$

Invariants in the form $\widehat{\varepsilon} = k\sqrt{I_{\varepsilon 2}}$, where k is constant, are used in the theories of thermal plasticity and creep.

Isotropic thermal expansion ε_T affects only the first strain invariant, the mechanical part of which is

$$I'_{\varepsilon I} = \theta - \theta_T = \theta - 3\varepsilon_T. \tag{1.29}$$

1.3.3.3 Interpretations of the Invariants of Stress and Strain Tensors

The first and second invariants of stress and strain tensors (or their combinations) possess several interpretations corresponding to their specific role in calculations of thermal strength of structures. It is worth noting that the multipliers of invariants influence those interpretations but do not change their nature in principle.

The First Invariants

It was shown above that the first stress invariant I_1 characterizes the average stress σ, while the first strain invariant $I_{\varepsilon I}$ is equal to the relative change in element volume θ, and $I_1 = K(\theta - \theta_T)$.

For a combined stress, a value $I_1 > 0$ specifies the volumetric tension, and $I_1 < 0$—volumetric compression. In case of isothermal loading ($\theta_T = 0$), tension or compression states can be described in either stresses or strains, whereas under the presence of thermal expansion they may differ. For example, under hydrostatic pressure $p = -\sigma > 0$ (compression affected by stresses), the body may be expanded (with $\theta > 0$) if its thermal expansion θ_T exceeds the mechanical change of volume σ/K.

The Second Invariant of Stress Tensor

Equation (1.21) for the second invariant of stress tensor I_2 can be transformed into a more convenient form

$$I_1^2 - 2I_2 = \sigma_x^2 + \sigma_y^2 + \sigma_z^2 + 2(\tau_{xy}^2 + \tau_{yz}^2 + \tau_{zx}^2) = \sigma_{ij}\sigma_{ij}. \tag{1.30}$$

In the theory of solids, the stress tensor σ_{ij} is decomposed into a spherical (mean) stress tensor σ and stress tensor deviator s_{ij}

$$\sigma_{ij} = \delta_{ij}\sigma + s_{ij}. \tag{1.31}$$

The expression

$$I_{2s} = s_{ij}s_{ij} = s_x^2 + s_x^2 + s_x^2 + 2(s_{xy}^2 + s_{xy}^2 + s_{xy}^2) \tag{1.32}$$

also represents the *second invariant of the stress tensor deviator*, which plays an important role in the deformation theory of solids and will be frequently used in what follows.

In applications of solid mechanics, the second invariant of the deviator is often used in the form $\sqrt{I_{2s}} = \sqrt{s_{ij}s_{ij}}$, which has sometimes different notation and name.

The quantity

$$\sigma_i = k_\sigma\sqrt{s_{ij}s_{ij}}, \tag{1.33}$$

with the appropriate factor k_σ is usually referred to as *the stress intensity*.

Equation (1.33) in explicit form is used with $k_\sigma = \sqrt{3/2}$ in two equivalent notations:

$$\sigma_i = \sqrt{\frac{3}{2}}\sqrt{(\sigma_x - \sigma)^2 + (\sigma_y - \sigma)^2 + (\sigma_z - \sigma)^2 + 2(\tau_{xy}^2 + \tau_{yz}^2 + \tau_{zx}^2)} \tag{1.34a}$$

or

$$\sigma_i = \frac{1}{\sqrt{2}}\sqrt{(\sigma_x - \sigma_y)^2 + (\sigma_y - \sigma_z)^2 + (\sigma_z - \sigma_x)^2 + 6(\tau_{xy}^2 + \tau_{yz}^2 + \tau_{zx}^2)} \tag{1.34b}$$

In case of a uniaxial stress under $\sigma_x = \sigma_0,\ \sigma_y = \sigma_z = 0,\ \tau_{ij} = 0$, the intensity of normal stresses in Eqs. (1.34a, b) yields $\sigma_i = \sigma_0$.

Sometimes *the intensity of tangential stresses* defined as

$$\tau_i = \frac{\sigma_i}{\sqrt{3}} \tag{1.35}$$

is utilized.

In case of pure shear, when $\tau_{xy} = \tau_{yx} = \tau_0$ and the other components of stresses in the coordinate system x, y, z are equal to zero, the intensity of tangential stresses becomes $\tau_i = \tau_0$.

The Second Invariant of Strain Tensor

Invariants of the strain tensor have the same general form as the invariants of the stress tensor.

The strain tensor ε_{ij} is decomposed into a spherical (the mean normal ε) strain tensor and strain deviator tensor e_{ij}

$$\varepsilon_{ij} = \delta_{ij}\varepsilon + e_{ij}. \tag{1.36}$$

The second invariant of the strain deviator tensor I_{2e} does not depend on thermal expansion and can be written analogously to invariant of the stress deviator I_{2s} in Eq. (1.32) as

$$I_{2e} = e_{ij}e_{ij} = e_x^2 + e_x^2 + e_x^2 + 2(e_{xy}^2 + e_{xy}^2 + e_{xy}^2). \tag{1.37}$$

The quantity

$$\varepsilon_i = k_\varepsilon\sqrt{e_{ij}e_{ij}}, \tag{1.38}$$

is named *the strain intensity.*

Defining the multiplier k_ε from the condition that the value ε_i is equal to the strain in the *uniaxial strain state,* when

$$\varepsilon_x = \varepsilon_0, \quad \varepsilon_y = \varepsilon_z = 0, \quad \gamma_{ij} = 2\varepsilon_{ij} = 0,$$

formulas for ε_i will yield relations similar to Eqs. (1.34a, b).

In most theoretical contributions and software packages, incorporating finite element analysis, it is usually assumed that the value ε_i coincides with the strain ε_0 in *uniaxial stress state* when $\varepsilon_x = \varepsilon_0,\ \varepsilon_y = \varepsilon_z = -\mu\varepsilon_0,\ \gamma_{ij} = 2\varepsilon_{ij} = 0$. In this case,

$$k_\varepsilon = \frac{1}{1+\mu}\sqrt{\frac{3}{2}}. \tag{1.39}$$

and by analogy with Eq. (1.34b), we obtain

$$\varepsilon_i = \frac{1}{(1+\mu)\sqrt{2}}\sqrt{(\varepsilon_x - \varepsilon_y)^2 + (\varepsilon_y - \varepsilon_z)^2 + (\varepsilon_z - \varepsilon_x)^2 + \frac{3}{2}(\gamma_{xy}^2 + \gamma_{yz}^2 + \gamma_{zx}^2)}. \tag{1.40}$$

Octahedral Stresses

Stresses σ_n and τ_n on a plane with normal making equal angles with the principal axes are also invariants. Such a plane can be considered as one of the eight sides of the octahedron surrounding the given element of a body. That is why the corresponding parameters are usually called "octahedral." Direction cosines of the octahedral normal are $n_1 = n_2 = n_3 = 1/\sqrt{3}$.

Therefore, the *octahedral normal stress* σ_n is equal to the mean stress

$$\sigma_n = \frac{1}{3}(\sigma_1 + \sigma_1 + \sigma_1) = \sigma. \tag{1.41}$$

The total stress acting on the octahedral plane $p_n = F_n/A_n$ is determined by the condition that the force F_n is equal to vector sum of its projections onto the principal directions, from where

$$p_n = \frac{1}{\sqrt{3}}\sqrt{\sigma_1^2 + \sigma_2^2 + \sigma_3^2}. \tag{1.42}$$

The *octahedral tangential stress* $\tau_n = \sqrt{p_n^2 - \sigma_n^2}$ is easily transformed to

$$\tau_n = \frac{1}{3}\sqrt{(\sigma_1 - \sigma_2)^2 + (\sigma_2 - \sigma_3)^2 + (\sigma_3 - \sigma_1)^2}. \tag{1.43}$$

The magnitude τ_n is connected with the intensity of normal stress Eq. (1.34b), and intensity of tangential stress Eq. (1.35), through

$$\tau_n = \frac{\sqrt{2}}{3}\sigma_i = \sqrt{\frac{2}{3}}\tau_i. \tag{1.44}$$

The stress intensity in the form of Eq. (1.34a) with $k_\sigma = 1/\sqrt{3}$ expressed in terms of the principal stresses as

$$\sigma_i = \sqrt{\frac{1}{3}\left[(\sigma_1 - \sigma)^2 + (\sigma_2 - \sigma)^2 + (\sigma_3 - \sigma)^2\right]}, \tag{1.45}$$

can be considered as a standard deviation of the principal stress values from the mean value.

This characteristic is sometimes regarded as one of the possible definitions of the stress intensity, see Malinin (1975).

1.3.3.4 Vector Representation of Stress and Strain Tensor Characteristics

Deformation process in a state of combined stress is sometimes more conveniently illustrated by means of vector representations of some tensor characteristics, as introduced by Prager (1937), Il'ushin (1963), and others.

To achieve this goal, a second-rank tensor' deviator is considered as a nine-dimensional vector $\mathbf{v}$ with components v_{ij} that allows using techniques of vector algebra in the subsequent analysis.

In solid mechanics, nine-dimensional stress vector $\boldsymbol{\sigma}_i$ with components $(3/2)^{0.5}s_{ij}$ and strain vector $\boldsymbol{\varepsilon}_i$ with components $(2/3)^{0.5}e_{ij}$ are commonly used. For simplicity, $\boldsymbol{\sigma}_i$ is referred to as a stress vector, and $\boldsymbol{\varepsilon}_i$—as a strain vector, though they characterize only deviators of the corresponding tensors. Magnitudes of these vectors are

$$|\boldsymbol{\sigma}_i| = \sqrt{\frac{3}{2}s_{ij}s_{ij}} = \sigma_i \quad \text{and} \quad |\boldsymbol{\varepsilon}_i| = \sqrt{\frac{2}{3}e_{ij}e_{ij}} = \varepsilon_i, \tag{1.47}$$

which for $\mu \approx 0.5$ coincide with definition of stress σ_i and strain ε_i intensities, respectively.

A projection of a vector $\mathbf{v}_i$ onto the direction of another vector $\boldsymbol{\omega}_i$ is

$$(\mathbf{v}_i)_\omega = v_i \cos\phi$$

where ϕ is an angle between these vectors.

According to the formula of scalar product of vectors

$$(\mathbf{v}_i \boldsymbol{\omega}_i) = v_i \omega_i \cos \phi = v_{ij} \omega_{ij}$$

the projection yields

$$(\mathbf{v}_i)_\omega = \frac{v_{ij}\omega_{ij}}{\omega_i}. \tag{1.48}$$

Condition of vector collinearity (parallelism) is

$$\frac{v_{ij}}{v_i} = \frac{\omega_{ij}}{\omega_i}. \tag{1.49}$$

Various vector representations in this book will be considered further.

1.3.4 Analysis of Structure Failure

Problems of continuum damage and structure failure under non-isothermal loading are widely discussed in specialized literature (see, e.g., Murakami 2012). In the present monograph, these problems are considered only within the frame of the phenomenological approach necessary for practical use in engineering.

1.3.4.1 Limit State and Load-Carrying Capacity of a Structure

Any mechanical structure (a machine part, a construction, etc.) can successfully perform its intended functions only to certain values of loads, temperatures, operating times, or some other external actions on the structure.

The mechanical structure should respond to the external actions by some combinations of internal parameters—stresses, strains, displacements, thermal expansion, structural changes, wear, and others, reflecting the physical and mechanical processes in the system. Condition at which the structure ceases to fulfill its functional requirements, i.e., loses its working capacity, is named a *limit state*.

The limit state can be connected with both damage of a structure and failure due to other reasons, e.g., deterioration of operational characteristics.

A special case of a limit state of the system, connected with mechanical damage, is referred to as *load-carrying* (*bearing*) *capacity* (Serensen et al. 1963), although it would be more precise to call such state a *loss* (*exhaustion*) of carrying capacity. Loss of mechanical load-carrying capacity can be a result of partial or total fracture of structure components or excessive changes of sizes or configuration of these components under operating conditions.

Further, we will refer to any reason leading to loss of load-carrying capacity as "structure failure," and the parameters of system corresponding to this event—parameters of load-carrying capacity. An example of such a parameter could be the critical rotation frequency of turbine disk at which it fails at spin up.

The present book contains analysis of some of the problems of losing load-carrying capacity in a mechanical structure, connected basically (though not exclusively) with its *temperature state*.

1.3.4.2 Local and Global Failure

Failure of a mechanical structure always begins with *material local damage* in the weakest area which can be an elementary volume, a fiber, and so forth. In some cases, the local damage leads to very rapid distribution of destruction in the entire structure resulting in its nearly immediate *global* collapse. For strength estimation of such systems, it is enough to consider the strength conditions of the weakest area.

In certain situations, undamaged parts of a mechanical structure are capable of sustaining additional loads, preserving the load-carrying capacity of a structure despite the local destruction underway.

The failure character depends on a variety of parameters: material properties, configuration of structure parts and their connection with other parts, loading and heating conditions, arising stresses and strains, temperature field of a body, and so forth. At elevated temperatures, the operating time plays essential role.

Generally, strength calculations should include conditions of both local material damage and load-carrying capacity of the mechanical structure as a whole.

In a construction, full fracture of which could lead to catastrophic consequences, a "weak link" is sometimes incorporated artificially. After its rupture, the machine will be stopped (or transferred into less severe loading regime)—i.e., the weak link is used as a safety lock in an electric chain.

Thermal expansion effect is quite often used in break switches (an example is given in Chap. 10).

Variable loads can cause *local cracks* due to low- or high-cycle fatigue (LCF or HCF), respectively.

The fatigue failure of a structure can occur at a wide range of nominal stresses and strains: LCF—basically within elastoplastic area and HCF—for elastic deformations. Temperature rise brings more peculiarity in development of the LCF.

Material HCF is discussed in manifold publication. The range of cyclic stress at high frequency oscillations is usually determined experimentally. The temperature influence on fatigue limit is not clearly determined unequivocal, although in most cases it decreases, as other material strength properties.

Non-isothermal LCF will be considered in Chap. 4.

At the beginning, LCF cracks grow without appreciable redistribution of stress over the structure and can be approximately calculated by methods of *fracture mechanics*. Redistribution of stresses due to crack growth reduces the number of cycles prior to failure.

Magnitudes of the damage parameters depend on the local temperature of the body. Growth of the cracks over the body changes the conditions of heat transfer that should be reflected in the change of a temperature field, and this, in turn, affects the speed of crack propagation. These effects are usually neglected in calculations.

1.3.4.3 Criteria of a Material Failure

The basic data on material strength are obtained through standard tension tests (less often—in torsion tests). In order to use these results for calculations of material failure at various combined SSS, some criteria of material failure are introduced. These criteria must match with experimental characteristics in a case of uniaxial tension. Since failure conditions for isotropic material cannot depend on coordinate system, only invariant characteristics of stress and/or strain tensors should be considered as such criteria.

The maximum principle stress σ_1, maximum tangential stress $\tau_{\max} = 0.5(\sigma_1 - \sigma_3)$, stress intensity (Mises' equivalent stress) σ_i, mean stress σ, and their combinations were proposed as "equivalent" stress characteristics σ_{eqv}, in different strength theories.

Invariants of strain tensor can also be utilized.

Application areas of material failure criteria will be discussed in the subsequent chapters.

1.3.4.4 Damage Models

In order to define the moment of the local material destruction, it is required to find the most stressed element and check it against some appropriate failure criterion.

Further modeling of the damage propagation resulting in the global collapse of a structure can be performed in several ways.

In case of gradually increasing loads, the modern computing facilities allow computation of the nonlinear elastoplastic SSS of the entire structure, taking into account not only altering forces and temperatures, but also a change in structure configuration.

Using the same failure criteria as for the first localized damage and excluding consistently the broken elements from the designed model, the development of damage progress over the structure is revealed. While approaching the state close to loss of structure capacity, the strains and displacements of non-fixed model elements increase dramatically. Moreover, accumulation of damage in material elements in the process of stress increase (Kachanov 1958) can be taken into account.

Calculations show that for structures of materials with enough plasticity, especially at high temperatures, visibly sharp increase of calculated strains and displacements occurs due to growth of plastic areas even if the excessively deformed elements are not excluded from designed model. It allows estimating the loss of a structural capacity without using criteria of material failure.

The accuracy of various failure criteria offered in the engineering literature for a combined SSS is still under research, but for sharp increase of calculated parameters the peculiarities of different failure criteria cease to play a significant role.

1.3.4.5 Strength and Durability Margins

A main goal of strength calculation—maintenance of reliable machine work—can be reached only provided that all operating parameters of the system will be less than their admissible values during a demanded lifetime. The operating parameters include the maximum values of stresses, strains, temperatures, duration of regimes, etc., under most severe working conditions. At this point, a possible scatter of material strength characteristics and industrial discrepancy of manufacturing must be taken into account.

A ratio of structure parameters at failure P_d and under operating conditions P_e is named the *safety factor*, or *strength margin* on this parameter $k = P_d/P_e$. In regard to time-dependent processes, the safety factors are usually named *durability margins*. As a rule in machine-building designs, it is necessary to use not just a single margin k but a combination of margins k_i, since each of those guarantees structural capacity only in relation to a separate characteristic.

For maintenance of structure integrity, all margins should not be less than their admissible values $k_i \geq [k_i]$. The appropriate values of margins $[k_i]$ should be established, based on the generalized operating experience of products—prototypes.

The maintenance of structural capacity and safety factors of mechanical structures at elevated temperature, including probability aspects, are considered in Chap. 9 in more detail.

References

Andrade EDC (1910) On the viscous flow of metals and allied phenomena. Proc R Soc Ser A 84: A567

Bailey RW (1935) The utilization of creep test data in engineering design. Proc Inst Mech Eng 131 (1):131–349

Balashov BF, Kozlov LA (1981) Methods of fatigue tests and results processing. In: Birger IA, Balashov BF (eds) Constructive strength of materials and details of gas-turbine engines. Mashinostroenie, Moscow (in Russian)

Birger IA (1957) Calculation of nonuniformly heated rods with variable elastic parameters. Trudy Instituta im. Baranova, No. 298, Oborongiz, Moscow (in Russian)

Birger IA, Dem'yanushko IV (1968) Theory of plasticity at nonisothermal loading. Ingenerny J Mechanika Tverdova Tela 6:70–77 (in Russian)

Boley BA, Weiner JH (1960) Theory of thermal stresses. Wiley, New York

Duhamel JMC (1837) Deuxième memoire sur les phénomènes thermo-mécaniques. J de l'Ecole Polytechnique 15(25):1–57

Dul'nev RA (1979) Experimental methods of research of thermo-fatigue characteristics. In: Constructive strength of materials and details of gaseous-turbine engines. CIAM, Proc. No. 835, Moscow, pp 497–516 (in Russian)

Gatewood BE (1957) Thermal stresses with applications to airplanes, missiles, turbines and nuclear reactors. McGraw-Hill Book Company Inc., New York
Il'ushin AA (1943) Some problems of the theory of plastic deformations. J Prikladnaya matematika i mekhanika YII(4):245–272 (in Russian)
Il'ushin AA (1963) Plasticity. Foundation of general mathematical theory. AN SSSR, Moscow (in Russian)
Kachanov LM (1958) About failure time in creep conditions. Izv.AN SSSR, ONT, No. 8 (in Russian)
Lebedev NN (1934) Thermal stresses in the theory of elasticity, PMM, vol 2(1). Moscow (in Russian)
Lebedev NN (1937) Thermal stresses in the theory of elasticity, ONTI. Leningrad-Moskva (in Russian)
Lemaitre J, Chaboche JL (1990) Mechanics of solid materials. Cambridge University Press, Cambridge
Lemaitre J, Chaboche JL (2004) Mecanique des materiaux solides, 2nd edn. Dunod, Paris
Malinin NN (1975) Applied theory of plasticity and creep, 2nd edn. Mashinostroenie, Moscow (in Russian)
Manson SS (1966) Thermal stress and low-cycle fatigue. McGraw-Hill Company, New York
Melan E, Parkus H (1953) Wärmespannungen Infolge Stationäre Temperatur-felder. Springer, Wien
Murakami S (2012) Continuum damage mechanics. Springer, Berlin
Noll W (1958) A mathematical theory of the mechanical behavior of continuous media. Arch Rat Mech Anal 2:197–226
Pisarenko GS, Novikov NV, Andreev LP et al (1969) About problem of investigation of metal mechanical characteristics at low temperatures under vacuum conditions. In: Thermostrength of materials and construction members, vol 5. Naukova Dumka, Kiev, pp 440–448 (in Russian)
Prager W (1937) Mechanique des solides isotropes au dela du domaine elastique. Memorial de Sciences Math, vol LXXXVII
Rabotnov YN (1966) Creep of construction elements. Physmatgiz, Moscow. Trans: (1969) Creep problems in structure members. North Holland, Amsterdam (in Russian)
Sedov LI (1973) Mechanics of continuum medium, vol 1, 2, 2nd edn. Nauka, Moscow (in Russian)
Serensen SV, Kogaev VP, Shneyderovich RM (1963) Load-capacity and strength calculation of machine structures, 1st edn (3rd edn, 1975). Mashinostroenie, Moscow (in Russian)
Shorr BF (1958) Effect of nonuniform heating under creep conditions on the change in a stress state. Dokl Akad Nauk SSSR 123(5):809–812 (in Russian)
Temis YM (1994) Applied solution methods of thermal plasticity problems. In: Frolov KV et al (eds) "Machinery, Encyclopedia". In: Kolesnikov KS et al (eds) Dynamics and machine strength, vol 1–3(B.1). Mashinostroenie, Moscow, pp 231–236 (in Russian)
Timoshenko SP, Goodier JN (1970) Theory of elasticity, 3rd edn. McGraw-Hill, New York
Touloukian YS et al (eds) (1975) Thermophysical properties of matter, vol 12. Thermal expansion. Plenum, NY, Washington
Troshchenko VT, Lebedev AA, Strijalo VA, Stepanov GV, Krivenjuk VV (2000) Mechanical behavior of materials at different load terms. Logos, Kiev (in Russian)
Valanis KC (1971) A theory of viscoplasticity without a yield surface. Arch Mech Stosow 23 (4):517–551
Volkov IA, Korotkikh YG (2008) State equations of viscoelasticplastic media with damages. Phyzmatgiz, Moscow (in Russian)
Wilshire B, Evans RW (1994) Acquisition and analysis of creep data. J Strain Anal Eng Des 29 (3):159–165
Zhou X (2008) Publications on thermal stresses. J Therm Stresses, 31:663–664 (Taylor & Francis)
Zienkiewicz OC, Taylor RL (2000) The finite element method, 5th edn. Butterworth-Heinemann, Oxford
Zubchaninov VG, Okhlopkov NL, Garanikov VV (2004) Experimental plasticity, B.2, Processes of combine loading. Tver's State Technical University, Tver (in Russian)

Chapter 2
Thermoelasticity

Abstract This chapter contains the foundations of the theory of thermoelasticity required for its implementation for calculating strength of structural members as well as for studying the following chapters of the monograph. It is demonstrated that thermoelastic deformations are independent of loading and heating sequence and are always reversible, including the case of temperature-dependent elasticity moduli and simultaneously varying temperatures and stresses. The "dynamic" elasticity moduli evaluating from measured resonance frequencies are more informative than the "static" ones originating from deformation diagrams and depending on loading rate. The governing equations in the theory of thermoelasticity for steady-state and transient non-uniform temperature fields are presented. Formation of temperature stresses in both statically determinate and indeterminate structures is described. Efficient methods for solving non-isothermal boundary value problems with temperature-dependent elasticity moduli are analyzed. The energy principles in the theory of elasticity usually underlying the FEM algorithms and specific features of the variational equations in thermoelasticity are subject to detailed discussion.

2.1 Foundations of Thermoelasticity

2.1.1 Thermal Stresses in an Elastic Body

2.1.1.1 Moduli of Elasticity

An elementary example of phenomenological description of a material property is the Hooke's law. Tensile tests of a rod demonstrated that its elastic elongation Δl can be defined as $\Delta l = Fl_0/EA_0$, where F is a longitudinal force applied to a rod, l_0 and A_0 are the initial length and cross-sectional area of a rod, respectively, and the constant E is a characteristic of the material named the *elasticity modulus* (also known as Young's modulus). Using the definition of stress $\sigma = F/A_0$ and relative

B.F. Shorr, *Thermal Integrity in Mechanics and Engineering*,
Foundations of Engineering Mechanics, DOI 10.1007/978-3-662-46968-2_2

deformation (strain) $\varepsilon = \Delta l / l_0$, we obtain a basic formula of material elastic deformation $\varepsilon = \sigma / E$.

The elasticity modulus basically depends on temperature, but the refined measurements show that it is also affected by loading conditions (tension or compression, increasing force or unloading, etc.). At elevated temperatures, there is a distinction between *dynamic* and *static* moduli depending on the loading speed. The deviation between those increases with the growth of temperature.

The dynamic modulus of elasticity $E_{dyn}(T)$ is evaluated from the resonance frequencies measured at standard tests at a given temperature (see Chap. 11). The accuracy of its definition can be rather high depending not only on the accuracy of processing of the vibrating signal, but also on an adopted dynamic model of the specimen.

The static modulus of elasticity can be determined as $E_{st,0}(T) = \tan \varphi_1$, where φ_1 is an angle of inclination of a tangent to any point of the initial part of the experimental curve $\sigma(\varepsilon, T)$ as shown by line 1 in Fig. 2.1.

The accuracy of modulus $E_{st}(T)$ definition is influenced by, at least, two factors.

- The initial part of a curve $\sigma(\varepsilon, T)$ is not quite rectilinear, and to eliminate subjectivity while evaluating $E_{st}(T)$ by a line 2, it should be calculated under some "contractual" conditions; and
- At elevated temperatures, the deformation curve $\sigma(\varepsilon, T)$ is influenced by the duration of loading which can lead to creep effects (line 3).

As a result, the values of static modulus $E_{st}(T)$, defined at higher temperatures and normal speeds of the static loading of a specimen, usually appear to be underestimated in comparison with the dynamic modulus $E_{dyn}(T)$. The area, representing ratio $E_{dyn}(T)/E_{st}(T)$ at various temperatures for selected four heat-resistant alloys, is demonstrated in Fig. 2.2 (Sizova's data in Birger and Shorr 1975).

For static calculations at high temperature, it is recommended to use the values $E_{dyn}(T)$.

The tests with unloading usually go fast, so creep does not occur even at high temperatures, and the modulus $E_{st}(T)$ defined on unloading line 4 in Fig. 2.1 differs slightly from the modulus $E_{dyn}(T)$.

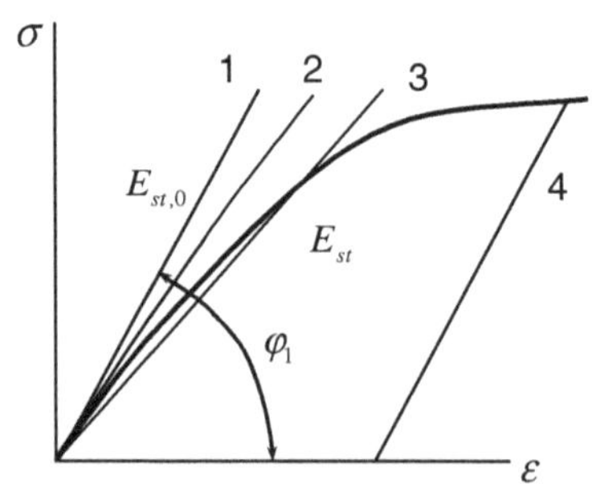

Fig. 2.1 Definition of the static modulus of elasticity

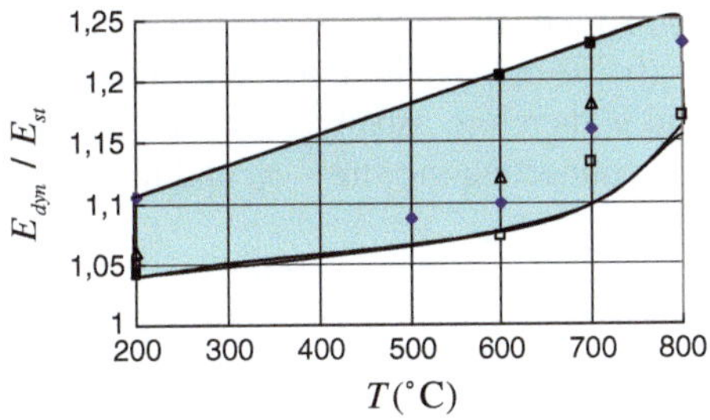

Fig. 2.2 Temperature dependence of the ratio of dynamic and static moduli of elasticity

Similar torsion tests define correlations between the temperature and the shear modulus $G(T)$.

2.1.1.2 Thermal Stresses

The terms corresponding to the temperature expansion have been introduced into the equations of elasticity by Duhamel (1837).

The classical linear theory of thermoelasticity is limited to small deformations and constant values of elasticity moduli E, G and coefficient of linear thermal expansion α_T. The theory is based on the following assumptions confirmed by experiments:

- When calculating the body stress–strain state (SSS), its temperature field is considered to be known on the basis of experimental measurements or calculations and does not change due to the stresses arising in the body. This approach is usually referred to as "the uncoupled theory of thermoelasticity."
- The elastic "mechanical" strain ε_σ induced by the action of the stress σ and temperature expansion of the material $\varepsilon_T = \alpha_T \Delta T$ may be summarized, resulting in a total strain $\varepsilon = \varepsilon_\sigma + \varepsilon_T$, according to the "the principle of strain superposition" (see Chap. 1). The total strain ε connected with displacements by geometric relations (1.14a) does not depend on material's mechanical characteristics.

The application of the linear theory remains valid within the moderate ranges of temperature variation.

The given stationary temperature field $T(x_i)$ differs from the initial one $T_0(x_i)$ as $\Delta T(x_i) = T(x_i) - T_0(x_i)$.

For the *uniaxial* (1D) stress state ($\sigma_1 = \sigma_0,\ \sigma_2 = \sigma_3 = 0$), we have

$$\varepsilon_1 = \frac{\sigma_0}{E} + \varepsilon_T, \quad \varepsilon_2 = \varepsilon_3 = -\mu\frac{\sigma_0}{E} + \varepsilon_T, \tag{2.1}$$

where μ is the Poisson's ratio. Its value for most of the engineering materials lies within the limits $\mu \approx (0.26\text{--}0.34)$. The Poisson's ratio μ typically varies slightly within temperature and is therefore usually assumed constant.

The inverse relation connecting the stress σ_0 and thermal expansion is given by

$$\sigma_0 = E(\varepsilon_1 - \varepsilon_T). \tag{2.2}$$

Alongside linear strains, a volume deformation is also of interest. The relative volume change of a small element of a body is

$$\theta = 3\varepsilon = \varepsilon_1 + \varepsilon_2 + \varepsilon_3, \tag{2.3}$$

where ε is the mean strain. Volume deformation may also be expressed as a sum of mechanical and thermal components $\theta = \theta_\sigma + \theta_T$, or

$$\theta = (1 - 2\mu)\varepsilon_\sigma + 3\varepsilon_T = \frac{\sigma_0}{3K_\theta} + 3\varepsilon_T, \tag{2.4}$$

where

$$K_\theta = \frac{E}{3(1 - 2\mu)} \tag{2.5}$$

is a *modulus of volumetric elasticity* (also known as a modulus of volumetric compression or a bulk modulus).

Thermal stresses in a homogeneous statically determinate uniformly heated system can emerge only due to restrictions in displacements along a boundary by the other constructive elements. In all other cases, thermal stresses occur when the temperature field is non-uniform.

If the connected parts of a structure are composed of materials with different characteristics of elasticity and thermal expansion, thermal stresses can be provoked by uniform heating of the whole structure.

Thermal Stresses in a Fixed Rod

As an example of arising thermal stresses, let us consider an elastic rod of length L with constant cross-sectional area A and fixed borders, heated in ΔT (Fig. 2.3). If one of the borders was free, the rod length would increase by $\Delta L = \alpha \Delta T L$ (Fig. 2.3a). If a rigid wall restricts expansion of a rod (Fig. 2.3b), the reaction force F causes its compression in $\Delta L_F = FL/EA$ (Fig. 2.3c).

Having utilized $\Delta L = \Delta L_F$, we find $F = E\alpha\Delta T A$ or $\sigma = F/A = -E\alpha\Delta T$ where the negative sign specifies compression.

The same result can be obtained, setting a rod strain to zero $\varepsilon = \sigma/E + \varepsilon_T = 0$ that leads to $\sigma = -E\varepsilon_T = -E\alpha\Delta T$.

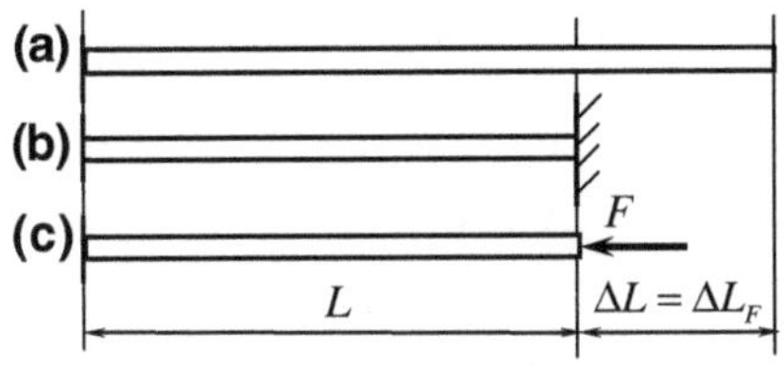

Fig. 2.3 Thermal stresses arising in a heated rod with fixed borders

2.1.2 Thermoelastic Strains

2.1.2.1 Temperature-Dependent Mechanical Properties

Under considerably non-uniform temperature field, it is necessary to treat the moduli of elasticity E, G, K and thermal expansion α_T as functions of temperature. For $\mu \approx$ const, the temperature variation of all elastic moduli is proportional to the change in the Young's modulus $E(T)$.

Different procedures for the thermal expansion coefficient defining in dependence on temperature $\alpha_T(T)$ have been described in Sect. 2.1.2.3.

Since Eqs. (2.1)–(2.5) correspond to any particular point of the body, they remain valid under non-uniform temperature field.

2.1.2.2 Independence of Thermoelastic Strains of the Order of Loading and Heating

An important feature of the linear theory of thermoelasticity is independence of a body SSS from the sequence of loading and heating. This applies also to mechanical characteristics depending on temperature. The influence of the sequence of loading and heating on energy relations is discussed below in Sect. 2.3.

Figure 2.4 demonstrates the deformation diagrams for a material at two temperatures T_1 and T_2, with corresponding elastic moduli E_1 and E_2. Under transition

Fig. 2.4 Independence of elastic strains from loading and heating order

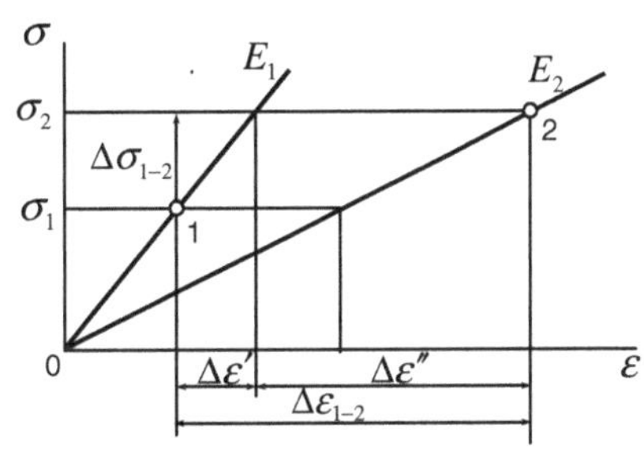

from one state $\sigma_1, \varepsilon_1, T_1$ to another state $\sigma_2, \varepsilon_2, T_2$, the total increment of elastic strain is

$$\Delta\varepsilon_{1-2} = \frac{\sigma_2}{E_2} - \frac{\sigma_1}{E_1}. \tag{2.6}$$

This transition can be achieved by various ways.

- By stress change $\Delta\sigma_{1-2} = \sigma_2 - \sigma_1$ at the initial temperature T_1 and strain increase

$$\Delta\varepsilon' = \frac{\Delta\sigma_{1-2}}{E_1}, \tag{2.7a}$$

followed by temperature change $\Delta T_{1-2} = T_2 - T_1$ at the new stress σ_2 with the additional increase in strain

$$\Delta\varepsilon'' = \sigma_2\left(\frac{1}{E_2} - \frac{1}{E_1}\right). \tag{2.7b}$$

- By temperature change $\Delta T_{1-2} = T_2 - T_1$ at the initial stress σ_1, and then change of the stress $\Delta\sigma_{1-2} = \sigma_2 - \sigma_1$ at a new temperature T_2 and corresponding increasing strain.
- By complete unloading $\Delta\sigma_{1-0} = -\sigma_1$ at the initial temperature T_1 and then by new full loading $\Delta\sigma_{0-2} = \sigma_2$ at the new temperature T_2. It is easy to check that all three ways will lead to the same value of the elastic strain due to Eq. (2.6).

2.1.2.3 Increments of Thermoelastic Strains

Under smooth variations of loads and temperature, equations of thermoelasticity can be differentiated with respect to operating or formal time t (see Chap. 1) and represented in differential form. This form is required, in particular, to solve contact problems with altering boundary conditions and in combination with plastic and creep strains.

Equation (2.1) yields

$$d\varepsilon_1 = \frac{d\sigma_0}{E} + \left(\frac{d\varepsilon_T}{dT} - \frac{\sigma_0}{E^2}\frac{dE}{dT}\right)dT, \tag{2.8}$$

where

$$\frac{d\varepsilon_T}{dT} = \frac{d(\alpha_T T)}{dT} = \alpha_T + T\frac{d\alpha_T}{dT}. \tag{2.9}$$

Instead of the modulus E, it is more convenient to use its reciprocal—the compliance coefficient

$$a(T) = [E(T)]^{-1}, \tag{2.10}$$

which has to be reconstructed beforehand from the temperature dependence $E(T)$.

Equation (2.8) then takes the form

$$d\varepsilon_1 = a d\sigma_0 + \left(\sigma_0 \frac{da}{dT} + \frac{d\varepsilon_T}{dT}\right) dT. \tag{2.11a}$$

During calculations, the differentials $d\sigma, dT, \ldots$, are replaced by small final increments $\Delta\sigma, \Delta T$, etc; then, Eq. (2.11a) yields

$$\Delta\varepsilon_1 = a(T)\Delta\sigma_0 + \sigma_0 \Delta a(T) + \Delta\varepsilon_T(T). \tag{2.11b}$$

Non-isothermal loading scenario with simultaneous changes of stresses, strains, and temperatures can have a sophisticated character. In Fig. 2.5, "a deformation chart" for the four-stage consecutive increase of stresses, strains, and temperatures (without thermal expansion) is shown. The solid and dotted broken lines assume that stepped temperature transitions occur at constant stresses or strain, respectively.

If a number of transitions are increased, the specified dependence becomes smooth, which is shown by dashed line. Under unloading according to any scenario, residual stresses do not arise.

Reproducing such non-isothermal elastic process at elevated temperatures experimentally is rather difficult because of possible occurrence of plastic strains and creep.

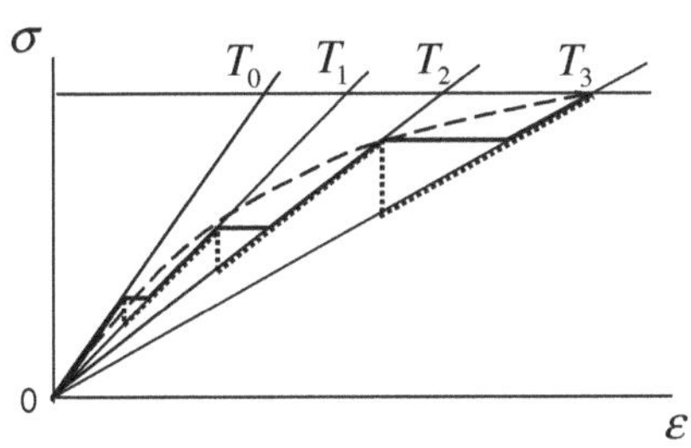

Fig. 2.5 A "deformation chart" at simultaneous changes of stresses, strains, and temperatures

2.2 The Theory of Thermoelasticity

2.2.1 Constitutive Equations

The equations of thermoelasticity are given in numerous books and are used in some software codes, incorporating final element method. Below, a concise description of the basic equations of thermoelasticity is presented with some comments useful for their correct application in engineering.

2.2.1.1 Equations of Combined Stress–Strain State

In case of a combined (multi-axial) SSS, the following basic equations of thermoelasticity yield:

$$\left.\begin{aligned} \varepsilon_x &= \frac{1}{E}\left[\sigma_x - \mu(\sigma_y + \sigma_z)\right] + \varepsilon_T, \\ \varepsilon_y &= \frac{1}{E}\left[\sigma_y - \mu(\sigma_z + \sigma_x)\right] + \varepsilon_T, \\ \varepsilon_z &= \frac{1}{E}\left[\sigma_z - \mu(\sigma_x + \sigma_y)\right] + \varepsilon_T, \\ \gamma_{xy} &= \frac{1}{G}\tau_{xy}, \gamma_{yz} = \frac{1}{G}\tau_{yz}, \gamma_{zx} = \frac{1}{G}\tau_{zx} \end{aligned}\right\}, \tag{2.12}$$

where $\sigma_x, \sigma_y, \sigma_z$ are normal (direct) stresses, $\tau_{xy}, \tau_{yz}, \tau_{zx}$ are tangential stresses, $\varepsilon_x, \varepsilon_y, \varepsilon_z$ are linear strains (elongations), $\gamma_{xy}, \gamma_{yz}, \gamma_{zx}$ are angular strains (shears), and $G = E/2(1 + \mu)$ is the shear modulus.

Equation (2.12) may be represented in other forms, using the expressions for the mean strain ε (see Eq. 2.3) and mean stress, namely

$$\sigma = \frac{1}{3}\left(\sigma_x + \sigma_y + \sigma_z\right), \tag{2.13}$$

resulting in

$$\varepsilon_x = \frac{1}{E}[(1 + \mu)\sigma_x - 3\mu\sigma] + \varepsilon_T. \tag{2.14}$$

Similar expressions can be obtained for the strain components $\varepsilon_y, \varepsilon_z$. Relations for the shear strains are independent of the mean stress.

The relative change of a small volume in the general case, according to Eq. (2.4), is equal to

$$\theta = \frac{\sigma}{K_\theta} + \theta_T, \tag{2.15}$$

where $\theta_T = 3\varepsilon_T = 3\alpha_T\Delta T = \beta_T\Delta T$.

In case of a hydrostatic pressure p, the mean stress is $\sigma = -p$, so that the bulk modulus K_θ characterizes the relative decrease of the elementary volume under compression.

Using index notation (see Chap. 1), Eq. (2.12) becomes

$$\varepsilon_{ij} = \frac{1}{E}\left[(1+\mu)\sigma_{ij} - 3\mu\delta_{ij}\sigma\right] + \delta_{ij}\varepsilon_T, \tag{2.16}$$

where for $i \neq j$ ($\delta_{ij} = 0$), we arrive at $\varepsilon_{ij} = \sigma_{ij}/2G = \gamma_{ij}/2$.

The constitutive Eq. (2.16) is also used in the form

$$\varepsilon_{ij} = \frac{1}{2G}\left(\sigma_{ij} - \delta_{ij}\sigma\right) + \delta_{ij}\left(\frac{\sigma}{K} + \varepsilon_T\right), \tag{2.17}$$

where $K = 3K_\theta$.

Inverse expressions for Eqs. (2.16) and (2.17) are given by

$$\sigma_{ij} = \frac{E}{1+\mu}\left[\varepsilon_{ij} + \frac{3\mu}{1-2\mu}\delta_{ij}\varepsilon + \frac{1+\mu}{1-2\mu}\delta_{ij}\varepsilon_T\right] \tag{2.18}$$

and

$$\sigma_{ij} = 2G(\varepsilon_{ij} - \delta_{ij}\varepsilon) + \delta_{ij}K(\varepsilon - \varepsilon_T). \tag{2.19}$$

Equations (2.16)–(2.19) are valid for both normal and tangential components of stresses and strains.

Thermoelasticity equations in the form (2.17) and (2.19) are used in the theories of thermoplasticity and creep. In these theories (see Chaps. 3 and 6), the stress tensor σ_{ij} is represented as a sum of spherical part $\delta_{ij}\sigma$ and deviator s_{ij}, and analogously for the strain tensor, namely

$$\begin{aligned} \sigma_{ij} &= \delta_{ij}\sigma + s_{ij}, \\ \varepsilon_{ij} &= \delta_{ij}\varepsilon + e_{ij}, \end{aligned} \tag{2.20}$$

where σ and ε are mean stress and strain, respectively.

Components of the deviators at $i \neq j$ coincide with the corresponding components of the tensors, and mean stresses associated with the mean strain and thermal expansion are given by Eq. (2.15). The deviator components are related to the equation

$$s_{ij} = 2Ge_{ij}, \tag{2.21}$$

arising from Eqs. (2.16) to (2.20).

2.2.1.2 Differential Form of the General Relations

Presentation of the general equations of thermoelasticity in differential form is necessary for theoretical study of thermoplasticity and creep problems. Thus, differentiating Eq. (2.17) gives

$$d\varepsilon_{ij} = 0.5a_G ds_{ij} + \delta_{ij}a_K d\sigma + \delta_{ij}d\varepsilon_T + 0.5s_{ij}da_G + \delta_{ij}\sigma da_K. \tag{2.22a}$$

In the inverse form,

$$d\sigma_{ij} = 2Gde_{ij} + \delta_{ij}Kd\varepsilon + 2e_{ij}dG + \delta_{ij}[(\varepsilon - \varepsilon_T)dK - Kd\varepsilon_T)], \tag{2.22b}$$

where the compliance coefficients are

$$a_G(T) = G^{-1}(T), \quad a_K(T) = K^{-1}(T). \tag{2.23}$$

Temperature increment of the elasticity modulus G is defined as

$$dG = \frac{dG}{dT}dT.$$

The latter may similarly be applied to other material characteristics.

Derivative dG/dT has to be obtained from the experimental curve $G(T)$ using numerical differentiation. Since the dependence $\mu(T)$ is often insignificant, then the functions $\bar{E}(T) = E(T)/E_0$, $\bar{G}(T) = G(T)/G_0$, and $\bar{K}(T) = K(T)/K_0$, where E_0, G_0, and K_0 refer to the normal temperature, are identical.

In the literature, equations, analogous to (2.22a), are sometimes expressed in terms of the modulus of elasticity $E(T)$, Poisson's ratio $\mu(T)$, and their derivatives in temperature. Thus, the differentiation of Eqs. (2.16) and (2.17) gives (Birger 1957):

$$d\varepsilon_{ij} = \frac{1}{E}\left[(1+\mu)d\sigma_{ij} - 3\mu\delta_{ij}d\sigma\right] - \left\{\frac{1}{E^2}\left[(1+\mu)\sigma_{ij} - 3\mu\delta_{ij}\sigma\right]\frac{dE}{dT} - \frac{\sigma_{ij} - 3\delta_{ij}\sigma}{E}\frac{d\mu}{dT} - \delta_{ij}\frac{d\varepsilon_T}{dT}\right\}dT.$$

2.2.2 Solution Methods

2.2.2.1 Constant Parameters of Elasticity

For the given surface and volume loads and temperature field, the constitutive equations in terms of Eqs. (2.16) or (2.17) together with Eqs. (1.12)–(1.14) of Chap. 1 form a complete system of equations of the linear theory of thermoelasticity, which has been considered by Duhamel (1837), Lebedev (1934), Maysel (1951), Boley and Weiner (1960), and many others. The material properties—the modulus of elasticity E and the Poisson's ratio μ are assumed to be constant for the whole body, whereas the displacements and deformations are considered to be small. If instead of the field ΔT one deals with a field of thermal expansion $\varepsilon_T = \alpha_T(T)\Delta T$, the nature of the problem does not change.

Common analytic approaches to the problems of linear thermoelasticity are mainly identical to the approaches of linear theory of isothermal elasticity and are presented in a series of monographs, for example, Timoshenko and Goodier (1970). A summary of methods for solving thermoelastic problems is presented in Boley and Weiner (1960).

A number of approximate analytical solutions obtained by methods of strength of materials, variational methods, etc., were provided by Melan and Parkus (1953), Gatewood (1957), Zhou (2008), and many others.

2.2.2.2 Variable Parameters of Elasticity

Analytical solutions, containing the variable parameters of elasticity, can be obtained for limited number of problems. General approaches are proposed in Birger (1957), Boley and Weiner (1960), and also in Birger and Shorr (1975). Calculation methods for unevenly heated pre-twisted rods are stated in Shorr (1960a) and Vorob'ev and Shorr (1983), for thin-walled rods—in Shorr (1960b), for stretched strings—in Birger and Mavljutov (1986), etc. (some examples see in App.).

In finite element method (FEM), the temperature dependence of the material elastic characteristics is used in the form of tables.

In calculations, solutions are usually established by successive approximations approach. These methods, in particular, the method of additional strains, have been developed mainly in relation to problems of thermoplasticity (see Chap. 3), but can also be applied to the problems of thermoelasticity.

The Method of Additional Strains

The method of additional strains allows consideration of an equivalent problem with constant elastic parameters together with the so-called additional strains (Birger and Shorr 1975 and others).

Along with the actual strains ε_{ij} of the original elastic body with variable values E and μ (or K, and G), let us consider "reduced" strains ε^r_{ij}, corresponding to the equivalent elastic body with constant values E_r and μ_r (or K_r and G_r) under the same loads and boundary conditions and at the same actual stresses σ_{ij}. The difference between these strains is referred to as the additional strain

$$\varepsilon_{ij}^{\circ} = \varepsilon_{ij} - \varepsilon_{ij}^{r}. \tag{2.24}$$

The strains ε_{ij}° at the moduli G_r and K_r must satisfy the equalities

$$s_{ij} = 2G(\varepsilon_{ij} - \delta_{ij}\varepsilon) = 2G_r\left(\varepsilon_{ij}^{r} - \delta_{ij}\varepsilon^{r}\right). \tag{2.25}$$

and

$$\sigma = K(\varepsilon - \varepsilon_T) = K_r(\varepsilon^r - \varepsilon_T). \tag{2.26}$$

Equations (2.24)–(2.26) yield

$$\varepsilon_{ij}^{\circ} = \left(1 - \frac{G}{G_r}\right)(\varepsilon_{ij} - \delta_{ij}\varepsilon) + \delta_{ij}\left(1 - \frac{K}{K_r}\right)(\varepsilon - \varepsilon_T). \tag{2.27a}$$

Usually, the difference in Poisson's ratio can be neglected, then

$$\varepsilon_{ij}^{\circ} \approx \left(1 - \frac{E}{E_r}\right)(\varepsilon_{ij} - \delta_{ij}\varepsilon_T). \tag{2.27b}$$

Due to Eq. (2.24), the stresses in the equivalent body will be the same as in the original one. After the strain ε_{ij}° is determined, Eq. (2.16) with the additional term ε_{ij}° in the right-hand side corresponding to the constant value E_r leads to the exact solution of the original elastic problem with variable modulus $E(T)$.

The strain ε_{ij}° is calculated using successive approximations

$$\left(\varepsilon_{ij}^{\circ}\right)^{(k)} \approx \left(1 - \frac{E}{E^r}\right)\left[(\varepsilon_{ij})^{(k-1)} - \delta_{ij}\varepsilon_T\right], \tag{2.28}$$

starting with the initial value $\left(\varepsilon_{ij}^{\circ}\right)^{(0)} = 0$ or $(\varepsilon_{ij})^{(0)} = \left(\varepsilon_{ij}^{r}\right)^{(0)}$.

To ensure rapid convergence of solutions, the reduced modulus E_r has to satisfy the condition $E_r(T_r) \geq E(T)$. Once calculated, stresses and strains corresponding to external loads and thermal expansion remain constant for every approximation. The self-equilibrated stress system and the corresponding elastic strains caused by the field of additional strains ε_{ij}° are calculated at each step.

Due to the development of the FEM, the majority of modern practical problems of thermoelasticity can be solved numerically by FEM (see, i.e., Zienkiewicz and Taylor 2000).

2.2.3 Interpretation of Calculation Results

Calculations of thermoelastic stresses and strains which take into account temperature-dependent moduli of elasticity can be used in a variety of ways, for example:

- To reveal the most stressed points of a structure and estimate the loads and temperatures at which plastic deformation in these points can start;
- To estimate a critical point of a structure of a poorly plastic material, where fracture may occur without visible plastic strain. In this case, the maximum principle tensile stress is often used as a criterion of material failure;
- To calculate deformations, which can sometimes essentially influence working capacity of a structure (e.g., in case of elimination of a backlash between stator and rotating parts of a turbomachine or due to distortion of a telescope lens profile);
- To calculate dynamic stress and strain distribution for natural oscillation modes of non-uniformly heated machine parts; and
- To calculate the ranges of cyclic stress defining structure endurance (HCF limit) at forced vibrations.

There are various ways applied in engineering to decrease the thermal stresses.

For example, in gas turbines disks, due to the hot rim at rather cold hub, considerable temperature difference along the radius leads to large tensile stresses at the central part of the disk. The temperature field of the disk is adjusted by heating the disk nave with hot air by the compressor.

Various compensating joints (also known as compensators) are used in the long pipelines in order to absorb longitudinal stresses along the pipes.

Frame structures, in particular, bridge supports, should be rigidly fixed only at one end in order to allow free expansion at heating, etc.

2.3 Energy Principles

2.3.1 Work and Internal Energy of Elastic Deformation

2.3.1.1 Deformation Work

Let us consider a small cubic element of a solid of side dx, hence the face area is given by $dA = dx^2$. The cubic element is affected by the principal stresses σ_k where index $k = 1, 2, 3$. Displacement speeds for each of the main directions $\dot{u}_k = du_k/dt$ vary along the length of the element as $\dot{u}_{k,k}dx$, so that the stresses σ_k in the absence of mass forces (when $\dot{\sigma}_{k,k} = 0$) perform the work per unit time

$$d\dot{W}_{in} = \left[(\sigma_k + \sigma_{k,k}dx)(\dot{u}_k + \dot{u}_{k,k}dx) - \sigma_k\dot{u}_k\right]dA = \sigma_k\dot{u}_{k,k}dxdA = \sigma_k\dot{\varepsilon}_k dV,$$

where dV is the volume of a small element of the body.

Here the time, as usually in statics problems, is understood as the parameter characterizing the scenario of the processes of loading and heating.

In general case work over entire element per unit time, i.e., power will be

$$\dot{W} = \sigma_{ij}\dot{\varepsilon}_{ij}. \tag{2.29a}$$

Having integrated Eq. (2.29a) in time at *constant stresses*, we obtain the work

$$W = \sigma_{ij}\varepsilon_{ij}. \tag{2.29b}$$

Note that the strain ε_{ij}, as well as its speed $\dot{\varepsilon}_{ij}$, includes temperature expansion. Therefore, the work W is equal to all mechanical work, which is done on the element from other parts of the body during loading and heating.

The total work of deformation in case of *proportional increase* of stresses and full strains is

$$W = 0.5\sigma_{ij}\varepsilon_{ij} = 0.5(\delta_{ij}\sigma + s_{ij})(\delta_{ij}\varepsilon + e_{ij}) = 0.5(\sigma\theta + s_{ij}e_{ij}). \tag{2.30}$$

It consists of work $W_\theta = 0.5\sigma\theta$ which occurs when the volume of a body changes and a work $W_s = 0.5s_{ij}e_{ij}$ that corresponds to the changes of a body shape.

2.3.1.2 Internal Mechanical Energy of Elastic Deformation

According to the law of energy conservation, the total work done on an element during a small step loading, in the absence of a heat supply, is equal to the increase of its internal mechanical energy

$$dP_{mech} = dW = \sigma_{ij}d\varepsilon_{ij}. \tag{2.31a}$$

The tensor of full strain increment $d\varepsilon_{ij}$ does not depend on the physical nature of the process, being a purely geometric characteristic of the deformable continuous medium.

Within a finite loading step transferring a body from state 1 to state 2, the work done is

$$P_{mech} = W = \int_1^2 \sigma_{ij}d\varepsilon_{ij}. \tag{2.31b}$$

The integral (2.31b) also does not depend on the physical nature of the process and the type of loading $\sigma_{ij} = f(\varepsilon_{ij})$. In the particular case, when the stress increases proportionally to strain, Eq. (2.31b) becomes

$$P_{mech} = 0.5\sigma_{ij}\varepsilon_{ij}. \tag{2.31c}$$

For *isothermal* loading of an elastic body ($\varepsilon_T = 0$), the internal energy is equal to the potential energy of the deformation P which, according to the equations of elasticity (2.17) and (2.19), can be presented in two equivalent forms:

$$P_{mech} = P = G\varepsilon_{ij}\varepsilon_{ij} + \frac{3}{2}(K - 2G)\varepsilon^2, \tag{2.32a}$$

or

$$P_{mech} = P = \frac{1}{4G}\sigma_{ij}\sigma_{ij} + \frac{3}{2}\left(\frac{1}{K} - \frac{1}{2G}\right)\sigma^2. \tag{2.32b}$$

Obviously, the potential energy of elastic deformation can also be expressed in terms of the Young's modulus E and Poisson's factor μ. The magnitude of P_{mech} is unequivocally defined by the elastic strain reached to the moment and does not depend on the loading path.

For *non-isothermal* loading of an elastic body ($\varepsilon_T \neq 0$) when the heat is supplied to the system, the potential energy of elastic deformation P includes a work done during the temperature expansion, depending on conditions of heating or cooling in loading process, and differs from internal mechanical energy P_{mech}. This fact shall be taken into account while treating the law of energy conservation. The problems of thermoelasticity concerning the energy were considered in detail by Boley and Weiner (1960), Birger (1959a, b), and others.

Basic features of energy relations of the uncoupled theory of thermoelasticity will be illustrated below on simple examples.

2.3.1.3 Work of Thermal Expansion

Let us consider three situations for an elastic rod being loaded with the stress σ_1 and heated with the thermal expansion $\varepsilon_T = \alpha_T \Delta T$. Here, we assume that $\alpha_T = \text{const}$ and $E = \text{const}$.

In the first process, a rod is stretched by stress σ_1, and then, it is heated up; in the second, on the contrary, the rod is first heated up and then stretched; and in the third situation, the heating and stretching processes occur simultaneously. In all three cases, the final quantity of the full strain will be identical and equal to $\varepsilon_1 = \sigma_1/E + \varepsilon_T$.

Work increment will be defined according to Eq. (2.32b) as

$$dW = \sigma_1(t)\left(\frac{d\sigma_1}{E} + d\varepsilon_T\right).$$

In order to find the total work W, it is required to integrate this equation in time parameter t.

In the first case, consisting of steps 0–1 and 1–2

$$W = \int_0^1 \sigma_1(t)\frac{d\sigma_1}{E} + \sigma_1\int_1^2 d\varepsilon_T = \frac{\sigma_1^2}{2E} + \sigma_1\varepsilon_T.$$

In the second case, there is no work done during the first step, resulting in the total work

$$W = \int_1^2 \sigma_1(t)\frac{d\sigma_1}{E} = \frac{\sigma_1^2}{2E}.$$

In the third case, under the assumption of identical rate of loading and heating, we arrive at

$$W = \int_0^2 \sigma_1(t)\left(\frac{d\sigma_1}{E} + d\varepsilon_T\right) = \frac{\sigma_1^2}{2E} + \frac{\sigma_1\varepsilon_T}{2}.$$

Thus, the independence of thermoelastic deformations of loading and heating order that has been specified above in Sect. 2.1.2.2, concerns only the estimation of total strain, but not the energy characteristics.

In a specific case of pure shear, the tangential stresses do not done the work under thermal expansion. This results in the identical amount of work at any sequence of loading and heating.

In papers on non-isothermal elasticity, the process of proportional heating and loading is usually assumed though the authors do not mention this explicitly. In this case, the work done on an element under uniaxial loading and heating can be presented in the following equivalent forms:

$$W = 0.5\sigma_1\varepsilon_1 = 0.5\left(\sigma_1^2/E + \sigma_1\varepsilon_T\right) = 0.5\left(E\varepsilon_1^2 - \varepsilon_T\varepsilon_1\right). \tag{2.33}$$

The work of the *constant* stress σ_1 at the thermal expansion ε_T will be further denoted as W_T and referred to as *the work of thermal expansion.* In the given example,

$$W_T = \sigma_1 \varepsilon_T. \tag{2.34}$$

It follows from comparison of Eqs. (2.33) and (2.34) that the potential energy of deformation, equal to work of elastic strain and expressed in terms of stresses, is defined by the difference

$$P = \frac{\sigma_1^2}{2E} = W - \frac{1}{2} W_T. \tag{2.35}$$

In the theories of thermoelasticity and thermoplasticity, the *additional work* of deformation is defined by the sum

$$P^* = W + \frac{1}{2} W_T. \tag{2.36}$$

It is obvious that $P + P^* = 2W = \sigma_1 \varepsilon_1$. Isothermal loading, characterized by $\varepsilon_T = 0$, yields $P = P^*$; for non-isothermal loading ($\varepsilon_T \neq 0$), these characteristics are different.

Having expressed the works W and W_T consecutively in terms of stresses and strains, and differentiating the obtained expressions accordingly by σ_1 and ε_1, we find that independently from thermal expansion

$$\frac{\partial P^*}{\partial \sigma_1} = \varepsilon_1, \quad \frac{\partial P}{\partial \varepsilon_1} = \sigma_1. \tag{2.37}$$

This justifies the introduction of those two functions.

Energy Balance

The following evident interpretation of maintaining energy balance is demonstrated in Fig. 2.6. The doubled work $2W = \sigma_1 \varepsilon_1$ done on an element is equal to the area of the rectangle consisting of four parts. Parts I and II, each of the area $P = 0.5\sigma_1^2/E$, are equal to the potential energy of elastic deformation of an element; parts III and IV,

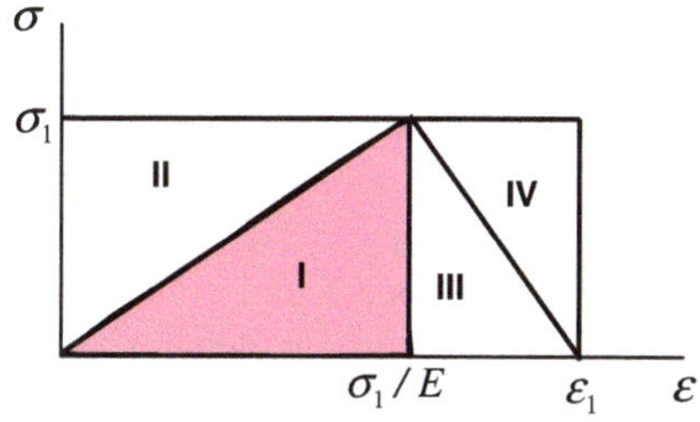

Fig. 2.6 Interpretation of energy balance under non-isothermal loading

each of the area $0.5W_T = 0.5\sigma_1\varepsilon_T$, are equal to half of the work of thermal loading. The sum of parts I or II and III or IV (in any combination) are equal to the total work W, which is in turn equal to the increase of the total internal energy of an element with heating taken into account.

It is obvious from Fig. 2.6 that for $\varepsilon_T \neq 0$, it is impossible to provide equality of the mechanical energy and work $P = W$ which also follows from the given above relations. But the sum of all parts of the work, shown in Fig. 2.6, gives the correct magnitude $2W$. Using the concept of additional work which compensates energy lack $W - P = 0.5W_T$ in P by increasing the energy in P^* at the same quantity $P^* = W + (W - P) = 2W - P$, we obtain in total the necessary energy balance. The magnitude of energy "lack" corresponds to the area of the part III, and additional work is equal to the sum of areas II + III + IV.

2.3.2 *Variational Equations of Isothermal Elasticity*

Variational methods based on energy relationships were widely applied in the "pre-computer" period for solving the problems of isothermal elasticity and then were extended to non-isothermal elasticity. Energy relationships are also often used in theoretical substantiations of FEM in software packages applied to non-isothermal problems, including nonelastic deformations. Therefore, it is useful to provide a brief demonstration of such approaches, taking into consideration the above-mentioned feature of energy relationships for both isothermal and non-isothermal cases.

2.3.2.1 Law of Mechanical Energy Conservation

The basic energy equation in isothermal theory connects the power (capacity) of all external forces applied to a body with the power (capacity) of the internal forces affecting all elements (see e.g., Kachanov 1948)

$$\int_S p_k \dot{u}_k dS + \int_V F_k \dot{u}_k dV = \int_V \sigma_{ij} \dot{\varepsilon}_{ij} dV, \tag{2.38}$$

where p_k, F_k are vectors of surface and volume forces, $\dot{u}_k$ is vector of displacements of the corresponding points, and σ_{ij} and $\dot{\varepsilon}_{ij}$ are tensors of stresses and strain speeds.

This formula makes use of the Eq. (2.29a) for work of internal forces. The integrals are taken over the body surface S and volume V.

Integration of the Eq. (2.38) in the time parameter t leads to

$$W_{S\Sigma} = W_{\Sigma}, \tag{2.39}$$

where

$$W_{S\Sigma} = \int_0^t \left(\int_S p_k \dot{u}_k dS + \int_V F_k \dot{u}_k dV \right) dt \tag{2.40a}$$

is the total work of external forces, and

$$W_{\Sigma} = \int_0^t \int_V \sigma_{ij} \dot{\varepsilon}_{ij} dV dt \tag{2.40b}$$

is the total work of internal forces.

The important feature of Eqs. (2.38) and (2.39) is that they rest only on the conditions of continuity and body equilibrium and therefore remain valid for any mechanical properties of a material.

In statics, assuming $\dot{u}_k dt = du_k$ and $\dot{\varepsilon}_{ij} dt = d\varepsilon_{ij}$, Eq. (2.38) is represented in the form

$$\int_S p_k u_k dS + \int_V F_k u_k dV = \int_V \sigma_{ij} \varepsilon_{ij} dV, \tag{2.41}$$

where the left-hand and right-hand sides of the equation correspond to the double value of the work of external and internal forces, respectively.

The total potential energy of elastic deformation of the body in case of $\varepsilon_T = 0$ is equal to the total work of internal forces, i.e., it is defined as

$$P_{\Sigma} = W_{\Sigma} = \int_V P dV = 0.5 \int_V \sigma_{ij} \varepsilon_{ij} dV. \tag{2.42}$$

Having expressed P according to Eqs. (2.16)–(2.19) in terms of strains or stresses, we arrive to the well-known relations

$$\frac{\partial P}{\partial \varepsilon_{ij}} = \sigma_{ij}, \quad \frac{\partial P}{\partial \sigma_{ij}} = \varepsilon_{ij}. \tag{2.43}$$

In the theory of elasticity, two variation types of Eq. (2.41) are considered.

2.3.2.2 Principle of Admissible Displacements

According to the principle of possible displacements, the kinematically possible (i.e., compatible to boundary conditions) displacements δu_k and deformations $\delta \varepsilon_{ij}$ vary, while external forces p_k, F_k and stresses σ_{ij} remain constant. Then, Eq. (2.41) yields

$$\int_S p_k \delta u_k dS + \int_V F_k \delta u_k dV = \int_V \sigma_{ij} \delta \varepsilon_{ij} dV. \tag{2.44a}$$

or

$$\delta W_{S\Sigma} = \delta W_{\Sigma}, \tag{2.44b}$$

where $\delta W_{S\Sigma}$ and δW_{Σ} are variations of the total work of external and internal forces, respectively.

For $\varepsilon_T = 0$, the magnitude δW_{Σ} is equal to the variation of the body potential energy of elastic deformation δP_{Σ}, expressed in terms of strains according to Eq. (2.32a).

Thus, $\delta W_{S\Sigma} = \delta P_{\Sigma}$ or

$$\delta (P_{\Sigma} - W_{S\Sigma}) = 0. \tag{2.45}$$

The difference $P_{\Sigma} - W_{S\Sigma}$ can be considered as the total potential energy of a body. Equation (2.45) shows that among kinematically admissible displacements of the points of the body, the real displacements correspond to minimum of the total potential energy.

2.3.2.3 Principle of Admissible Changes of a Stress State

According to the principle of possible changes of a stress state, statically possible (i.e., compatible to equilibrium and boundary conditions) stresses $\delta \sigma_{ij}$ and external forces δp_i, δF_i vary, while displacements u_i and strains ε_{ij} remain constant. Then,

$$\int_S u_i \delta p_i dS + \int_V u_i \delta F_i dV = \int_V \varepsilon_{ij} \delta \sigma_{ij} dV \tag{2.46a}$$

or

$$\delta W^*_{S\Sigma} = \delta W^*_{\Sigma}. \tag{2.46b}$$

Equations (2.46a, b) show that variations of "additional" works of external $\delta W^*_{S\Sigma}$ and internal δW^*_{Σ} forces are equal.

For $\varepsilon_T = 0$, variation δW_{Σ}^* standing in the right-hand side of Eqs. (2.46a, b) is equal to a variation of potential energy δP_{Σ}. Analysis similar to that carried out above shows that among statically possible stresses, the real stresses correspond to equality of variations of potential energy of a body and additional work of external forces.

2.3.3 *Variational Equations of Thermoelasticity*

Variational equations of thermoelasticity incorporating the temperature terms were considered in Kachanov (1948), Maysel (1951), and Argyris (1953) and were further developed by Hemp (1954), Birger (1959a), and others.

2.3.3.1 Potential Functions of Thermoelasticity

In general, the *work of thermal expansion* represented in case of uniaxial extension and heating by Eq. (2.34) is defined as

$$W_{T\Sigma} = \int_V \sigma_{ij}\delta_{ij}\varepsilon_T dV. \tag{2.47}$$

Temperature expansion ε_T does not vary, as the temperature field is assumed to be known, but with variation of stress σ_{ij}, the work of thermal expansion varies.

The basic energy Eq. (2.39) for $\varepsilon_T \neq 0$ may be presented as

$$W_{S\Sigma} = W_{\Sigma} = 0.5\left(P'_{\Sigma} + P''_{\Sigma}\right). \tag{2.48}$$

The functions P'_{Σ} and P''_{Σ} are typically quoted as "potential functions of thermoelasticity" (Birger 1959b) defined as

$$P'_{\Sigma} = W_{\Sigma} - 0.5W_{T\Sigma} = 0.5\int_V \sigma_{ij}\left(\varepsilon_{ij} - \delta_{ij}\varepsilon_T\right)dV, \tag{2.49a}$$

$$P''_{\Sigma} = W_{\Sigma} + 0.5W_{T\Sigma} = P'_{\Sigma} + W_{T\Sigma} = 0.5\int_V \sigma_{ij}\left(\varepsilon_{ij} + \delta_{ij}\varepsilon_T\right)dV. \tag{2.49b}$$

Expressing the function P'_{Σ} in terms of the strains by means of Eqs. (2.19), we obtain from Eq. (2.49a)

$$P'_\Sigma == \int_V P' dV = \int_V \left[G\varepsilon_{ij}\varepsilon_{ij} + \frac{3}{2}(K - 2G)\varepsilon^2 - 3K\varepsilon\varepsilon_T \right] dV, \tag{2.50}$$

hence

$$\frac{\partial P'}{\partial \varepsilon_{ij}} = \sigma_{ij}. \tag{2.51}$$

Similarly, the function P''_Σ may be expressed in terms of the stresses through Eq. (2.17)

$$P''_\Sigma = \int_V P'' dV = \frac{1}{2}\int_V \left[\frac{1}{2G}\sigma_{ij}\sigma_{ij} + 3\left(\frac{1}{K} - \frac{1}{2G}\right)\sigma^2 + 3\sigma\varepsilon_T \right] dV, \tag{2.52}$$

resulting in

$$\frac{\partial P''}{\partial \sigma_{ij}} = \varepsilon_{ij}. \tag{2.53}$$

For $\varepsilon_T = 0$, the potential functions of thermoelasticity are $P' = P$, $P'' = P^*$. However, for $\varepsilon_T \neq 0$, these functions contain not only elastic, but also the strains including thermal expansion.

Further, Eq. (2.47) together with Eqs. (2.48) and (2.49a) varies either in strains or stresses.

2.3.3.2 Non-isothermal Principle of Admissible Displacements

According to the principle of admissible displacements, let us consider the variations of kinematically possible displacements δu_i and strains $\delta \varepsilon_{ij}$ under fixed external forces and stresses σ_{ij}. Here, the variation of work of thermal expansion is $\delta W_{T\Sigma} = 0$ and $\delta P'_\Sigma = \delta P_\Sigma$. Then,

$$\delta W_{S\Sigma} = \delta W_\Sigma = \delta P'_\Sigma = 0.5 \int_V \sigma_{ij}\delta\varepsilon_{ij} dV$$

or

$$\delta(P_\Sigma - W_{S\Sigma}) = 0. \tag{2.54}$$

This equation looks similar to Eq. (2.45) and states the condition for minimum of the full potential energy, whereas the stress in the expression for potential energy P_Σ is connected with the full deformations by Eq. (2.19).

2.3.3.3 The Non-isothermal Principle of Admissible Changes of Stress States

As in case of $\varepsilon_T = 0$, we consider the variation of statically admissible stress $\delta\sigma_{ij}$ and external forces δp_i, δF_i, while displacements u_i and strains ε_{ij} remain constant. The variation of additional work of external forces is defined as

$$\delta W_{S\Sigma} = \int_0^t \left(\int_S u_i \delta p_i dS + \int_V u_i \delta F_i dV \right) dt, \qquad (2.55)$$

and the variation of the work of thermal expansion is given by

$$\delta W_{T\Sigma} = \int_V \delta_{ij} \delta\sigma_{ij} \varepsilon_T dV. \qquad (2.56)$$

The stress variation $\delta\sigma_{ij}$ in Eq. (2.52) yields

$$\delta W_{S\Sigma} = \delta W_\Sigma = \delta P''_\Sigma = 0.5 \int_V \varepsilon_{ij} \delta\sigma_{ij} dV$$

or

$$\delta\left(P''_\Sigma - W_{S\Sigma}\right) = 0, \qquad (2.57)$$

Thus, among statically admissible thermoelastic stresses, the real stresses satisfy the equality of variations of additional work (or potential thermoelasticity function P''_Σ of a body) and additional work of external forces.

Under heating, in the absence of external forces when $W_{S\Sigma} = 0$, Eqs. (2.54) and (2.57) enable to determine the thermal SSS. If the state is expressed in strains, the condition of potential energy minimum $\delta P_\Sigma = 0$ is used; if the state is expressed in stresses, then the condition of a minimum of additional work $\delta P''_\Sigma = 0$ is imposed.

Thermoelasticity energy correlations are used in a number of works, see for example, Birger (1959b) and Boley and Weiner (1960).

References

Argyris JH (1953) Thermal stress analysis and energy theorems. Great Britain Air Ministry, Aero. Research Council, p 16

Birger IA (1957) Calculation of nonuniformly heated rods with variable elastic parameters. Trudy Instituta im. Baranova No. 298, Oborongiz, Moscow (in Russian)

Birger IA (1959a) Variational methods in structural mechanics of turbomachines. Oborongiz, Moscow (in Russian)

Birger IA (1959b) Variational equations in theory of thermoelasticity and thermoplasticity. In: Birger IA (ed) Strength and reliability of machine-building structures. Selected works, Ufimsky gosudarstvenny Tekhnologichesky Universitet, Ufa, 1998, pp 65–74 (in Russian)
Birger IA, Mavljutov RR (1986) Strength of materials. Nauka, Moscow (in Russian)
Birger IA, Shorr BF (eds) (1975) Thermal strength of machine parts. Mashinostroenie, Moscow (in Russian)
Boley BA, Weiner JH (1960) Theory of thermal stresses. Wiley, New York
Duhamel JMC (1837) Deuxième memoire sur les phénomènes thermoméсaniques. Journal de l'Ecole Polytechnique 15(25):1–57
Gatewood BE (1957) Thermal stresses with applications to airplanes, missiles, turbines, and nuclear reactors. McGraw-Hill Book Company Inc, New York
Hemp WS (1954) Fundamental principles and theorems of thermoelasticity. Aircraft Eng 26 (302):126–127
Kachanov LM (1948) Mechanics of plastic media. Gostechizdat, Moscow (in Russian)
Lebedev NN (1934) Thermal stresses in the theory of elasticity, vol 2(1). PMM, Moscow (in Russian)
Maysel VM (1951) Temperature problem of the theory of elasticity. Izv. AN USSR, Kiev (in Russian)
Melan E, Parkus H (1953) Wärmespannungen infolge stationäre Temperatur-felder. Springer, Wien
Shorr BF (1960a) To theory of pre-twisted unevenly heated rods. Izv AN SSSR Mekhanika i Mashinostroenie 1:102–112 (in Russian)
Shorr BF (1960b) On the theory of thin-walled pre-twisted rods. Izv AN SSSR Mekhanika i Mashinostroenie 5:74–79 (in Russian)
Timoshenko SP, Goodier JN (1970) Theory of elasticity, 3rd edn. McGraw-Hill, New York
Vorob'ev YuS, Shorr BF (1983) Theory of pre-twisted rods. Naukova Dumka, Kiev (in Russian)
Zhou X (2008) Publications on thermal stresses. J Therm Stresses 31:1146–1147
Zienkiewicz OC, Taylor RL (2000) The finite element method, 5th edn. Butterworth-Heinemann, Oxford

Chapter 3
Thermal Plasticity: Unidirectional Proportional Loading

Abstract This chapter deals with the problems of the theory of thermal plasticity in which, under proportional change of stress tensor components, the plastic deformation increases at a loading and heating, and does not vary at unloading and cooling. The temperature range in which creep can be neglected is considered. Analysis of results of material tests under uni- and multi-axis stress at various temperatures shows that, with reference to the specified brand of the problems, non-isothermal variants of the theory of plastic flow and the deformation theory correctly enough present evolution of plastic deformations. The beginning of the plastic flow or its renewal after the intermediate unloading happens when the stress intensity reaches a generalized non-isothermal surface of yielding. Unlike an isothermal loading, the conditions and rate of plastic strain evolution depend on a relationship of rates of the temperature and stress intensity; in particular, plastic strains can increase under reduce of stresses, but increase of temperature. Rational methods of step-by-step calculation of elastoplastic deformations are stated in application to the non-isothermal loading. As the initial approach, the elastic solution is used at small plastic strains, and the solution for ideally plastic body—at major strains. Instances of such solutions are demonstrated. It is shown that the growth of temperatures at combined stress leads to diminish of a short-term carrying capacity of a structure because of the change of a material strength characteristics, while a role of temperature stresses induced by a non-uniform heating distinctly weakens when plastic strains considerably exceed the thermal expansion.

3.1 Isothermal Plasticity

Plastic deformation of a solid occurring in a temperature field, which could be non-uniform over a structure but remains stationary in time, will be named "isothermal plasticity."

B.F. Shorr, *Thermal Integrity in Mechanics and Engineering*,
Foundations of Engineering Mechanics, DOI 10.1007/978-3-662-46968-2_3

3.1.1 Basic Aspects of Thermal Plasticity

3.1.1.1 Statement of the Problem

In some engineering problems, a structure is rapidly enough heated to required temperature and then affected by gradually increasing loads which initiate the occurrence and monotonic increase of plastic deformations up to failure of a structure. The loading can be interrupted by unloading but if herewith the plastic strain does not vary, the process should be considered as unidirectional.

Such problems are the main issue of this chapter.

Cyclic loading and heating of the machine parts accompanied by repetitive plastic deformation in opposite directions are discussed in Chap. 4. The most general approach to description of material behavior under arbitrary loading is provided in Chap. 5.

3.1.1.2 Proportional Loading

There is a class of unidirectional problems where the outer forces affecting a body increase synchronously and proportionally to each other, whereas the temperature field is stationary. The components of a combined stress in various elements of a body also change synchronously, but in general, not proportionally. However, in many practically important cases, the variation of the stress components can be considered as proportional.

If all components of the stress tensor in a given point increase synchronously, the loading is named "*proportional.*" This typically for statically determinate systems, however, is also possible in more general cases. For example, consider a turbine disk heated at a small rotational speed. After the stationary (usually non-uniform) temperature field is established, the disk will be spun to a given rotational speed. During the trail, the stress distribution over the disk varies, but correlations between stress components in particular points remain constant almost until fracture.

For calculation of strains in the body affected by monotonic proportionally increasing stresses, the simplest approaches of the theory of thermal plasticity can be used.

Proportional loading does not often occur in practice, but in the presence of prevailing unidirectional loading, the deviation from such conditions can be insignificant. If all components of the stress tensor steadily increase, the development of plastic deformation may not strongly differ from conditions of proportional. This allows considering the processes as "quasi-proportional" using the calculation approaches based on experimental data for proportional loading.

3.1.2 Diagrams of Elastoplastic Deformation

3.1.2.1 Tensile Diagram

Basic experimental data for revealing thermoplastic properties of a material are obtained by tensile tests, during which the heated specimens are stretched at constant rate of increase of their length.

The tension force F and the changing gauge length L of the specimen are controlled. Their values are used to calculate a normal stress $\sigma = F/A_0$ and longitudinal strain $\varepsilon = (L - L_0)/L_0$ referred to the initial cross-sectional area A_0 and initial gauge length L_0. The temperature field within this length must be maintained uniform and constant.

Total deformation $\varepsilon(\sigma, T)$ of the initially unloaded specimen, heated from temperature T_0 up to the controlled temperature T, contains thermal expansion given by $\varepsilon_T(T) = \alpha_{Tm}(T - T_0)$, so

$$\varepsilon(\sigma, T) = \varepsilon_\sigma(\sigma, T) + \varepsilon_T(T). \tag{3.1}$$

Excluding thermal expansion, diagrams can be represented as a nonlinear dependence between the stress σ and "mechanical" part of elastoplastic strain $\varepsilon_\sigma(\sigma, T) = \varepsilon(\sigma, T) - \varepsilon_T(T)$ at various temperatures. Further, the subscript (σ) in a description of a mechanical part of total deformation ε_σ will be applied only when necessary.

In Fig. 3.1a, the stress–strain curves $\sigma[(\varepsilon), T]$ are presented due to Zakharova's tests (1966) for a heat-resistant nickel alloy at various constant temperatures T. Along with this typical diagram, the curves $\varepsilon = f[(T), \sigma]$ at various constant stresses σ can be plotted as it is shown in Fig. 3.1b. Diagram $\sigma = f[(T), \varepsilon]$ at

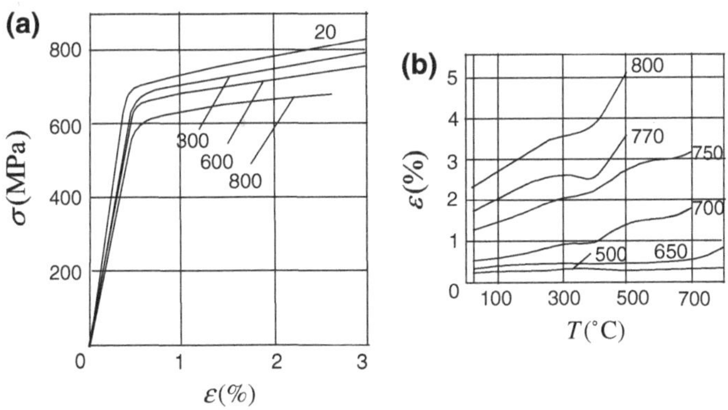

Fig. 3.1 Two types of elastoplastic diagrams with **a** constant values of temperatures T, °C, and **b** constant values of stresses σ, MPa shown for each curve

various constant strains ε can be obtained analogously. Below, the stress–strain curves $\sigma[(\varepsilon), T]$ are designated as $\sigma(\varepsilon, T)$.

In Fig. 3.2, typical deformation diagrams $\bar{\sigma}(\varepsilon, T)$ of a sufficiently plastic titanic alloy are shown in relative form, with stress taken as a ratio $\bar{\sigma} = \sigma/\sigma_{B,0}$ where $\sigma_{B,0}$ is the maximum value of stress (ultimate strength) at normal temperature T_1. Elastic unloading to $\bar{\sigma} = 0$ occurs after the specimen rupture.

Extensive test results for some materials were analyzed by Lode (1926) followed by many other researchers (Fridman 1948, 1974; Kachanov 1948, 1969; Bridgman 1964, and others). The tests conducted on these materials under gradually increasing tension at constant temperature revealed the following typical features:

- The stress increases along with strain but not proportionally. When the stress exceeds some threshold depending on a temperature T, the relation $\varepsilon(\sigma)$ becomes essentially nonlinear: Strain increases significantly faster than stress. Initial parts of the stress–strain curves $\varepsilon(\sigma, T)$ in the standard form $\sigma(\varepsilon, T)$ are shown on an enlarged scale in Fig. 3.3a. Curve 1 demonstrates a flat part of yielding (usually named flow platform), within which the strain grows under almost constant stress, which is typical for pure metals. This stress can be considered as a *trueyield point* σ_y. At the same time, the derivative $E_t = d\sigma/d\varepsilon$ (tangential modulus) on curve 2 decreases monotonously as stresses increase, which is usual for the majority of structural materials.

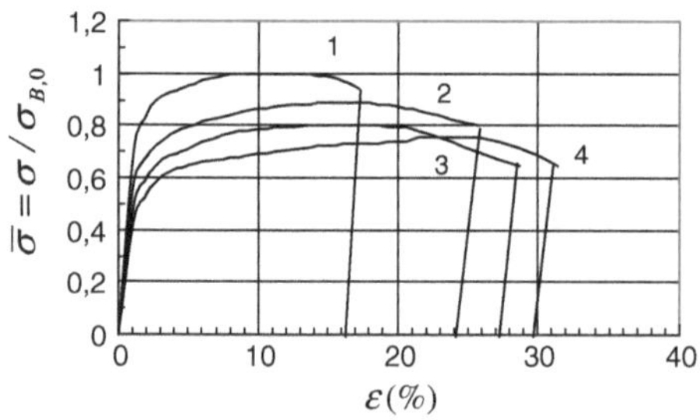

Fig. 3.2 Typical elastoplastic diagrams for a sufficiently plastic alloy at different temperatures: $T_4 > T_3 > T_2 > T_1$

Fig. 3.3 Initial part of elastoplastic diagrams at constant temperature **a** real curves, **b** idealized curve

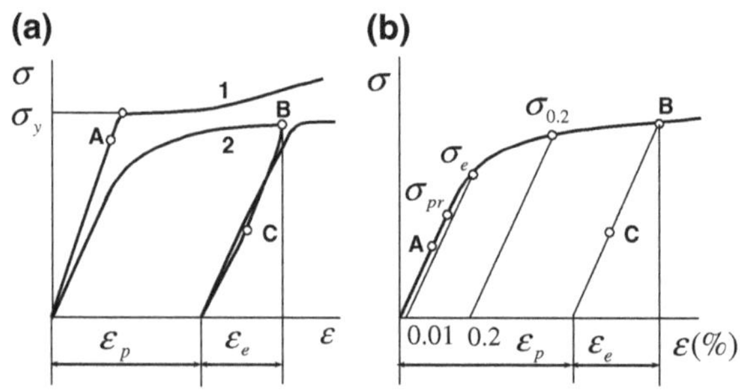

- For a diagram with a well-defined yield point (curve 1), unloading from a point $A(\sigma < \sigma_y)$ results in zero strain, i.e., deformation is completely reversible (elastic). Dependence $\varepsilon(\sigma)$ in this area is linear to the reasonable accuracy, and the modulus of elasticity $E(T) = \sigma/\varepsilon$ for a given temperature is constant.
  For the majority of other materials (curve 2) if unloading starts from point B, the full strain $\varepsilon(\sigma, T)$ decreases by the *elastic* part $\varepsilon_e(\sigma, T) \approx \sigma/E(T)$. The remaining plastic part of a strain so-called as *residual* strain is given by as follows:

$$\varepsilon_p(\sigma, T) = \varepsilon(\sigma, T) - \varepsilon_e(\sigma, T),$$

  Experiments reveal some insignificant differences in slopes for the unloading and subsequent loading lines, shown at point C in Fig. 3.3a. For practical purposes, it is possible to use $\varepsilon_e(\sigma, T) \approx \sigma/E(T)$ in both cases and accept the initial part of the deformation diagram in idealized form as shown in Fig. 3.3b.
- The entire experimental curve $\varepsilon(\sigma)$ can be used as a description of material properties for gradually increasing loads at constant temperatures. However, for loads varying in quantity and direction, and moreover, for varying temperatures, it is necessary to divide a diagram into elastic and plastic parts. These parts are separated by the initial yield point, which can be estimated in various ways.
  A stress at the point where the curve $\sigma(\varepsilon, T)$ starts considerably deviate from the straight line is termed *the limit of proportionality.* At the point named *the limit of elasticity* σ_e, unloading leads to insignificant residual deformation, e.g., $\varepsilon_p(\sigma_e) = 0.01\ \%$. In engineering texts, the *yield point* $\sigma_y = \sigma_{0.2}$ is defined as a stress inducing residual strain of value $\varepsilon_p(\sigma_{0.2}) = 0.2\ \%$. It may be used as a comparative characteristic of materials. The specified points are demonstrated in Fig. 3.3b.
  All these points may be conventionally combined into one, since which it is necessary to take into account plastic yielding. This stress will be quoted as the *initial yield point* σ_{y0} (in some works, an equivalent term, initial proportionality limit, is used). Though the quantity of initial yield point is not precisely defined on diagrams, the rate of the strain development varies rapidly at the border of the specified areas (usually 1.5–2 order), as seen in Figs. 3.1 and 3.2. That allows distinguishing these areas confidently enough.
  The magnitude of the initial yield point σ_{y0} is also used for the simplified bilinear approximation of the deformation diagram.
- If the stress is reduced to zero and then *repeated* loading is applied in the former direction at the same temperature, the dependence $\varepsilon(\sigma, T)$ remains linear to a good approximation, until the stress returns to the value from which the previous unloading has been started. Therefore, the points of deformation curves may be considered as current ("instant") yield points $\sigma_y(\varepsilon_p, T)$. Since these yield points exceed the initial one, $\sigma_y(T) > \sigma_{y0}(T)$, such behavior of a material is referred to *hardening* and used in engineering for expansion of operation area, in which deformation of a machine part under unidirectional loading remains elastic.

Other definitions of the term "hardening" will be given in the next two chapters. The effect of hardening also appears during repeated loading after incomplete unloading (point C in Fig. 3.3).

If stress grows further, deformation increases along the curve almost coinciding with the initial one.

- It is worth separating the elastoplastic problems into two area: the area of relatively small plastic deformations (theoretically at $\varepsilon_\sigma \ll 1$ but practically for $\varepsilon_\sigma < 0.15-0.2$), and the area of the large plastic deformations. At the failure point, plastic deformations can reach tens, and under some conditions of "superfluidity," hundreds of percents. The area of the large plastic deformations is of particular interest for technological processes.

Machine parts usually operate in conditions corresponding to the area of small plastic strains; therefore, the change in dimensions of a structure may be neglected. Some features of deformation curves at the large plastic deformations will be considered below in connection with the short-term strength calculation.

An unloading followed by subsequent loading allows the total strain $\varepsilon(\sigma, T)$ to be presented as the sum of two parts: the recovered after unloading (elastic) and residual (plastic) ones

$$\varepsilon(\sigma, T) = \varepsilon_e(\sigma, T) + \varepsilon_p(\sigma, T). \tag{3.2}$$

In Fig. 3.4, initial portion of deformation curves $\bar{\sigma}(\varepsilon, T)$ for a titanic alloy at several constant temperatures are presented. Black dots mark initial yield points σ_{y0}, and white dots correspond to values of $\sigma_{0.2}$. Stresses $\bar{\sigma}$ are related to ultimate stress at normal temperature $\sigma_{B,0}$. Under increasing temperature, values of σ_{y0} and $\sigma_{0.2}$ decrease, and nonlinear plastic deformation starts at the stress σ_{y0}, which is essentially less than the yield stress $\sigma_{0.2}$.

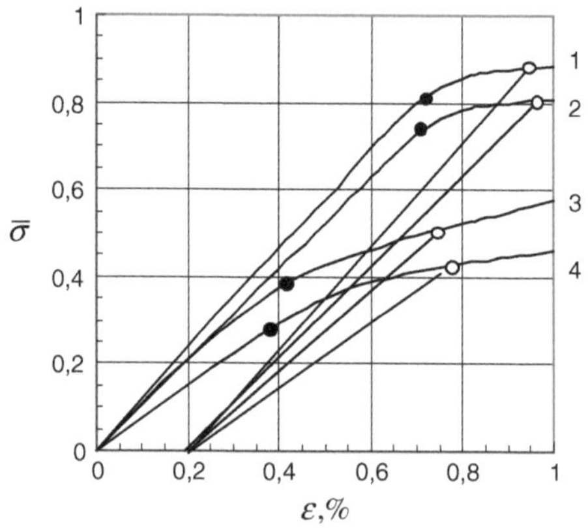

Fig. 3.4 Initial portions of deformation diagrams for a titanium alloy at different temperatures: *1T* = 20 °C, *2T* = 100 °C, *3T* = 300 °C, *4T* = 500 °C

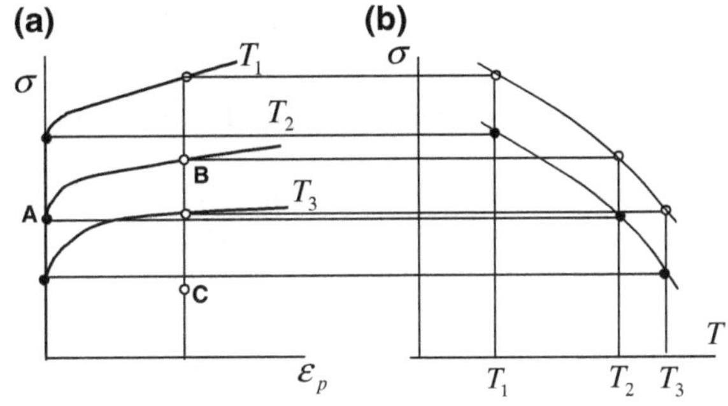

Fig. 3.5 Diagrams $\sigma(\varepsilon_p, T)$ and $\sigma(T, \varepsilon_p)$ at different temperatures

At moderated temperatures, specific for each material, the strain values $\varepsilon(\sigma, T)$ corresponding to given stress and temperature practically do not depend on the duration of their action. Thus, elastoplastic deformation can be considered irrespective of the creep process considered in Chaps. 6 and 7.

Usually, at relatively low speed of specimen's static loading ($d\varepsilon/dt \leq 10^{-4}$ s^{-1}), the dependence $\varepsilon(\sigma, T)$ is not affected by the value of the speed.

The curves $\sigma(\varepsilon, T)$ can be transformed into the curves $\sigma(\varepsilon_p, T)$ excluding the elastic deformations, as shown in Fig. 3.5a. Points B and C in Fig. 3.5a at temperature T_2 correspond to the same points in Fig. 3.3b.

The curves $\sigma(T, \varepsilon_p)$ corresponding to the constant value of ε_p can be obtained from the curves $\sigma(\varepsilon_p, T)$. In Fig. 3.5b, a case $\varepsilon_p = 0$ is demonstrated with black dots, and white dots depict the current value of ε_p.

3.1.2.2 Compression and Torsion Diagrams

Under compression, the cross-sectional area of a specimen gradually increases. Therefore, deformation diagrams for tension and compression incorporating large plastic deformation essentially differ. Under small deformations, the distinction of elastoplastic deformation curves for tension and compression is usually negligible.

Torsion tests allow determining deformation diagrams for shear strain $\gamma(\tau, T)$ versus tangential stress. The thin-walled tubular specimens are typically used for these tests. The tangential stress is given by $\tau = M/2\pi r_m^2 h$, where M is gradually increasing torsion moment, r_m is an average radius, and h is a thickness of the tube wall. The shear strain is given by $\gamma = \Delta\varphi\, r_m/L$, where $\Delta\varphi$ is the angle of a specimen twist within the length L. Initial portion of these diagrams have the same qualitative character as for tension tests, but at the large plastic strains, they may differ Thus, there is no neck in a specimen as its cross-sectional area remains almost constant under torsion.

3.1.3 Generalized Deformation Diagram

3.1.3.1 Plasticity Under Combined Stress States

Throughout many years, a number of researchers have carried out extensive experimental investigations on the study of materials' elastoplastic behavior under the combined stress states (Lode 1926; Schmidt 1932; Odqvist 1933, and others). Historical analysis of the specified works can be found in Il'ushin (1948), Johnson and Mellor (1962), and other publications. Data on mechanical properties of various structural materials under combined stress are provided in Lebedev et al. (2003) and in others handbooks.

Diagrams of elastoplastic deformation for combined stress are usually obtained by trials of thin-walled tubular specimens heated to the required temperature under gradual proportional increase of a tensile force, a torque, and interior pressure in various combinations. In the course of trials, the following quantities are measured: normal (direct) stresses σ_1 and σ_2 in the axial and circumference directions, respectively, linear ε_1 and ε_2 strains in the same directions, tangential stress τ, and shear strain γ in a sample's cross section. These quantities are represented as components of stress and strain tensors, $\sigma_{11} = \sigma_1$, $\sigma_{22} = \sigma_2$, $\sigma_{12} = \tau$ and $\varepsilon_{11} = \varepsilon_1$, $\varepsilon_{22} = \varepsilon_2$, $\varepsilon_{12} = 0.5\gamma$. Other tensors components are assumed be zero.

As material behavior in an isotropic body must not depend on a coordinate system, the deformation diagrams for combined stress (named *generalized* diagrams) should represent dependence between some *invariants* of stresses and strains (see Chap. 1).

3.1.3.2 Maximum Tangential Stress as a Yield Criterion

As plastic deformation occurs through shear (sliding) of material particles, the tangential components of stress tensor should be responsible for the strain development at any stress states. The maximum tangential stress τ_{max} is the simplest invariant which could influence the occurrence and evolution of plastic deformation.

Tresca (1868) and Saint-Venant (1871) suggested that a material plastic flow (yielding) occurs when the maximum tangential stress τ_{max} reaches some critical value—*tangential yield point* τ_y. This idea is also connected with Coulomb's name. In case of ideal plasticity, a quantity $\tau_y = \text{const}$, while in common case, generalized deformation diagram can be presented as a function $\tau_y(\gamma_{max})$. Reliability of such presentation of deformation diagrams in case of proportional loading was confirmed experimentally by Davis (1955) and a number of other researchers.

The tangential stresses acting in corresponding planes are expressed in terms of the principal normal stresses $\sigma_1 \geq \sigma_2 \geq \sigma_3$ through

$$\tau_{12} = 0.5|\sigma_1 - \sigma_2|, \quad \tau_{23} = 0.5|\sigma_2 - \sigma_3|, \quad \tau_{31} = 0.5|\sigma_3 - \sigma_1|. \tag{3.3}$$

The maximum tangential stress is $\tau_{\max} = 0.5|\sigma_1 - \sigma_3|$. Similarly, the shears $\gamma_{12}, \gamma_{23}, \gamma_{32}$ and their maximum value $\gamma_{\max} = |\varepsilon_1 - \varepsilon_3|$ are defined through the principal linear strains $\varepsilon_1 \geq \varepsilon_2 \geq \varepsilon_3$.

Dependence $\tau_y(\gamma_{\max}, T)$ can be directly obtained by using results of torsion tests. Current tangential yield points $\tau_y(\gamma, T)$ are the points lying on a deformation curve.

3.1.3.3 Stress Intensity as a Yield Criterion

Theoretically, the disadvantage of the yield criterion $\tau_{\max}$ is that it does not incorporate the possible influence of tangential stresses τ_{12} and τ_{32}, while the potential energy of shape change P_s depends on all those stresses. The latter follows from the first term of Eq. (2.32b) of Chap. 2 that can be transformed to

$$P_s = \frac{1}{3G}\left(\tau_{12}^2 + \tau_{23}^2 + \tau_{31}^2\right). \tag{3.4}$$

This disadvantage is eliminated if the criterion of plastic deformation is taken in the form of stress intensity σ_i as offered by Huber (1904) and Mises (1913).

Equations (1.35a and 1.35b) of Chap. 1 for stress intensity presented in principal stresses are as follows:

$$\sigma_i = \sqrt{\frac{3}{2}}\sqrt{(\sigma_1 - \sigma)^2 + (\sigma_2 - \sigma)^2 + (\sigma_3 - \sigma)^2} \tag{3.5a}$$

or

$$\sigma_i = \frac{1}{\sqrt{2}}\sqrt{(\sigma_1 - \sigma_2)^2 + (\sigma_2 - \sigma_3)^2 + (\sigma_3 - \sigma_1)^2}. \tag{3.5b}$$

It follows from Eqs. (3.3) and (3.5b)

$$\sigma_i = \sqrt{2}\sqrt{\tau_{12}^2 + \tau_{23}^2 + \tau_{31}^2} = \sqrt{6GP_s}$$

Equation (3.5a) is commonly used in theoretical papers, while Eq. (3.5b) is widely adopted in engineering applications. In a number of works, and particularly in the FEM codes, a value of stress intensity σ_i in the form of Eq. (3.5b) is referred to as *equivalent stress* or "*von Mises stress.*" We shall designate it as σ_{eqv}.

Similar formulas may be written for strains with appropriate multipliers (see Chap. 1).

Strain intensity can be expressed in two equivalent forms:

$$\varepsilon_i = \sqrt{\frac{2}{3}}\sqrt{(\varepsilon_1 - \varepsilon)^2 + (\varepsilon_2 - \varepsilon)^2 + (\varepsilon_3 - \varepsilon)^2}, \tag{3.6a}$$

or

$$\varepsilon_i = \frac{\sqrt{2}}{3}\sqrt{(\varepsilon_1 - \varepsilon_2)^2 + (\varepsilon_2 - \varepsilon_3)^2 + (\varepsilon_3 - \varepsilon_1)^2} \tag{3.6b}$$

which in case of tension becomes

$$\varepsilon_{i0} = \frac{2(1+\mu)}{3}\varepsilon_0. \tag{3.7}$$

where μ is a varying factor of lateral contraction, which coincides with the Poisson's ratio $\mu = \mu_e$ in the elastic area.

A relative "mechanical" change of element's volume in case $\varepsilon_T = 0$ is defined by the geometrical relation $\theta = 3\varepsilon$. Experiments show that after the plastic deformations have occured, the volume change remain elastic. Therefore, in case of tension,

$$(1 - 2\mu_e)\sigma_0/E = (1 - 2\mu)\varepsilon_0,$$

that yields

$$\mu(\bar{\varepsilon}) = 0.5\,[1 - (1 - 2\mu_e)\bar{\varepsilon}], \tag{3.8}$$

where $\bar{\varepsilon}(\sigma_0) = \varepsilon_{0e}/\varepsilon_0 = \sigma_0/E\varepsilon_0(\sigma_0)$.

Behavior of the factor μ with increase of plastic strain $\bar{\varepsilon}_p$ taken in the form $\bar{\varepsilon}_p = \varepsilon_p/\varepsilon_e = (1 - \bar{\varepsilon})/\bar{\varepsilon}$, for some typical values of the Poisson's ratio μ_e is displayed in Fig. 3.6.

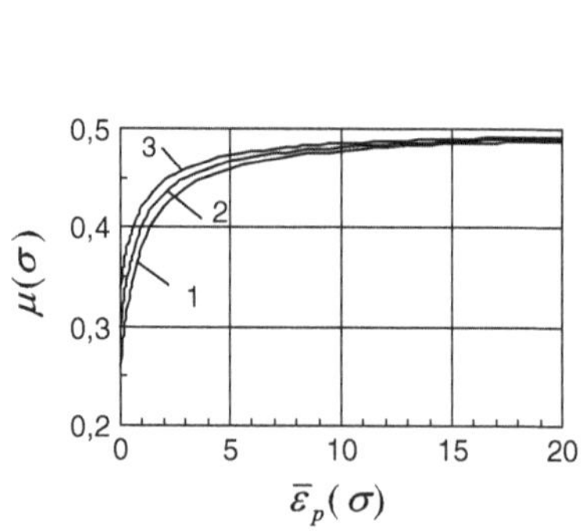

Fig. 3.6 Lateral contraction factor μ versus relative plastic strain; *1* $\mu_e = 0.26$, *2* $\mu_e = 0.3$, *3* $\mu_e = 0.34$

With growth of the ratio $\bar{\varepsilon}_p$, the factor μ tends to a limit $\mu = 0.5$ and the strain $\varepsilon_{i0} \rightarrow \varepsilon_0$ that allows neglecting the difference between curves $\sigma_i(\varepsilon_i)$ and $\sigma_0(\varepsilon_0)$ at $\bar{\varepsilon}_p \gg 1$. Dependence $\mu(\bar{\varepsilon})$ shown in Fig. 3.6 is confirmed by test measurement (see, i.e., Fleming and Temis 2002).

Similar to the stress intensity σ_i, strain intensity ε_i in the form of Eq. (3.6b) is usually referred to as *equivalent strain*, or "*von Mises strain*" ε_{eqv}.

According to Lebedev et al. (2003), the generalized deformation diagrams $\sigma_i(\varepsilon_i, T)$ of a titanic alloy at temperatures 20 °C (curve 1) and 300 °C (curve 2) are presented in Fig. 3.7.

The diagrams are obtained by loading the tubular samples in the axial z and circumferential θ directions with various ratios $K = \sigma_z/\sigma_\theta$ of the stresses in these directions. The principal stresses σ_1 and σ_2 have been taken as $\sigma_1 = \sigma_z$, $\sigma_2 = \sigma_\theta$ for $\sigma_z \geq \sigma_\theta$ and $\sigma_1 = \sigma_\theta$, $\sigma_2 = \sigma_z$ for $\sigma_z \leq \sigma_\theta$; in both cases, it is assumed $\sigma_3 = 0$.

Experimental points show that the deformation diagrams at the combined stress can be presented in the generalized form $\sigma_i(\varepsilon_i, T)$. Analogous data were obtained by Marin and Yu (1956) applying to investigation of biaxial stresses produced by subjecting tubular specimens to internal pressure and axial tension. It is also confirmed by many other contributions.

Justification of diagrams $\sigma_i(\varepsilon_i, T)$ and $\tau_{\max}(\gamma_{\max})$ shows that in some cases, mainly at the beginning of plastic deformation, a relation between intensities σ_i and ε_i satisfies experimental data slightly better. However, in other cases, mainly in the area of large plastic strains prior to failure, a relation between $\tau_{\max}$ and $\gamma_{\max}$ appears more reliable. Accounting for the scatter of experimental characteristics, it is possible to use one or another diagram, observing only the rule: at calculations of safety factors (see Chap. 8), the generalized diagrams should be constructed identically.

At unloadings and repeated loadings, the material behavior is described in the generalized diagram in the same way as has been shown in Fig. 3.3a.

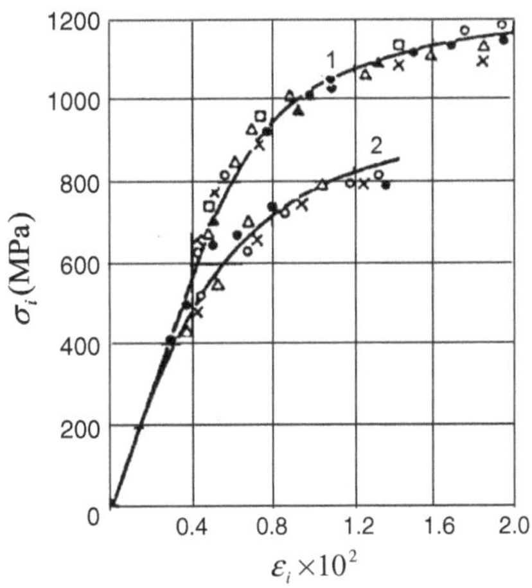

Fig. 3.7 Generalized elastoplastic diagram of a titanium alloy at different temperatures and plane stress states: *1T* = 20 °C, 2 *T* = 300 °C; *solid line* (—) $K = \infty$, *open circle* (○) $K = 2$, *filled circle* (●) $K = 1$, times (×) $K = 0.5$, *open triangle* (Δ) $K = 0$

Odqvist's Parameter

The idea of representing the generalized deformation diagram in the form of dependence between invariants was proposed by Odqvist (1933). Neglecting elastic strains, he introduced the equation $\tau_y = F(I)$, where τ_y is a varying yield point in torsion and (in our notations)

$$I = \int_{t_0}^{t} (\dot{e}^p_{ij}\dot{e}^p_{ij})^{0.5} dt \approx \sum \left(de^p_{ij}de^p_{ij}\right)^{0.5} = \sqrt{3/2}\sum d\varepsilon^p_i. \quad (3.9)$$

The quantity $\varepsilon^p_\Sigma = \sum d\varepsilon^p_i$ is called Odqvist's parameter. In case of proportional unidirectional loading, the quantity ε^p_Σ coincides with the current value of strain intensity ε_i.

3.1.4 Proportional Elastoplastic Loading

3.1.4.1 Branches of Loading Theory

In case of proportional (simple) loading, an orientation of tensor stress in each point of a body remains invariable during all process. In vector representation (see Chap. 1), this means that the direction cosines $\sqrt{3/2}s_{is}/\sigma_i$ and $\sqrt{3/2}ds_{ij}/(d\sigma)_i$ of the stress vector and its increment have equal values. Then

$$\frac{ds_{ij}}{(d\sigma)_i} = \frac{s_{ij}}{\sigma_i}, \quad (3.10)$$

where $\sigma_i = \left(3s_{ij}s_{ij}/2\right)^{0.5}$ and $(d\sigma)_i = (3ds_{ij}ds_{ij}/2)^{0.5}$ are magnitudes of stress vector and its increment, respectively.

Proportional loading is assumed here unidirectional.

The theory of proportional elastoplastic loading has two main branches:

- *The deformation theory of plasticity* proposed by Hencky (1925) and Nadai (1931) for ideal plasticity and developed by Il'ushin (1943, 1948) as *the theory of small elastoplastic deformation* for hardening materials;
- *The theory of plastic flow* (yielding) established in different formulations by Saint-Venant (1871), Mises (1913), Prandtl (1920), Reuss (1930), Kachanov (1948, 1969), and others.

3.1.4.2 Deformation Theory

Strains evolution at monotonous proportional increase of stresses without unloading, the so-called *active* loading, is completely described by generalized deformation diagram $\sigma_y = \sigma_i = f(\varepsilon_i)$. In this case, there is no need to separate strains on elastic and plastic parts and consider specific conditions of plastic strains arising.

It is assumed that deviators' components of strain and stress tensors are proportional

$$e_{ij} = \lambda s_{ij}, \quad \lambda > 0, \tag{3.11}$$

i.e., these tensors presented in vector forms are coaxial $\boldsymbol{\varepsilon}_i \| \boldsymbol{\sigma}_i$.

A scalar λ depends on increasing strain intensity $\varepsilon_i = \left(2e_{ij}e_{ij}/3\right)^{0.5}$ (or on work of shape change $W_s = \int s_{ij} de_{ij}$) and constant temperature.

Presenting stress intensity due to Eq. (3.5a) as $\sigma_i = \left(3s_{ij}s_{ij}/2\right)^{0.5}$, we obtain $\lambda = 3\varepsilon_i/2\sigma_i$ and

$$e_{ij} = \frac{3\varepsilon_i}{2\sigma_i} s_{ij}. \tag{3.12}$$

Equation for strain and stress vectors coaxiality in the form identical to Eq. (3.12) was obtained by Hencky (1925).

Experimental curves $\sigma_y = \sigma_i(\varepsilon_i, T)$ like those shown in Fig. 3.7 may be used to obtain a relation

$$E_s(\varepsilon_i, T) = \frac{\sigma_y}{\varepsilon_i}, \tag{3.13}$$

where $E_s(\varepsilon_i, T)$ is a varying *secant modulus,* which is a function of strain intensity ε_i (or W_s) and temperature T.

Then

$$\sigma_y = \sigma_i = E_s(\varepsilon_i, T)\varepsilon_i. \tag{3.14}$$

From Eqs. (3.10) and (3.12), it follows

$$\frac{de_{ij}}{(d\varepsilon)_i} = \frac{e_{ij}}{\varepsilon_i}, \tag{3.15}$$

i.e., during the overall active process, the orientation of stress and strain tensors and their increments in each point of the body is identical and remains constant.

The constitutive equation of the deformation theory of thermoplasticity including mean stress and thermal expansion is

$$\varepsilon_{ij} = \frac{3}{2E_s}(\sigma_{ij} - \delta_{ij}\sigma) + \delta_{ij}\left(\frac{\sigma}{K} + \varepsilon_T\right). \tag{3.16}$$

Equation (3.16) becomes identical to Eq. (2.16b) of the thermoelasticity provided that the elasticity modulus $E(T)$ is changed for a reduced modulus $E_{red}(\varepsilon_i, T)$ depending on the strain

$$E_{red}(\varepsilon_i, T) = \frac{2(1+\mu)}{3}E_s(\varepsilon_i, T) \approx E_s(\varepsilon_i, T). \tag{3.17}$$

The value of E_{red} can be determined by successive approximations.

3.1.4.3 Solution Methods

For calculation of isothermal plastic deformations by the method of successive approximations, the solution of a corresponding thermoelastic problem with necessary addition terms is usually used. The principle of "elastic solution" was offered by Il'ushin (1943). Since then more convenient ways for practical application were developed.

Method of Variable Elasticity Parameters (Birger 1964)

The sequence of calculation due to the given method is illustrated by Fig. 3.8a. Having carried out the first "elastic" calculation at $E_*^{(0)} = E$, we obtain values $\sigma_i^{(1)}$ and $\varepsilon_i^{(1)}$ (point B), find on the deformation curve the value of the current yield stress $\sigma_y^{(1)}$ at the same strain $\varepsilon_i^{(1)}$, and then calculate the secant modulus $E_s^{(1)} = \sigma_y^{(1)}/\varepsilon_i^{(1)}$ and corrected one $E_*^{(1)}$. Calculation comes to an end when for each plastically

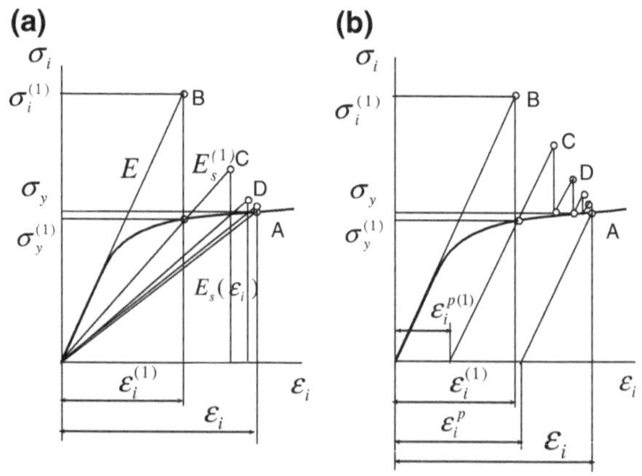

Fig. 3.8 Calculation procedure of deformation theory: **a** method of variable elasticity parameters, **b** method of additional deformation parameters

deformed element of a body the difference $\sigma_i^{(n)} - \sigma_y^{(n)}$ becomes less than the given value (point A in Fig. 3.8a).

Method of Additional Deformation Parameters

As in the theory of thermoelasticity (see Sect. 2.2.2.2), the constitutive Eq. (3.16) of thermoplasticity can be written in the form of Eq. (2.17) with constant values of the elasticity modulus but with additional plastic strain ε_i^p or related additional stress.

A real value of ε_i^p is defined as a difference between full strain ε_i and elastic strain σ_i/E at identical real stress $\sigma_i = \sigma_y$, corresponding to the deformation diagram (Fig. 3.8b)

$$\varepsilon_i^p = \varepsilon_i - \frac{\sigma_y(\varepsilon_i^p)}{E}. \tag{3.18}$$

The first approximate value of $\varepsilon_i^{p(1)}$ can be obtained using $\varepsilon_i^{p(1)} = \varepsilon_i^{(1)} - \sigma_y^{(1)}(\varepsilon_i^{(1)})/E$ where $\varepsilon_i^{(1)}$ is a strain to the elastic solution. A difference $\sigma_i^{(1)} - \sigma_y^{(1)}$ for the same strain $\varepsilon_i^{(1)}$ can be named additional stress. A quantity $\varepsilon_i^{p(1)}$ is "specific" additional strain (Birger 1963; Shorr 1974) unlike the theoretical additional strain corresponding to elastic stress $\sigma_i^{(1)}$. Using the latter for real deformation diagrams might be inefficient and even impossible.

The further calculation procedure is illustrated by Fig. 3.8b.

Convergence problems of various computation procedures of successive approximations in application to the "elastic solutions" of thermoplasticity was considered by Temis (1980).

Method of Additional Volume Deformations (Shorr 1960)

If the solution of the thermoplasticity problem for an incompressible body is known (see, i.e., Sokolovsky 1969), compressibility can be considered by the method of additional volume deformations by consecutive approximations. We will define the resultant thermal expansion as follows:

$$\varepsilon_T^* = \varepsilon_T + (1 - 2\mu)\sigma/E \tag{3.19a}$$

Having solved the problem in thermoplasticity setting $\varepsilon_T^{*(1)} = \varepsilon_T$, we find distribution of stresses $\sigma_{ij}^{(1)}$ and their mean values $\sigma^{(1)}$ and calculate a new value

$$\varepsilon_T^{*(2)} = \varepsilon_T + (1 - 2\mu)\sigma^{(1)}/E \tag{3.19b}$$

and so on.

It has been shown that this calculation procedure converges even in the case of the purely elastic problems and for elastoplastic problems provides very fast convergence.

3.1.4.4 Theory of Plastic Yielding

Considering a loading process in increments, full strain tensor components are represented as sum of elastic and plastic parts and temperature expansion

$$d\varepsilon_{ij} = d\varepsilon_{ij}^e + d\varepsilon_{ij}^p + \delta_{ij} d\varepsilon_{ij}^T. \tag{3.20}$$

Dependence of the elastic strain increment on stresses and temperature has been considered in Chap. 2.

Increments of plastic strain components $de_{ij}^p = d\varepsilon_{ij}^p$ are assumed to be proportional to the corresponding components of the stress deviator

$$d\varepsilon_{ij}^p = s_{ij} d\lambda. \tag{3.21}$$

where a scalar $d\lambda$ depends on increasing hardening parameter $\chi(\varepsilon_\Sigma^p)$ or $W_s^p = \int s_{ij} de_{ij}^p$ and temperature T.

The value of plastic strain increment intensity is

$$(d\varepsilon^p)_i = \sqrt{\frac{2}{3} d\varepsilon_{ij}^p d\varepsilon_{ij}^p} = d\lambda \sqrt{\frac{2}{3} s_{ij} s_{ij}} = \frac{2}{3}\sigma_i d\lambda, \tag{3.22}$$

and

$$d\varepsilon_{ij}^p = \frac{3 s_{ij}}{2\sigma_i} (d\varepsilon^p)_i. \tag{3.23}$$

According to Eqs. (3.21)–(3.23), plastic deformations grow in the direction, in which the stress tensor acts (the vectors are coaxial, $d\boldsymbol{\varepsilon}_i \| \boldsymbol{\sigma}_i$). This is analogous to viscous liquid flow. Therefore, this approach is sometimes called "the theory of plastic flow."

Under loading when $d\sigma_i = d\sigma_y$, a rate of plastic strain intensity (an increment $d\varepsilon_i^p$) can be determined from the generalized deformation diagram using current value of plasticity modulus E_p

$$d\varepsilon_i^p = \frac{d\sigma_i}{E_p(\varepsilon_i^p, T)}. \tag{3.24}$$

where

$$\frac{1}{E_p} = \frac{1}{E_t} - \frac{1}{E} \tag{3.25}$$

or

$$E_p = \frac{EE_t}{E - E_t}. \tag{3.26}$$

If the plastic strains considerably exceed elastic ones, then $E_t \ll E$ and $E_p \approx E_t$.

In case of isothermal proportional loading, calculation results derived by means of deformation theory and theory of plastic yielding completely coincide. In case of non-isothermal proportional loading, the latter allows analyzing heating and cooling influence on strain alteration for material stressed state (see below).

In case of non-proportional loading, both approaches do not take into account increase of unsteady plastic strain rate, but the theory of yielding qualitatively reflects a various direction of elastic and plastic strain increments (see Chap. 5).

3.2 Non-isothermal Plasticity

In common cases, outer loads and temperature of a body do not change synchronously and they may either increase or decrease. For instance, many parts of machines and energy plants at the beginning of their work are subjected to rapid loading by internal pressure or centrifugal forces. The temperature induced by gradual heating raises slower. During unloading, a rapid decrease of pressure and rotational speed take place, followed by a gradual cooling.

3.2.1 Uniaxial Non-isothermal Stressed State

3.2.1.1 Non-isothermal Loading and Unloading

Figure 3.9a demonstrates that a non-isothermal stress state σ_0, T in a point A can be reached either by heating from normal temperature T_0 to T at $\sigma_0 = 0$ and then loading to stress $\sigma_0 = \sigma_y^A$ (curve OA) or by loading to stress $\sigma_0 = \sigma_y^A$ and then heating to temperature T (the pass OFA). In both cases, temperature and stress induce increase of plastic strain, and an "active" non-isothermal loading takes place.

The elementary assumption used in the theory of thermoplasticity states that under active non-isothermal loading, the resulting full ε_0^A and plastic ε_{p0}^A strains are identical for any sequence of loading and heating.

During further active loading, the plastic strain ε_{p0}^B at a point B referring to temperature $T' = T + \Delta T$ and stress $\sigma_0' = \sigma_0 + \Delta\sigma_0$ can be obtained, as shown in Fig. 3.9a, by transition from point A either through the path ACB (at first stress is increased in $\Delta\sigma_0$, and then heating in ΔT occurs), or through the trajectory ADB (heating is followed by additional loading). The results are assumed to be identical.

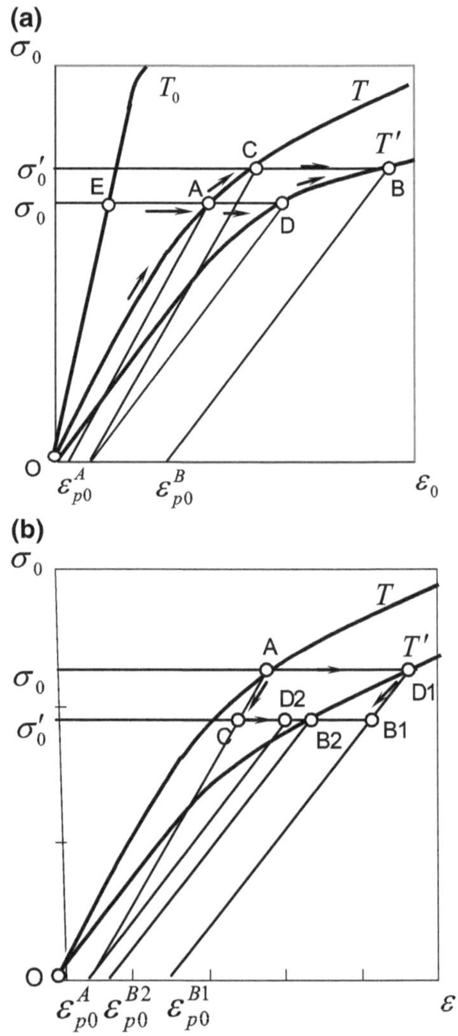

Fig. 3.9 Transition from one state of temperature-stressed conditions into another one **a** $\Delta\sigma_0 > 0$ and $\Delta T > 0$; **b** $\Delta\sigma_0 < 0$ and $\Delta T > 0$

Under simultaneous stress and temperature reduction ($\Delta\sigma_0 < 0$ and $\Delta T < 0$), the plastic strain remains invariable, and the sequence of cooling and unloading does not usually play any role.

Different situations occur if stress and temperature change in opposite directions, for example, stress decreases, $\Delta\sigma_0 < 0$, while temperature rises, $\Delta T > 0$ (Fig. 3.9b). If transition from point A with σ_0 and T to point B1 with σ_0' and T' occurs through the line AD1B1 (i.e., a body is first heated to $T' = T + \Delta T$ at $\sigma_0 = \text{const}$, and then stress decreases to $\sigma_0' < \sigma_0$) then the plastic strain increases at heating to the value ε_{p0}^{B1} and remains the same during unloading. However, if unloading to $\sigma_0' < \sigma_0$ occurs at T = const (through line *AC*), followed by the subsequent heating (line CD2B2), the plastic strain grows to ε_{p0}^{B2} only along the site D2B2. In some cases, depending on the character of the deformation curves at different temperatures, plastic strain does not increase at all (when points B2 and D2 in Fig. 3.9b will coincide).

Plastic strain development under processes at $\Delta\sigma_0 > 0$ and $\Delta T < 0$ is also indeterminable.

As steps of $\Delta\sigma_0$ and ΔT decrease, the points B1 and B2 approach each other. Considerable distinction in calculated results is shown for significant changes in stresses and temperatures. Usually, smooth changes of stress and temperature can be presented as functions $\sigma_0(t)$ and $T(t)$, depending on some formal quantity *t*. Then, a real loading/heating trajectory for given material characteristics $E_t = \partial\sigma_y/\partial\varepsilon_0$ and $E_T = \partial\sigma_y/\partial T$ will be a function of relation of derivatives $d\sigma_0/dt$ (or $d\varepsilon_0/dt$) and dT/dt.

3.2.1.2 Surface of Non-isothermal Plastic Deformation

Since plastic yielding does not depend on a way deformation occurs during *active non-isothermal,* a continuous function in the form $\varepsilon_{p0} = f(\sigma_0, T)$ can be introduced. This function geometrically corresponds to some surface of non-isothermal plastic deformation (Birger and Dem'yanushko 1968; Birger and Shorr 1975, and others). Similar to the stress–strain diagrams for isothermal deformation, this surface can be

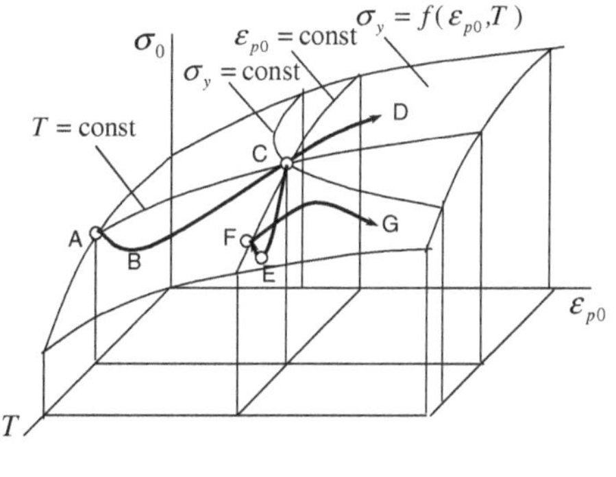

Fig. 3.10 Surface of non-isothermal plastic deformation

presented in the form $\sigma_y = f(\varepsilon_{p0}, T)$ as shown in Fig. 3.10. Lines of intersection of the planes $\sigma_y = \text{const}$, $\varepsilon_{p0} = \text{const}$, and $T = \text{const}$ with this surface are also demonstrated. The usual deformation diagrams shown above in Fig. 3.4 represent the lines of intersection of the planes $T = \text{const}$ with the deformation surface.

Plastic deformation starts at some initial temperature T_{in} at a point A where the stress σ_0 reaches the value $\sigma_{y0}(T_{in})$. Then, the "representing" point with varying coordinates σ_0, T_0 stays on a deformation surface during overall active loading, moving in the direction of the plastic strain ε_{p0} increase, for example, along the line ABCD.

In this case, $\sigma_0 = \sigma_y$ and $d\sigma_0 = d\sigma_y$, where $d\sigma_y$ is the full differential

$$d\sigma_y = \frac{\partial \sigma_y}{\partial \varepsilon_{p0}} d\varepsilon_{p0} + \frac{\partial \sigma_y}{\partial T} dT. \tag{3.27}$$

A plasticity modulus $E_p = \partial \sigma_y / \partial \varepsilon_{p0}$ represents a usual derivative along the curve $\sigma_y(\varepsilon_{p0})$ at $T = \text{const}$, and a "temperature" modulus $E_T = -\partial \sigma_y / \partial T$—a derivative along the line $\varepsilon_{p0} = \text{const}$ lying on a deformation surface.

The modulus E_T can be defined by curves $\sigma_y(T)$ at constant values ε_{p0} presented above in Fig. 3.5b. For structural materials, the specified derivatives always have opposite signs.

If in any point B at $\sigma_0 = \sigma_y$, simultaneous change of stress and temperature occurs such that the full stress increment is less than the yield point increment $d\sigma_0 < d\sigma_y$, the representing point departs from the yield surface, for example, along the line CE, that corresponds to elastic unloading at $\varepsilon_{p0} = \text{const}$. During this process, the plastic deformation remains constant, until the representing point will again reach the yield surface (point F in Fig. 3.10)—certainly, if it occurs.

Under isothermal loading, the renewed plastic flow always continues the interrupted process. Unlike this, the renewed non-isothermal process, on the contrary, is bonded with the interrupted one only by the fact that their representing points lie on the general yield surface. However, the representing points themselves, as well as the directions of their further movement along the yield surface, in a common case do not coincide (compare curves FG and CD).

3.2.2 Multi-axial Non-isothermal Stressed State

3.2.2.1 Generalized Elastoplastic Characteristics

A concept of a surface of non-isothermal plastic deformation presented in Fig. 3.10 is naturally generalized on a combined stress state by substituting stress σ_i and plastic strain ε_i^p intensities instead of stress σ_0 and plastic strain ε_{p0} in case of uniaxial loading. So it can be assumed that for each material, a *generalized surface of non-isothermal plastic deformation* can be established in the form of $f(\sigma_y, \varepsilon_i^p, T) = 0$ and that conditions of plastic strain proportional evolution may be defined by changing the position of a representing point σ_i, ε_i^p, T relatively to this surface.

In the present chapter, we consider a *non-isothermal proportional loading* at a combined stress state. This process is characterized by the simultaneous change of stress and temperature under condition that any of the stress tensor components in the given point of a body do not change the sign (but the signs of stress tensor components may be different). Under the specified condition, a process allows the possibility of proportional partial or full unloading to $\sigma_i = 0$. Moreover, all changes of the stress and strain tensor components must occur smoothly.

Note, that in the more general statement (see Chap. 5), proportional loading can occur at simultaneous changes of signs (but not magnitudes) of the ratios s_{ij}/σ_i for all points of a body, i.e., loading might be not active.

Theoretical analysis of non-isothermal proportional loading is based on the same assumptions as taken in the theory of plastic flow, see Eqs. (3.20)–(3.24).

From the data presented in Chap. 2, it follows that an increment of total deformation work done on elementary volume $dW = \sigma_{ij} d\varepsilon_{ij}$ is formed of the work of volume change $dW_\theta = \sigma d\theta$, which is always defined by elastic deformation and the work of shape change $dW_s = s_{ij} de_{ij} = \sigma_i d\varepsilon_i$.

The work dW_s may be presented as a sum of the stress works upon elastic and plastic strain increments. In case of proportional loading

$$dW_s = dW_s^e + dW_s^p = \sigma_i (d\varepsilon_i^e + d\varepsilon_i^p). \tag{3.28}$$

Work of plastic deformation for the entire history of loading and heating is resulted from integration of the second term of Eq. (3.28)

$$W_s^p = \int_0^t \sigma_i d\varepsilon_i^p. \tag{3.29}$$

The quantity W_s^p, as well as Odqvist's parameter ε_Σ^p that is often called the *accumulated plastic strain,* to a certain degree reflects development of material hardening and damaging through the non-isothermal proportional loading. According to the tension diagram, the quantity ε_Σ^p coincides with the current plastic strain ε_p, increasing within range $0 \le \varepsilon_\Sigma^p \le \delta$, and the maximum quantity W_s^p coincides with the work accumulated in the material till the moment of rupture,

The impossibility of using these quantities as general criteria of material hardening and damaging accumulation is discussed in Chaps. 4 and 5.

3.2.2.2 Non-isothermal Loading and Unloading

Sign-constant loading together with the concept of generalized plastic strain surface must satisfy the following conditions (Birger and Dem'yanushko 1968; Kachanov 1969):

Plastic strain can appear if a representing point belongs to this surface, i.e.,

$$\sigma_i = \sigma_y(\varepsilon_i^p, T). \tag{3.30}$$

As in case of tension, a process of loading/unloading under constant temperature is determined by the *increment of stress intensity* which is given by

$$d\sigma_i = d\left[(3s_{ij}s_{ij}/2)^{0.5}\right] = \frac{3s_{ij}ds_{ij}}{2\sigma_i} \tag{3.31}$$

for active loading, $d\sigma_i = (d\sigma)_i = (3ds_{ij}ds_{ij}/2)^{0.5}$.

Under *active or neutral non-isothermal loadings,* the representing point remains on the generalized plastic strain surface, hence:

$$d\sigma_i = d\sigma_y(\varepsilon_i^p, T), \tag{3.32a}$$

where due to Eq. (3.27)

$$d\sigma_y = \frac{\partial\sigma_y(\varepsilon_i^p, T)}{\partial\varepsilon_i^p} d\varepsilon_i^p + \frac{\partial\sigma_y}{\partial T}(\varepsilon_i^p, T)dT, \tag{3.32b}$$

or

$$d\sigma_y + E_T dT = E_p d\varepsilon_i^p. \tag{3.32c}$$

Moduli $E_p(\varepsilon_{p0}, T) = \partial\sigma_y/\partial\varepsilon_i^p$ and $E_T(\varepsilon_{p0}, T) = -\partial\sigma_y/\partial T$ are determined from 1D deformation diagrams at various temperatures.

An increment of plastic strain intensity is

$$d\varepsilon_i^p = \frac{1}{E_p}(d\sigma_i + E_T(\varepsilon_i^p, T)dT), \tag{3.33a}$$

that yields due to Eq. (3.23)

$$d\varepsilon_{ij}^p = \frac{3s_{ij}}{2\sigma_i E_p}[d\sigma_i + E_T(\varepsilon_i^p, T)dT]. \tag{3.33b}$$

Under active non-isothermal loading, $d\varepsilon_i^p > 0$ and hence Eq. (3.33a) yields

$$d\sigma_i + E_T(\varepsilon_i^p, T)dT > 0. \tag{3.34}$$

Neutral non-isothermal loading is characterized by $d\varepsilon_i^p = 0$ and therefore

$$d\sigma_i + E_T(\varepsilon_i^p, T)dT = 0 \tag{3.35}$$

This situation is possible if stress and temperature increments have different signs. In this case, the stress can grow by the same magnitude as the yield point, with the latter increasing due to temperature decay (and visa versa).

Under *non-isothermal unloading,* the representing point departs from the plastic strain surface, $d\sigma_i < d\sigma_y$ at $d\varepsilon_i^p = 0$ and Eq. (3.32c) yields

$$d\sigma_i + E_T(\varepsilon_i^p, T)dT < 0. \tag{3.36}$$

Non-isothermal conditions of plastic deformation essentially differ from the conditions in case of constant temperature. Plastic deformation can develop not only under stress increase, but also under its reduction if temperature growth considerably lowers the yield stress. On the other hand, unloading at essential temperature reduction might begin under stress increase. For the same reasons, neutral loading is possible at stress changes.

A partial derivative of strain under plastic yielding at $d\sigma = 0$

$$\beta = \left.\frac{\partial \varepsilon_0}{\partial T}\right|_{d\sigma_0 = 0} = \left.\frac{\partial \varepsilon_y}{\partial T}\right|_{d\sigma = 0} \tag{3.37}$$

can be defined from the experimental curves $\varepsilon_0 = f(\sigma_0, T)$ shown in Fig. 3.1b using $\varepsilon_i = \varepsilon_y$. As

$$\left.\frac{\partial \varepsilon_i}{\partial T}\right|_{d\sigma = 0} = \frac{\partial \varepsilon_i^p}{\partial T} + \frac{\partial \varepsilon_i^e}{\partial T} = \frac{\partial \varepsilon_i^p}{\partial T} - \frac{\sigma}{E^2}\frac{dE}{dT},$$

then

$$d\varepsilon_i^p = \frac{\partial \varepsilon_i^p}{\partial \sigma_i} d\sigma_i + \frac{\partial \varepsilon_i^p}{\partial T} dT = \frac{d\sigma_i}{E_p} + \left(\beta + \frac{\sigma_y}{E^2}\frac{dE}{dT}\right) dT, \tag{3.38}$$

where $E_p = \partial \sigma_i / \partial \varepsilon_i^p$.

Equation (3.38) allow obtaining experimental characteristics of thermal plasticity using deformation diagrams $\varepsilon_0(\sigma_0, T)$ and $E(T)$, available in reference books.

3.2.2.3 Solution Methods

The basic equation of the non-isothermal plastic flow theory, Eq. (3.33b), can be presented (Birger 1964) as

$$d\varepsilon_{ij}^{p} = [F_{\sigma}(\sigma_i, T)d\sigma_i + F_T(\sigma_i, T)dT]s_{ij}, \tag{3.39}$$

where

$$\begin{aligned} F_{\sigma}(\sigma_i, T) &= \frac{3}{2\sigma_i E_p}, \\ F_T(\sigma_i, T) &= F_{\sigma}(\sigma_i, T)E_T. \end{aligned} \tag{3.40}$$

Under unloading is $F_{\sigma} = F_T = 0$.

In order to formulate the constitutive equations for the theory of thermal plasticity, Eq. (3.39) must be summarized with Eq. (2.22a) of the thermoelasticity presented in Chap. 2. As a result, we obtain an expression

$$\begin{aligned} d\varepsilon_{ij} = {} & 0.5a_G ds_{ij} + \delta_{ij}a_K d\sigma + \delta_{ij}d\varepsilon_T + 0.5s_{ij}da_G + \delta_{ij}\sigma da_K \\ & + (F_{\sigma}d\sigma_i + F_T dT)s_{ij} \end{aligned} \tag{3.41}$$

As well as in case of thermoelasticity, the total system of the differential equations of thermal plasticity includes Eqs. (1.12)–(1.14) of Chap. 1 for the given surface and volume forces and known temperature field. Loads and temperature changes should satisfy the restrictions referred to the active loading.

Under step-by-step integration of the system of equations of thermal plasticity, the elastoplastic stress/strain state of the body at the beginning of each step is assumed to be known. This allows to verify conditions, mentioned in Eqs. (3.30) and (3.34), in every point of the body. In those points where they are satisfied, the representing "state point" lies on the generalized surface of plastic deformation and the plastic yielding will continue. In case of simultaneous reduction of all external forces and temperatures, the whole body will be elastically unloaded.

In the points where the condition (3.30) is not satisfied at the beginning of the current integration step, deformations were elastic. The further character of the process will depend on stress and temperature increments.

For varying temperatures, the transition from curve $\sigma_y(T)$ to curve $\sigma_y(T + \Delta T)$ occurs in a way that has been shown above in Fig. 3.9.

The method of additional deformations is more reliable since it is based on the solution of the linear elastic problem. The terms containing temperature-dependent elasticity parameters can be included in the additional strains. Then, the primary problem of thermoelasticity with constant values of elasticity parameters is solved.

The plastic strain increments presented in the form of Eqs. (3.33b) or (3.39) are considered as the increments of additional strains.

Computation in each step is carried out by successive approximations.

3.2.2.4 Additional Remarks

In a number of works on thermoplasticity (Prager 1956; Boley and Weiner 1960, etc.), a surface of plastic flow (yielding) in 6D space $f(s_{ii}, \varepsilon_{ij}^p, T) = 0$ instead of a surface of non-isothermal plastic deformation in 3D space $F(\sigma_i, \varepsilon_i^p, T) = 0$ is considered. As such, the equation of compatibility

$$df = \frac{\partial f}{\partial s_{ij}} ds_{ij} + \frac{\partial f}{\partial \varepsilon_{ij}^p} d\varepsilon_{ij}^p + \frac{\partial f}{\partial T} dT = 0, \tag{a}$$

a condition of irreversibility $s_{ij} d\varepsilon_{ij}^p > 0$ and a condition of stability $ds_{ij} d\varepsilon_{ij}^p > 0$ should be satisfied.

In Birger and Shorr (1975), it was shown that for the yield surface given by

$$f = \frac{3}{2} s_{ij} s_{ij} - \sigma_y^2(\varepsilon_i^p, T) = 0 \tag{b}$$

the specified relations, together with Boley and Weiner's (1960) deviations, result in the basic equation of non-isothermal plasticity Eq. (3.33b), which has been obtained above.

Experiments with neutral loading revealed an increase of some strain components $\Delta\varepsilon_{ij}^p$ though theoretically $d\varepsilon_{ij}^p = 0$. Assuming that this phenomenon does not occur as a result of continued active loading at finite stress increment $\Delta\sigma_i$ (see Chap. 5), it can be described by accepting

$$d\varepsilon_{ij}^p = s_{ij} d\lambda_1 + \lambda_2 ds_{ij}. \tag{c}$$

A magnitude of λ_2 could be found from the tests including gradual rotation of a stress vector. Then for the neutral loading

$$\lambda_2 = \frac{3 ds_{ij} d\varepsilon_{ij}^p}{2(d\sigma)_i^2}. \tag{d}$$

In Birger and Shorr (1975), it was shown that this approach yields

$$d\varepsilon_{ij}^p = \frac{3}{2}\left[\left(\frac{1}{E_t} - \frac{1}{E_s}\right)\frac{s_{ij}}{\sigma_i} d\sigma_i + \left(\frac{1}{E_s} - \frac{1}{E}\right) ds_{ij}\right], \tag{e}$$

describing both proportional and neutral loading.

Equation (e) represents a special case of equations of the generalized deformation theory of non-isothermal plasticity (Birger 1964; Birger and Dem'yanushko 1968)

$$d\varepsilon_{ij}^{p} = (F_{\sigma} d\sigma_i + F_T dT)s_{ij} + F_s ds_{ij}. \tag{f}$$

where parameters F_{σ} and F_T are defined by Eq. (3.40) and $F_s = \lambda_2$.

Conditions of applicability of Eqs. (e) and (f) require experimental studying.

3.2.3 *Strength and Durability at Unidirectional Non-isothermal Loading*

3.2.3.1 Generalized Deformation Diagrams

A generalized deformation diagram $\sigma_i(\varepsilon_i)$ describes a material behavior until destruction. The most correct presentation on such features gives trials at *torsion* of a thin-walled tubular sample in which the volume and area of cross section practically do not change until the destruction by shearing. Therefore, the deformation diagram at torsion $\tau(\gamma)$ differs from the primary test diagram $M(\varphi)$, bonding a torque with the twist angle φ, only in the axes parameters. This diagram usually demonstrates monotonous behavior up to specimen's failure without neck formation.

The stress and strain intensities under torsion of the tubular sample (i.e., at pure shear) are $\sigma_i = \tau\sqrt{3}$ and $\varepsilon_i \approx \gamma_{xz}/\sqrt{3}$ (at $\mu \to 0.5$), respectively (see Chap. 2). This allows easy transformation of the torsion diagrams $\tau(\gamma)$ into the generalized diagrams $\sigma_i(\varepsilon_i)$ as it is demonstrated in Fig. 3.11.

As a generalized diagram coincides with the extension one, the above procedure shows how the latter might look without neck occurrence.

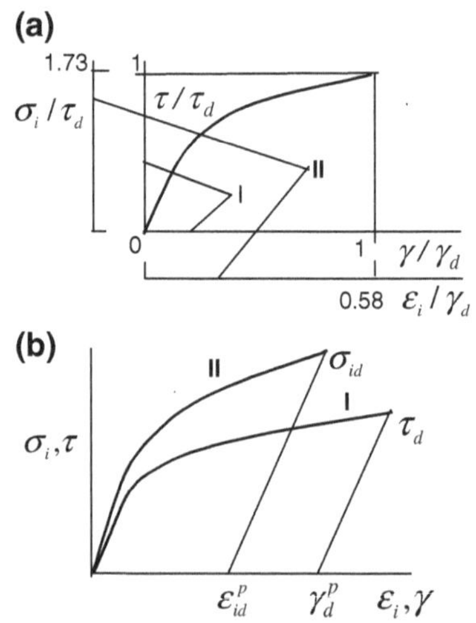

Fig. 3.11 Transformation of diagram $\tau(\gamma)$ into $\sigma_i(\varepsilon_i)$ **a** by scale change, **b** in common scale

A typical generalized deformation diagram $\sigma_i(\varepsilon_i)$ is presented in Fig. 3.12. Specimens of not too brittle material rupture when stress intensity σ_i reaches a true limit (breaking) strength σ_d at the given temperature T. A corresponding limit for the plastic strain intensity is $\varepsilon_i = \varepsilon_d$, and a true residual strain after full unloading is $\varepsilon_i = \varepsilon_{p,d}$.

Some heat-resistant alloys are mildly plastic, and their short-term damage under tension occurs without visible necking. In this case, a generalized diagram almost completely coincides with the experimental diagram for a specimen in tension.

On the other hand, tensile diagrams of many structural materials operating at elevated temperatures have a more common view shown above in Fig. 3.2. Extension of a sample with constant strain rate leads to a neck formation. Features of such diagrams were widely covered in the literature, in particular, the "neck effect" that can be considered as a loss of the stability of the rod's cylindrical shape (see Panovko and Gubanova 1964, and also Chap. 10).

As necking takes place, the sample's stress state becomes nonhomogeneous along its length which leads to further increase in stress and deformation in the necking zone. Reduction of the cross-sectional area in the neck from A_0 to A_d causes the measured load falling from its maximum value F_B to the less value F_d despite the fact that strains continue growing until full rupture. A nominal stress follows the load $F(\varepsilon)$ which results in reduction of magnitude $\sigma'_d = F_d/A_0$ in relation to the ultimate strength $\sigma_B = F_B/A_0$ and "true" stress $\sigma''_d = F_d/A_d$. A nominal strain ε'_d and residual elongation δ_k correspond to the rapture stress σ'_d (Fig. 3.12).

A quantity of δ_k, usually provided in the reference books, depends on the chosen base of strain measurement including the neck area. Therefore, the value δ_k does not completely reflect plastic properties of a material.

3.2.3.2 Reduction of a Rod Cross Section in Tension

Unlike value of δ_k, a relative residual reduction of a specimen's cross-sectional area $\psi = 1 - A_d/A_0$ does not depend on the gauge length at strain measurement that makes it a more objective characteristic of material plastic behavior under uniaxial

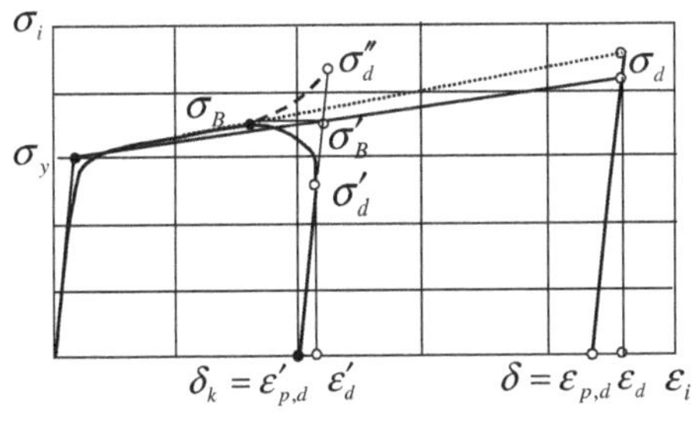

Fig. 3.12 Definition of real breaking stress σ_d, total strain ε_d, and residual plastic strain $\varepsilon_{p.d} = \delta$

extension. The quantity ψ may be computed by measuring the neck residual diameter after rupture and also presented in the reference books.

If the tensile load $F(\varepsilon)$ is measured during the test up to the moment of rupture, an average stress over the neck cross section at this instant could be defined as

$$\sigma''_d = \frac{F_d}{A_d} = \sigma'_d \frac{1}{1-\psi}. \tag{3.42}$$

The value σ''_d is usually treated as a "true" breaking stress.

Direct measurement of maximum strain in a neck at rupture is practically impossible. Therefore, the values of "true" strain ε'_d and plastic (residual) strain $\delta_k = \varepsilon'_{p,d}$ at rupture (as material characteristic) should be estimated indirectly.

Let us consider a volume $V' = L'_0 A_0$ of a small region L'_0 at a neck throat and assume that such volume during the neck formation, until rupture at the length L'_d, remains constant, e.g., $L'_0 A_0 = L'_d A_d$. Then

$$\varepsilon'_d = \frac{L'_d}{L'_0} - 1 = \frac{A_0}{A_d} - 1 = \frac{\psi}{1-\psi}, \tag{3.43a}$$

or with the assumption of large strains

$$\varepsilon''_d = \ln\left(\frac{L'_d}{L'_{00}}\right) = \ln\left(\frac{1}{1-\psi}\right) = \ln\,(1+\varepsilon'_d). \tag{3.43b}$$

Equation (3.43b) is sometimes used in computations of materials' cyclic durability at LCF (see Chap. 4) as a characteristic of plastic properties.

If there are no torsion trials, then at $\psi > 0.05$, it is useful to calculate the "real" damage parameters of a material σ''_d and ε''_d by Eqs. (3.42) and (3.43a, b), respectively, yet in the absence of a visible neck.

If the plastic strains considerably exceed elastic ones, the complete diagram may be represented as consisting of two parts. The elastic portion within stresses $\sigma \le \sigma_y$ has the elasticity modulus E, and the plastic one—the tangential modulus $E_{t\infty}$, which can be obtained as

$$E_{t\infty} \approx \frac{\sigma_d - \sigma_y}{\delta}, \tag{3.44}$$

as shown in Fig. 3.12.

If a real rupture stress quantity σ_d is not measured during the tensile test, the modulus $E_{t\infty}$ can be approximately estimated as

$$E_{t\infty} \approx \frac{\sigma'_B - \sigma_y}{\delta_k}. \tag{3.45a}$$

Elongating the line with this slope till the real rupture strain ε_d, one can estimate the value σ'_B

$$\sigma''_d \approx \sigma_y + E_{t\infty}\delta_k. \tag{3.45b}$$

This definition of value σ''_d is more meaningful than a value $S = \sigma_B/(1-\psi)$ which is often taken as a "real" ultimate stress limit.

3.2.3.3 Short-Term Damage Under Multi-axial Stressed State

In case of specimen's extension, the rupture of sufficiently plastic material occurs by a plane inclined 45° to the specimen's axis associated with the maximum tangential stress.

As criteria of short-term strength of structures under multi-axial stress, the value of breaking stress σ_{id} or plastic strain ε^p_{id} intensities may be used. These quantities are assumed to be equal to the corresponding "true" breaking values of stress σ_d and plastic strain δ under uniaxial extension. Therefore, the reliability of stress σ_d and strain δ definition from elastoplastic strain diagrams has important value.

For this purpose, it is useful to find the breaking intensities σ_{id} and ε_{id} from the torsion test results if they are known.

Adequacy of damage criteria in stress intensities obtained by tension and torsion was revealed by Serensen et al. (1963) though their identity was not close enough.

For materials with neck forming under tensile tests, it is necessary to take into account a possible significant deviation of failure characteristics depending on the factor ψ and it shall be taken $\varepsilon_{id} = \delta$ and $\sigma_{id} = \sigma_d$.

Tests show that under combined stress the diagrams $\sigma_i(\varepsilon_i, T)$ or $\tau_i(\gamma_i, T)$ may be used till failure. However, in some cases, a maximum tangential stress value

$$\tau_{\max} = 0.5(\sigma_1 - \sigma_3) \tag{3.46}$$

corresponds slightly better to the short-term failure criterion.

Non-uniformly heated bodies, all elements of which are not able to freely extend in any directions at heating, will experience pure thermal stresses without outer loads. Calculations show that at the occurrence of plastic deformations, the field of thermal stresses becomes level and their value decreases. But as these stresses are self-balanced in the body, there always exists elastic regions where stresses have opposite signs and they are less than yield points at appropriate temperatures.

The presence of the elastic regions at any non-uniform temperature field makes the standard methods for estimation of a structure loading capacity unacceptable. The occurrence of local damages can additionally change temperature non-uniformity that in turn affects loading capacity.

A structure made of brittle materials (glass, ceramics) fractures practically immediately after a local crack arises. As for calculation of damage evolution in a non-uniformly heated structure, it demands special analysis leaning upon the methods of fracture mechanics.

3.2.4 Numerical Examples

3.2.4.1 A Rod Exposed to Simultaneous Loading and Heating

Let a rod of rectangular cross section with width b be exposed to simultaneous loading of increasing tensile force F and gradual heating. Active loading and heating stop when the average tension stress reaches $\sigma_m = 200$ MPa, and non-uniformity of a temperature field become 200 °C. A temperature distribution along the rod width (for half of the section at $0 \leq \xi \leq 1$ where $\xi = 2x/b$) is shown in Fig. 3.13a. Deformation diagrams of the material corresponding to tension and compression are assumed to be identical. Creep is neglected due to the short period of the overall process.

Results of elastioplastic computation of the stresses in the rod (see Appendix 1) are shown in Fig. 3.13b.

Elastic computation (curve 1) shows the total stresses from tensile force and non-uniform heating that considerably surpass yield points of material, both in the colder zone at a rod axis where $\sigma_y \approx 160$ MPa and at hotter edges where $\sigma_y \approx 120$ MPa. Plastic strains arising under active loading were calculated using the deformation

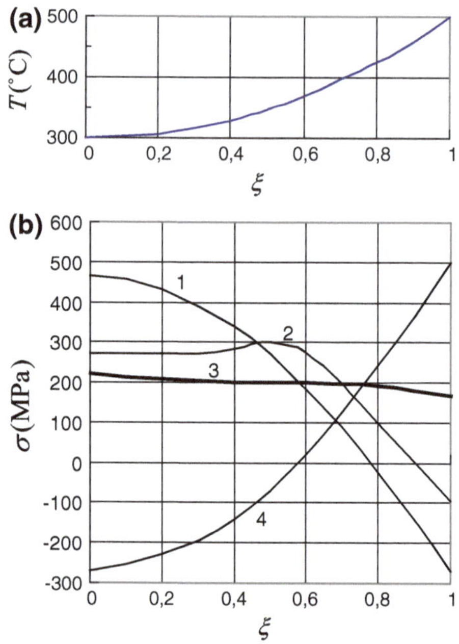

Fig. 3.13 Elastic and elastoplastic stresses under tension of a non-uniformly heated rod

theory by the method of variable elasticity parameters specified above. Curve 2 corresponds to the first approach, bold curve 3—to the final solution.

It is visible that plastic deformation results in almost entire alignment of the stresses over the section. The mean stress does not change as it is defined by the equilibrium condition of the rod. The maximum tension stress which has remained in the colder central part exceeds the mean stress only by 10 %. In the considered temperature range, the material strength does not practically change; therefore, the minimum *local strength margin* (see Chap. 8) corresponds to the region of the maximum stress, and at $\sigma_B = 490$ MPa is equal to $k_{\min} = \sigma_B/\sigma_{\xi=0} = 2.23$, while the value calculated for the mean stress is $k_m = 2.44$.

In Fig. 3.13b, the residual stress distribution after unloading and cooling of a rod to the normal temperature is also displayed (curve 4). Here, the residual stresses are assumed to be elastic. In reality, the maximum residual stresses exceed a yield point at normal temperature ($\sigma_{pr} \approx 230$ MPa) and plastic strains ought to arise also after unloading. Computations of sign-varying plastic strains will be considered in Chaps. 4 and 5.

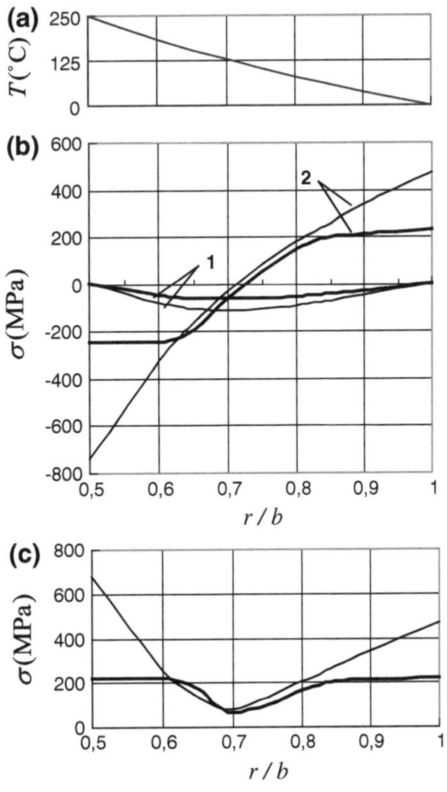

Fig. 3.14 Elastic and elastoplastic thermal stresses in a hollow cylinder

3.2.4.2 Pure Thermal Stresses

As an example, we consider a long circular cylinder in the axially symmetric temperature field $T(r/b)$ presented in Fig. 3.14a. The ratio of the cylinder's internal radius to external one is $a/b = 0.5$ and the material yield point is $\sigma_y = 220$ MPa.

In Fig. 3.14b, distributions of radial 1 and tangential 2 elastic stresses and in Fig. 3.14c elastic stress intensity along the cylinder thickness are shown by thin curves. For given conditions, two plastic zones arise in the non-uniformly heated cylinder—near cold external and hot internal surfaces. Between these zones, there remains an elastic region. In the same figures, distributions of the corresponding elastoplastic stresses computed by a method of additional strains (Shorr 1960) are shown with bold lines.

Calculation of the elastoplastic stress distribution under given temperature field allows defining the minimum value of the local strength margin of a structure.

References

Birger IA (1963) Method of additional deformation in problems of theory of plasticity. Izv.AN SSSR. Mechanika i Mashinostroyenie, No. 1, pp 47–56 (in Russian)

Birger IA (1964) Theory of plastic flow at non-isothermal loading. Izv. AN SSSR. Mechanika i Mashinostroyenie, No. 1, pp 193–196 (in Russian)

Birger IA, Dem'yanushko IV (1968) Theory of plasticity at nonisothermal loading. Injenerny Jurnal. Mechanika Tverdova Tela 6:70–77 (in Russian)

Birger IA, Shorr BF (eds) (1975) Thermal strength of machine parts. Mashinostroenie, Moscow (in Russian)

Boley BA, Weiner JH (1960) Theory of thermal stresses. Wiley, New York

Bridgman PW (1964) Collected experimental papers of P.W. Bridgman. Vols I–VII. Harvard University Press, Cambridge

Davis EA (1955) Combined tension-torsion tests with fixed principal directions. J Appl Mech 22 (3):411–415

Fleming WJ, Temis JuM (2002) Numerical simulation of cyclic plasticity and damage of an aluminium metal matrix composite with particulate SiC inclusions. Int J Fatigue 24:1079–1088

Fridman YB (1948) Mechanical properties of metals. Oborongiz, Moscow (in Russian)

Fridman YB (1974) Mechanical properties of metals. In: Deformation and damage, vol 1, 3rd edn. Mashinostroenie, Moscow (in Russian)

Hencky H (1925) On slow steady flows in plastic bodies. ZAMM, Bd.25, H2:115–124

Huber MT (1904) Die spezifische Formanderungsarbeit als Mass der Anstrengung eines Materials. Lemberg

Il'ushin AA (1943) Some problems of the theory of plastic deformations. J Prikladnaya matematika i mekhanika YII(4):245–272 (in Russian)

Il'ushin AA (1948) Plasticity (in Russian). OGIZ, Moscow-Leningrad

Johnson W, Mellor PB (1962) Plasticity for mechanical engineers. Van Nastrand Company LTD, London

Kachanov LM (1948) Mechanics of plastic media. Gostechizdat, Moscow (in Russian)

Kachanov LM (1969) Foundation on theory of plasticity, 2nd edn. Nauka, Moscow (in Russian)

Lebedev AA Koval'chuk BI, Giginjak FF, Lamashevsky VP (2003) Mechanical behavior of construction materials at combined stress state. In: Lebedev A (ed), 3rd edn. Izdatel'sky Dom "In Yure", Kiev

Lode W (1926) Versuche über den Einfluss der mittleren Hauptshpannung auf das Fliessen der Metalle Eisen, Kupfer und Nickel. Zeitschr. für Physik, B.36, S.913

Marin J, Yu LW (1956) Biaxial plastic stress-strain relations of a mild steel for variable stress ratios. Trans ASME, Paper No. 54-A-243, pp 499–509

Mises R (1913) Meckanik der festen Körper im plastisch-deformablen Zustand, Gottinger Nachr., Math.-Phys. Klasse, H.4:582–592

Nadai A (1931) Plasticity. McGraw-Hill Book Co., NJ

Odqvist FKG (1933) Die Verfestigung von flusseisenähnlichen Körpern, ZAMM, B.13:360

Panovko YG, Gubanova II (1964) Stability and vibrations of elastic systems. Nauka, Moscow (in Russian)

Prager W (1956) Plastic design and thermal stresses. British Welding J 3(8):355–359

Prandtl L (1920) Über die Härte plastischer Körber, Gött. Nachr., mathphys K1:71

Reuss A (1930) Berücksichtung der elastichen Formänderungen in der Plastizitätstheorie . Zeitschr. für angew. Math. u. Mech, vol 10

Saint-Venant B (1871) About establish equatons of inner displacements arising in plastic solids after elastic limits. J.d. Math. pures et appl. ser.II 16:308

Schmidt R (1932) Über die Zusammenhang von Spannungen und Form-änderungen im Verfestigungsgebiet. Ing. Archive, Bd.III, H.3:215–235

Serensen SV, Kogaev VP, Shneyderovich RM (1963) Load-ability and strength calculation of machine structures (in Russian), 1st edn (3rd edn., 1975), Mash-inostroenie, Moscow

Shorr BF (1960) To calculation of no uniformly heated cylinders in elastic-plastic domain. Izv. AN SSSR, Mekhanika i Mashinostroenie 6:57–62 (in Russian)

Shorr BF (1974) Method of additional functions in mechanics problems of nonisothermally deforming bodies. Thermal stresses in constrution elements, vol 14. Naukova Dumka, Kiev, pp 13–18 (in Russian)

Sokolovsky VV (1969) Theory of plasticity, 3rd edn. Vysshaya shkola, Moscow (in Russian)

Temis YM (1980) Method of successive approximations with error correction in geometrically nonlinear elastic problems. In: Proceedings of applied problems of strength and plasticity. Algoriphmic solution procedures in elasticity and plasticity Problems. Izd.Gor'kovskogo Gos. Universiteta. Go'kiy, vol 16, pp 3–10 (in Russia)

Tresca H (1868) Memoire sur l'ecoulement des corps solides. Mem. pres. par div. sav. a l'Acad. de l'Inst. imp. de France, 18:739–799; (1872); 75–135

Zakharova TP (1966) Accumulation of plastic strains at non-isothermal loading. In: Trans Strength and dynamics of aviation engines, vol 4. Mashinostroenie, Moscow, pp 201–213 (in Russian)

Chapter 4
Thermal Plasticity: Cyclic Loading and Heating

Abstract Cyclic durability of machine parts is related to material resistance against low cycle fatigue (LCF) under periodic (cyclic) loads and temperatures within the range from 10 to 10^4 cycles and above. This chapter deals with experimental and theoretical foundations of computational methods for estimating both isothermal and non-isothermal cyclic durability of engineering structures. Typical examples of material cyclic tests under varying loads and peculiarities of cyclic non-isothermal deformation diagrams are considered. In the last case, the heating and cooling half-cycles are not identical due to a difference in the occurrence of elastic and plastic zones. The role of material the so-called cyclic adaptability is emphasized. This phenomenon leads to cycle stabilization that ensures to a great extend of the long-term working capacity of machines. Stabilized cycles can be both simple and consisting of a set of half-cycles of different stress and temperature. It is shown that the number of cycles before LCF failure under arbitrary programs of periodic loading and heating can be calculated from Coffin–Manson's curves by averaging plastic strain intensity magnitude over whole deformation process. Analysis of the factors influencing the thermal fatigue of machine parts coming from thermocyclic tests is presented along with efficient methods for calculating stresses and strains corresponding to synchronous non-isothermal cycles. The possibility for summation of cyclic damage is demonstrated for evaluating durability of variable cycles associated with unsteady loading and heating or unstable material parameters. It is shown that LCF can be evaluated using the normalized accumulated plastic strain related to its magnitude at failure. The absolute accumulated plastic strain (Odqvist's parameter) at cyclic loading varies over a wide range and cannot be a criterion for cyclic damage accumulation and failure.

B.F. Shorr, *Thermal Integrity in Mechanics and Engineering*,
Foundations of Engineering Mechanics, DOI 10.1007/978-3-662-46968-2_4

4.1 Cyclic Elastoplastic Processes in Solids

4.1.1 Cyclic Loadings and Unloadings

Specific elastoplastic behavior of materials subjected to proportional unidirectional loading has been studied in Chap. 3. By doing so, the plastic strains were assumed to be sign-constant under unloading. In this chapter, another special case is considered when stresses change periodically (cyclically) and plastic strain in each half-cycle develops in opposite direction.

4.1.1.1 Cyclic Loading

In some works, a small number (sometimes two–three) of sign-varying loadings are treated as "cyclic." Here, we shall consider thermoplasticity problems associated with large number of cycles (until $\sim 10^5$) which is typical for the operation of many parts of machines and power plants. The study of these processes demands special experimental and theoretical approaches.

Since duration of thermocyclic processes at high temperatures can be rather large, plastic deformations, creep, and accumulation of a material damage might be considered jointly. However, the operating regimes of short-term change of loadings and temperatures quite often alternate with long-term regimes in which loads and temperature change very slowly. This allows analyzing the thermoplasticity problems of cyclic loading and heating separately of creep and long-term durability. The latter will be discussed in Chap. 7.

A number of cycles associated with repeated working conditions of machine parts (start, operating modes, shutdown, etc.) usually have orders from 10^2 to 10^5. This loading is named *low cycle*, in contrast to *high cycle*, caused by vibration when the number of cycles may greatly exceed those values and reach 10^8 and above. In both cases, material fails due to fatigue, but at low cycle fatigue (LCF), visible plastic deformation may occur in each cycle, while for high cycle fatigue (HCF), calculated macro-strains do not go beyond their elastic limit. Temperature increase influences both LCF and HCF, but LCF is more affected by high temperatures.

The classification of fracture modes has important diagnostic meaning because it allows revealing the causes of material's failure and suggesting the actions to eliminate them.

4.1.1.2 Unloading with Variable Sign of Stresses

In Fig. 4.1, an initial portion of deformation diagram $\sigma(\varepsilon)$ is presented. In case of tension/compression and sign-varying torsion, initial parts of deformation diagrams are identical; therefore, notations σ and ε can reference here to any uniaxial stress state.

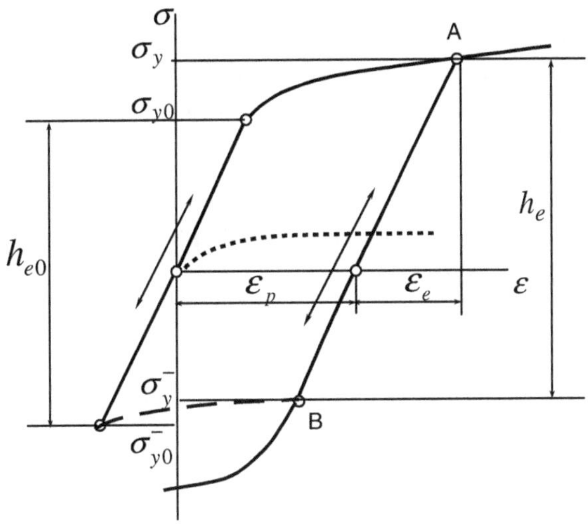

Fig. 4.1 Unloading with change of stress sign

If the unloading has been started at point A and will continue with the opposite stress sign, then dependence $\sigma(\varepsilon)$ remains almost rectilinear till the stress σ_y^- is reached (point B). As far as this point, the value of elasticity modulus is approximately the same as at initial loading.

The stresses $\sigma_y^- < 0$ represent a new "instant" yield points referred to the opposite stress sign. Within limits $\sigma_y^- \leq \sigma \leq \sigma_y$ (between points A ↔ B), the strains $\varepsilon_e = \sigma/E$ behave elastically, while a plastic strain remains constant. The area of elastic deformations $h_e = \sigma_y - \sigma_y^- = \sigma_y + |\sigma_y^-|$ depends on a material structure variation during the deformation process and generally is not constant. Provided $|\sigma_{y0}^-| = \sigma_{y0}$, the initial size of the elastic area at $\varepsilon_p = 0$ is $h_{e0} = 2\sigma_{y0}$.

Enlargement of the elastic area (as is shown in Fig. 4.1 where $h_e > h_{e0}$) specifies the *isotropic strain hardening* of a material, narrowing relates to the *isotropic strain softening,* and constant width corresponds to the *stable state*. This definition of the term "hardening" differs from that given in Chap. 3 where a width of the elastic area is not recognized. Terms *work hardening* and *work softening* are also sometimes used.

As the plastic deformation increases, position of elastic area characterized by coordinates of its middle line also changes (see dotted curve in Fig. 4.1). Motion of the elastic area always occurs in a direction of current increasing stress because it is caused by the appearance of material *strain anisotropy*. Usually, the inequality

$$\left|\sigma_y^-\right| < \sigma_y. \tag{4.1}$$

takes place, in spite of possible isotropic increase of elastic area in the width.

Bauschinger (1881) experimentally studied how repeated sign-varying loading affects the yield points of steels. The reduction of the "instant" yield value after change of a stress sign is named the "Bauschinger's effect."

Essential difference of the initial site of deformation diagram from the curves of subsequent half-cycles was shown by Arutjunjan's tests (1964).

4.1.1.3 Idealized Cycles with Sign-Varying Stresses

In an idealized cycle, the initial yield points in tension and compression are assumed to be identical. The deformation process after the stress sign has been changed is displayed in Fig. 4.2. The increase of stress value is attended by plastic yielding in the opposite direction (along the secondary curve BCEG). Unloading proceeds now in the reverse direction. In Fig. 4.2, positions of the unloading states both from the primary yield curve (to point B) and from the secondary one (to points D, F, or H) are shown with black points.

Let us consider various situations that can be resulted by the stress sign alteration.

- Under stress increase from point B toward points C → E → G, the value of earlier reached plastic strain decreases. When the stress passes point E, the plastic strain becomes zero and then will again start to increase in magnitude, but in the opposite direction.
- The curve of plastic loading in the opposite direction B → C → E → G after the first unloading differs from the initial curve of loading in the direction of point A.
- In case of unloading in point E and repeated stress increase toward point F, the development of plastic deformation generally differs from that corresponding to the first loading.

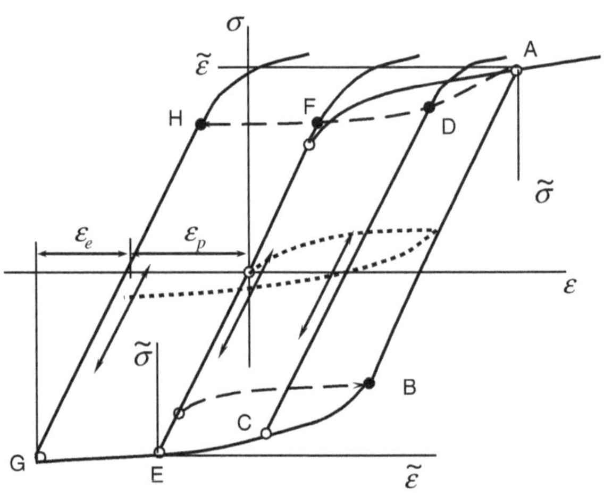

Fig. 4.2 Loadings and unloadings with alteration of stress and plastic strain signs

- For the material behavior shown in Fig. 4.2, the width of the elastic area continues increasing with growth of absolute value of plastic strain $|\varepsilon_p|$ irrespective of its sign. The change of elastic area width at the material isotropic hardening or softening does not depend on a load direction.
- For quite stable state of a material, the width of the elastic area is constant, and at $|\sigma_{y0}^{-}| = \sigma_{y0}$ is equal to the double value of the initial yield point

$$h_e = \sigma_{y0} + \left|\sigma_{y0}^{-}\right| = h_{e0} = 2\sigma_{y0}. \tag{4.2}$$

  This relation has been established by Masing (1926), and it is often used for the approximate analysis of sign-varying elastoplastic deformation.
  Alteration of the elastic area width in the process of plastic deformation shows a material unstable state.
- Position (a coordinate) of the average line of the elastic area shown by a dotted curve follows up the stress direction. Thus, the central point of the elastic area can take a symmetric position between border points C and D, as well as in the unloaded state. Under unloading from these points and next loading in opposite direction, the Bauschinger's effect does not occur. For the repeated changes of stress sign, the specified effect repeats in the opposite directions.

This specifies that the Bauschinger's effect depends not only on a material property, but also on conditions of loading and unloading.

4.1.1.4 Full and Half-Cycles

Under repeated loading from point E to point A, the deformation curve produces a closed loop—a *strain cycle*. The cycle can be presented as two *half-cycles* (A → E and E → A in Fig. 4.2) each of which begins at a point where the loading changes its direction. This point is taken as the origin of «tracking coordinates» $\tilde{\sigma}, \tilde{\varepsilon}$ used in the majority of design methods.

We will number half-cycles by index $k = 0, 1, 2, \ldots$, and parameters of a half-cycle k will be designated with the upper index (k). In some cases when there is no need to distinguish the half-cycles, the same designations will be utilized in respect of the number of full cycles.

The initial half-cycle begins from the point $\varepsilon_p = 0$ and may be regarded as incomplete. It qualitatively differs from the subsequent half-cycles which begin in the presence of residual stresses from the previous loading. Therefore, we subscribe the initial half-cycle with index $k = 0$ and count subsequent half-cycles, beginning from the first *full* half-cycle $k = 1$ (i.e., from the points A in Figs. 4.1 and 4.2).

Beginning from the first half-cycle $k = 1$, the dependence $\sigma(\varepsilon)$ for a typical cycle may be represented in terms of tracking coordinates $\tilde{\sigma}, \tilde{\varepsilon}$ by two mirror identical curves $\tilde{\sigma}(\tilde{\varepsilon})$, named as *diagram of cyclic deformation* (Arutjunjan 1964; Manson 1966). In Fig. 4.3a, this diagram is displayed by curves AE and EA.

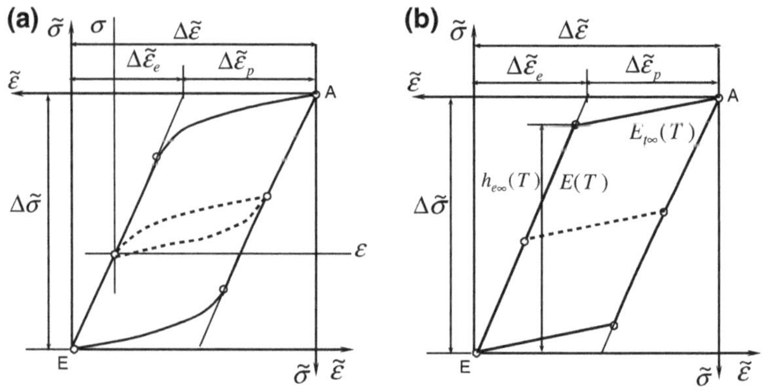

Fig. 4.3 Parameters of a stable deformation cycle (**a**) and its bilinear approximation (**b**)

At the stable state of a material, its cyclic deformation diagrams in tracking coordinates $\tilde{\sigma}(\tilde{\varepsilon})$ are the same for all half-cycles $k \geq 1$ over the entire operating period. At the non-stable material state, and also during asynchronous non-isothermal processes, each half-cycle should be considered separately.

4.1.1.5 Different Sign-Varying Cycles

Cyclic tests are mostly carried out at tensile/compression of short-length cylindrical specimens with solid cross section. Branches of tension and compression curves for some materials slightly differ, mainly because of distinctions in yield stresses. This sometimes leads to different character of the initial half-cycles in tension and compression. However, subsequent half-cycles for many materials become almost identical (Shneyderovich 1968).

If the differences of tension and compression cycle characteristics hold true, unidirectional shifting of stress and strain ranges may be observed with increasing number of cycles.

Cyclic tests for shear are carried out by sign-variable torsion of thin-walled tubular specimens. The tests of some materials under repeatedly varying torsion (i.e., Roytman and Fridman 1947) showed that in this case, the basic features of materials behavior remain the same as for tensile/compression tests.

Cyclic tests of tubular specimens under simultaneous or consecutive application of twisting moment, internal pressure, and axial force are conducted for research purposes (see Chap. 5).

On the basis of available experimental data, the cyclic deformation diagrams in tracking coordinates $\tilde{\sigma}(\tilde{\varepsilon})$ obtained at rod axial tension/compression tests or torsion of thin-walled tubular specimens can be represented for cyclic multi-dimensional proportional isothermal loading using the dependences $\tilde{\sigma}_i(\tilde{\varepsilon}_i)$ between intensities of stress σ_i and strain ε_i. These diagrams for different points of a body can vary due to uneven temperature distribution over the body.

4.1.2 Isothermal Cyclic Loading

4.1.2.1 Stable Cycles

Long-term cyclic work of a machine part is possible only at a stable state of a material.

A stable cycle can be realized at constant range (the doubled amplitude) of stresses $\Delta\sigma = \sigma^+ - \sigma^-$ or total strains $\Delta\varepsilon = \varepsilon^+ - \varepsilon^-$. In the first case, the loading is named *soft*, and in the second one—*rigid*. The cycles may be symmetric with stresses $\sigma^- = -\sigma^+$ or $\varepsilon^- = -\varepsilon^+$, or asymmetric with $\sigma^- \neq -\sigma^+$ or $\varepsilon^- \neq -\varepsilon^+$.

Dependence $\tilde{\sigma}(\tilde{\varepsilon})$ for many materials has two almost linear parts—elastic with the modulus E and plastic with the minimal tangential modulus $E_t = E_{t,\min}$ significantly smaller than E. Between them, there is a nonlinear transient curve in which the tangential modulus $E_t = d\tilde{\sigma}/d\tilde{\varepsilon}$ quickly decreases from E to $E_{t,\min}$. Both E and $E_{t,\min}$ depend only on temperature, while E_t is a function of temperature T and strain $\tilde{\varepsilon}$.

Under the great plastic strains, the role of the nonlinear part becomes negligible, which allows the function $\tilde{\sigma}(\tilde{\varepsilon})$ to be represented in the approximate bilinear form shown above in Fig. 4.3b.

Increments of stresses and strains in the cyclic coordinates $\tilde{\sigma}(\tilde{\varepsilon})$ are always positive: $\Delta\tilde{\sigma} > 0$, $\Delta\tilde{\varepsilon} > 0$, and $\Delta\tilde{\varepsilon}_p \geq 0$.

Further, we shall note the difference between signs of increments $\Delta\varepsilon_p$ and $\Delta\tilde{\varepsilon}_p$ only in case of need.

A half-cycle at any constant temperature T, described by a curve $\tilde{\sigma}_i(\tilde{\varepsilon}_i, T)$, may be named *simple*. A cycle whose both half-cycles are simple and entirely described by the same diagram $\tilde{\sigma}_i(\tilde{\varepsilon}_i, T)$ we name also *simple*. The cycles 3–4 in Fig. 4.4a, b are examples of simple ones.

The cyclic diagrams $\tilde{\sigma}_i(\tilde{\varepsilon}_i, T)$ for each simple half-cycle and static diagrams $\sigma_i(\varepsilon_i, T)$ look alike. Therefore, the calculation methods stated in Chap. 3 can be used for determining the strain and stress ranges $\Delta\varepsilon_i$ и $\Delta\sigma_i$ for the given ranges of outer loads. To this goal, the stress and strain components σ_{ij} and ε_{ij} in the formulas of Chap. 3 must be replaced by their finite increments $\Delta\sigma_{ij}$ and $\Delta\varepsilon_{ij}$.

4.1.2.2 Cycle Stabilization

For many materials, the first several half-cycles are not identical in spite of constant loading conditions, but then, they quickly become stabilized, as shown in Fig. 4.4a for soft and in Fig. 4.4b for rigid loading cycles of a heat-resistant nickel–chrome alloy (Dul'nev in Birger and Shorr 1975). In these figures, the indices of sequential half-cycles k are displayed.

When the strains at stabilized cycles become overall elastic, *full adaptability* (*shakedown*) may be spoken about. The problem of full adaptability was studied theoretically in works of Melan (1938), Koiter (1960), Kachanov (1969), and others.

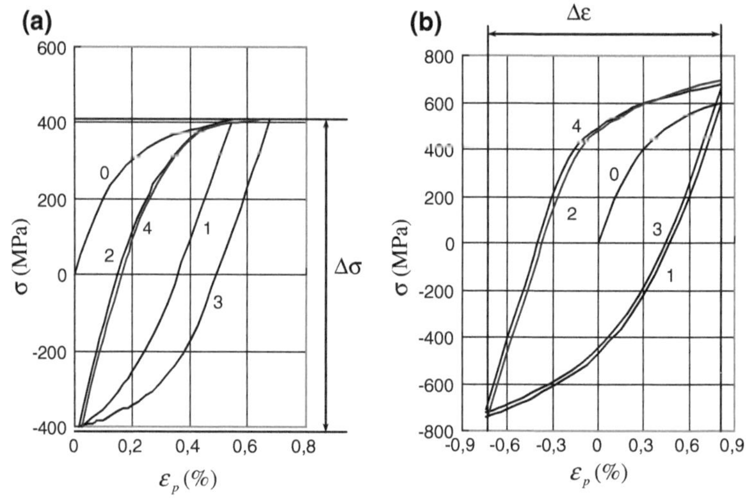

Fig. 4.4 Stabilization of deformation cycles for a CrNi alloy: **a** $T = 800\,°\mathrm{C}$, soft loading; **b** $T = 700\,°\mathrm{C}$, rigid loading

In other cases, the development of plastic deformation under constant load and temperature ranges does not stop, but stresses and strains within each half-cycle identically alter in opposite directions so that within the whole cycle, the plastic strain does not increase. This effect can be treated as *partial adaptability*. In conditions of adaptability, damage in material continues to accumulate, so the term "adaptability" reflects only the deformation characteristics of cyclic processes. The further discussion on this effect is given in Chap. 5.

4.1.2.3 Unstable Cycles

Under unstable state of a material, even and odd half-cycles remain different during all process and the loops become open, but their evolution runs usually gradually. In this case, a loop width $\Delta\tilde{\varepsilon}_p$ and work ΔW_p for a cycle may be considered as variable quantities. The character of loops evolution is schematically shown in Fig. 4.5a, b for soft and in Fig. 4.5c, d for rigid loadings.

Elasticity moduli usually alter at repeated cycles insufficiently, though in some tests, their viewable changes were noted and ought to be taken into account for accumulated plastic strain calculation at large cycle number (Temis 1989, 1994, 2005).

In the case of cyclically hardening materials, the loop width $\Delta\tilde{\varepsilon}_p$ gradually decreases, aspiring to the stabilized size (Fig. 4.4a). Be the material's hardening

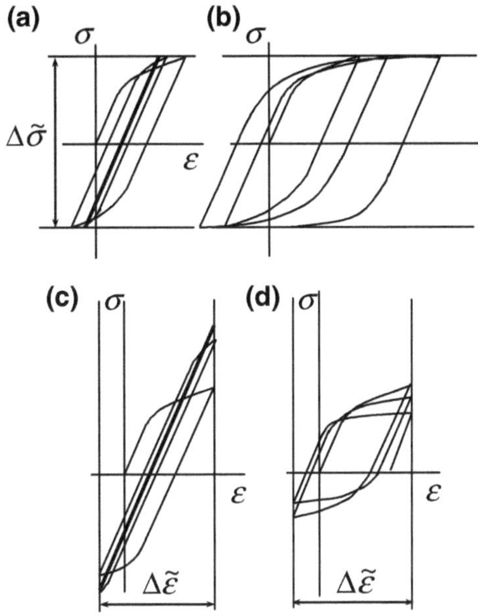

Fig. 4.5 Different cycle types: **a**, **b** soft loading, **a** hardening material, **b** softening material; **c**, **d** rigid loading, **c** hardening material, **d** softening material

unlimited, the loop width $\Delta\tilde{\varepsilon}_p \to 0$ (Fig. 4.5a, c). For cyclically softening materials, the stabilization can also occur (Fig. 4.4b) and in the limiting case (at the ratio $\Delta\tilde{\varepsilon}_p/\Delta\tilde{\varepsilon} \to 1$), the loop width comes nearer to full strain range (Fig. 4.5b, d).

Sometimes, a material can initially harden and then soften (see Shneyderovich 1968).

4.1.2.4 Asymmetrical Cycles

Cyclic tests are sometimes carried out using asymmetric loading cycles, usually "zero-to-maximum stress," with asymmetry factor $r = \sigma_{\min}/\sigma_{\max} = 0$. For rigid cycles, a role of asymmetry is usually insignificant, but at soft cycles, in particular for zero-to-tension stress, an unilateral strain increase after each cycle may happen in a form of "ratchet motion" (Chaboche and Nouailhas 1989).

A one-side increase of plastic deformation may be also connected with creep (Troshchenko et al. 2000).

4.2 Isothermal Low Cycle Fatigue (LCF)

The final goal of computations of machine parts durability at cyclic loading and heating is to find a number of cycles to failure $N = N_d$ depending on strain (or stress) and temperature ranges $\Delta\varepsilon$, $(\Delta\sigma)$, and ΔT, which in common case might be variable. This problem is called *computation of cyclic durability or a low cycle fatigue* (*LCF*). The procedure defines the number of cycles prior to the local fracture in the most stressed point of the body and also the number of cycles to failure of the whole body.

For the computation of cyclic durability, it is necessary to have experimental data of a material LCF, while it is subjected to the operation conditions of the machine parts. It is also necessary to calculate the distribution of cyclic stress and strain ranges over the body under the set cyclic loads and temperatures, taking into account their change during the work.

4.2.1 Experimental Curves of Cyclic Durability

4.2.1.1 Primary LCF Curves

Cyclic tests of specimens come to an end to destruction from LCF. The number of destructive cycles N_d is sometimes called the *cyclic durability,* in contrast to the *time durability* measured in hours. A dependence of cycle number N_d against full strain range $N_d(\Delta\varepsilon)$ is usually obtained by rigid-cyclic tension/compression tests. If the stress ranges $\Delta\sigma$ are inspected during testing, then the plastic strain ranges $\Delta\varepsilon_p = \Delta\varepsilon - \Delta\sigma/E$ and the function $N_d(\Delta\varepsilon_p)$ can be experimentally obtained.

Soft loading tests are less popular. If the ranges of the total strains $\Delta\varepsilon$ are supervised, then ranges of the plastic strains $\Delta\varepsilon_p$ are easy to be calculated and dependence $N_d(\Delta\sigma)$ can be reconstructed into the function $N_d(\Delta\varepsilon_p)$. For stable states of a material, correct computation of curves $N_d(\Delta\varepsilon_p)$ by both methods should yield close results. To destroy a sample at a small number of cycles and considerable residual deformations, soft loading is more preferable.

Failure at asymmetric cycles can have quasi-static character similar to the damage at short-term strength tests.

The specified pattern persists for elevated constant temperatures, but the cyclic durability decreases with temperature increase.

Typical dependences $\Delta\varepsilon(N_d, T)$ for three heat-resistant alloys at high temperatures are shown in Fig. 4.6 (Dul'nev in Birger and Balashov 1981). Curve 1 meets the wrought alloy at T = 750 °C, and curves 2 and 3 meets the two cast alloys at T = 950 °C.

As temperature rises, the influence of the static stress component on number of the destruction cycles N_d significantly increases.

4.2.1.2 Generalized LCF Curves

Manson (1966), according to Smith et al. (1963), presents the results of LCF tests at normal temperature for a significant number of structure materials (steels, a nickel

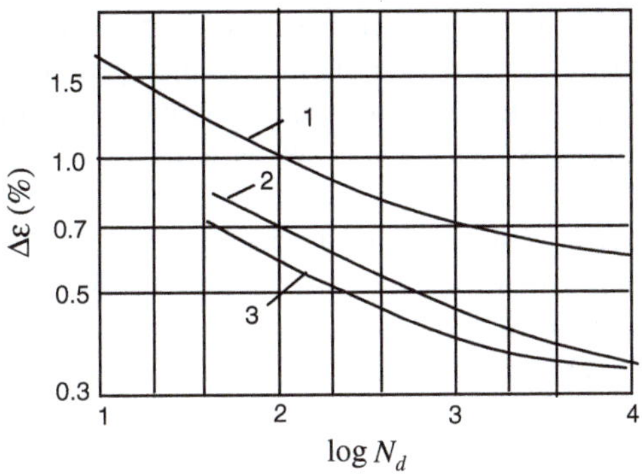

Fig. 4.6 Dependence $\Delta\varepsilon(N_d)$ for three heat-resistant alloys at different temperatures

alloy, as also titanium, beryllium, and aluminum). The total strain ranges $\Delta\varepsilon$ versus a number of cycles to failure N_d were plotted in a double logarithmic coordinates as shown in generalized form in Fig. 4.7. In the graph, an area, in which points for most tested materials are compactly enough located, is painted, but an out-of-order curve 1 for one of the steels is also displayed.

Qualitatively, similar dependences were obtained by many other researchers under testing of different materials, including testing at high temperatures. However, the number of cycles to damage N_d for present-day alloys within the same strain ranges is noticeably higher than in Manson's data.

Three parts can be allocated at curves $\Delta\varepsilon(N_d)$ presented in Fig. 4.7 and in other publications on LCF. At a small number of cycles, approximately to $N_d \approx 10^2$, the range of the total strains may exceed 10 % and be close to the plastic strain $\Delta\varepsilon \approx \Delta\varepsilon_p$ since $\Delta\varepsilon_p \gg \Delta\varepsilon_e$. Within this part, the failure character is close to the static. Here, a small reduction of strain range $\Delta\varepsilon$ increases the number of cycle destruction 10 and above times.

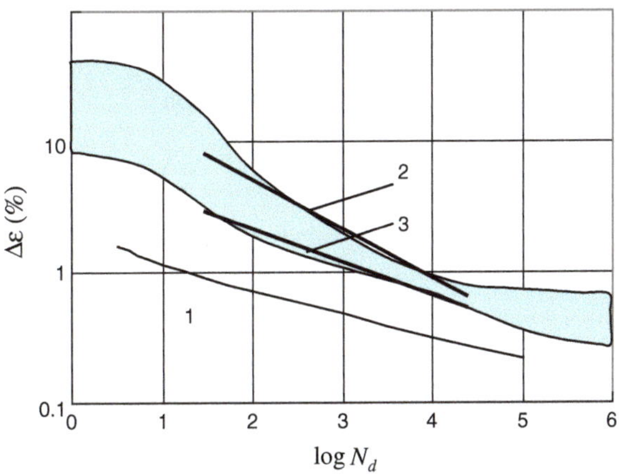

Fig. 4.7 Generalized dependence $\Delta\varepsilon(N_d)$ for some materials (due to Manson 1966)

At the second, the basic part (approximately from $N_d \approx 10^2$ to $N_d \approx 10^4 - 10^5$), the ratio $\Delta\varepsilon_p/\Delta\varepsilon_e$ of plastic strain range to the elastic one decreases but nevertheless exceeds unit. The destruction comes at smaller plastic strains and has more local character since it occurs as a result of the formation of fatigue micro-cracks. Therefore, stress concentrators render great influence on the development of low cyclic damage. The dependence $\log \Delta\varepsilon = f(\log N_d)$ for each material does not appreciably deviate from a straight line that facilitates its linear approximation, shown by bold lines 2 and 3.

The third site (beginning from $N_d \approx 10^5$), in which ranges of full strain become almost elastic, $\Delta\varepsilon \approx \Delta\varepsilon_e$, is related to the area of HCF.

The curves at elevated temperatures, shown above in Fig. 4.6, are qualitatively similar to the generalized curve in Fig. 4.7, but the area of the second part, where linear approximation of dependence $\log \Delta\varepsilon = f(\log N_d)$ is possible, considerably decreases.

This once more shows the need for experimental definition of LCF curve of a material within the temperature range in which it is assumed to be used.

4.2.2 Approximate Formulas of Cyclic Durability

For estimation of the cyclic durability of machine parts with lack of fair experimental data, various approximate formulas were proposed. A short analysis of some approximates is presented below.

4.2.2.1 Coffin–Manson's Approximations

From Figs. 4.6 and 4.7, a dependence between number of cycles (or half-cycles) to failure N_d and total strain range in a cycle $\Delta\varepsilon$ in the field of numbers $N_d \approx 10^2 - 10^5$ may be approximated in logarithmic coordinates by a straight line

$$\log \Delta\varepsilon = A - m \log N_d, \tag{4.3a}$$

to which corresponds a power function

$$\Delta\varepsilon N_d^m = C. \tag{4.3b}$$

Using experimental values $N_d^{(1)}$ and $N_d^{(2)}$ for two ranges $\Delta\varepsilon^{(1)}$ and $\Delta\varepsilon^{(2)}$, the index of power m may be defined by the formula

$$m = \frac{\log(\Delta\varepsilon^{(1)}/\Delta\varepsilon^{(2)})}{\log(N_d^{(2)}/N_d^{(1)})}$$

and constant C—by Eq. (4.3b).

Here, $m > 0$ and $A = \log C$. At cyclic loading, these coefficients depend on the test temperature.

The similar approximate dependence between the number of cycles N_d and the range of plastic strain $\Delta\varepsilon_p$,

$$\Delta\varepsilon_p N_d^{m_p} = C_p \tag{4.4}$$

was proposed by Coffin (1954) and Manson (1966).

Some other approximating formulas for estimation of cycle number N_d in dependence of the ranges $\Delta\varepsilon$ or $\Delta\varepsilon_p$ were also offered.

Experiments referred to the third site of the fatigue curve show that dependence of elastic strain ranges $\Delta\varepsilon_e$ on the number of failure cycles N_d can also be approached in the form of power function

$$\Delta\varepsilon_e N_d^{m_e} = C_e. \tag{4.5}$$

Increase of cycle's number N_d with reduction of strain range $\Delta\varepsilon_e$ proceeds more quickly, than in the field of elastoplastic deformations. For many materials, it is possible to consider $N_d \to \infty$ at $\Delta\varepsilon_{-1} = 2\sigma_{-1}/E$, where σ_{-1} is an endurance (fatigue) limit taken usually 2×10^7 (or sometimes more).

Since Manson (1966), Eq. (4.5) is extended to the entire area of LCF in which the range of total strain is represented as a sum of the ranges

$$\Delta\varepsilon = \Delta\varepsilon_e + \Delta\varepsilon_p, \tag{4.6}$$

as it has been shown in Fig. 4.3.

It follows from the Eqs. (4.4)–(4.6) that

$$\Delta\varepsilon = C_p N_d^{-m_p} + C_e N_d^{-m_e}. \tag{4.7}$$

Equation (4.7) is supposed to be true both for elastoplastic and purely elastic areas of cyclic deformation.

Analysis of experimental data reveals that exponents in N_d change for various materials in rather narrow limits, namely $m_p \approx 0.5{-}0.6$ and $m_e \approx 0.06{-}0.16$.

4.2.2.2 Other Formulas

Extending Eq. (4.7) conditionally to cyclic deformation at small number of half-cycles $N_d = 1$ or $N_d = 10$, when $\Delta\varepsilon_e \ll \Delta\varepsilon_p$, the constant C_p may be presented as

$$C_p \approx \delta^{m_C} = \left[\ln\left(\frac{1}{1-\psi}\right)\right]^{m_C}. \tag{4.8}$$

The value δ is assumed to correspond with a "true" relative residual contraction of a specimen cross-sectional area under rupture (see Chap. 3).

Using the cyclic test results for torsion when destruction occurs without neck formation, the value δ can be defined by measured residual relative shear strain $\delta = 2\gamma_p$.

If a factor C_p is defined from a curve of initial half-cycle at $N_d = 1$, then $m_C = 1$, and for definition C_p by the 10th half-cycle, $m_C = m_p$.

Applying Eq. (4.7) to cyclic deformation with a great number of cycles $N_d \approx 10^6 - 10^7$ when $\Delta\varepsilon_e \gg \Delta\varepsilon_p$, the constant C_e can be connected with the endurance limit σ_{-1} in a form

$$C_e \approx \frac{2\sigma_{-1}}{E}, \tag{4.9a}$$

For an asymmetric cycle with mean stresses σ_m, Birger and Mavljutov (1986) offered to take

$$C_e \approx \frac{2\sigma_{-1}}{E}\left(1 - \frac{\sigma_m}{\sigma_{-1}}\right)\left(\frac{N_0}{N_d}\right)^{1/m}, \tag{4.9b}$$

where σ_{-1} is a limit of endurance for a cycle number $N_0 = 10^7$.

Under LCF tests of nominally identical specimens in the same conditions, a scatter of fatigue life can reach 100 % and more. This scatter is taken into account by specification of safety factors (see Chap. 8). It has been accented above that the approached formulas are recommended to be used only for qualitative or comparative estimations; otherwise, durability margins should increase even more. Tests show that the actual number of cycles N_d before destruction, taking into account the result's dispersion, can be significantly smaller than the number of cycles calculated by Eqs. (4.7)–(4.9a).

4.2.3 Failure Conditions for Cyclic Loading

4.2.3.1 Local and Global Failures

In operating conditions of a mechanical structure, the stress state is as a rule non-uniform. Then, the magnitude N_{d0} can be considered as a criterion for appearance of the LCF first local crack (usually in zones of stresses concentration). A cyclic durability margin (see Chap. 8) might be defined in relation to this cycle number. However, the structure can in some cases keep its work ability until a gradual growth of the crack results in the "avalanche-like" destruction. Corresponding evaluation can be carried out by methods of fracture mechanics with obligatory experimental confirmation.

At simultaneous action of variables and constant loads, deformation cycles become asymmetrical. Thus, the accumulation of fatigue damage is attended by gradual increase of plastic strain in the direction of prevailing constant loads. At the elevated temperatures, it is promoted by creep (see Chap. 7).

4.2.3.2 Synchronous Isothermal Cycles

In a simplest operation case, all external loads simultaneously (synchronously) change from zero values in the inherent state to the maximum values at the end of the *initial half-cycle* $(k = 0)$. If a temperature field does not change during loading and thermal stresses are absent, the calculated stresses and strains in all points of a structure increase in absolute size from zero values according to the curves $\sigma_i = f^{(0)}(\varepsilon_i, T)$ of usual unidirectional static deformation diagrams, since the loading is proportional. Quantities of $\sigma_i^{(0)}$, $\varepsilon_i^{(0)}$, and $\varepsilon_{pi}^{(0)}$ at the end of the initial half-cycle can be computed by the methods stated above in Chap. 3.

Computation of the *first half-cycle* of unloading $(k = 1)$, which finishes with the reduction of external loads to inherent zero values, can be fulfilled by the same methods, but with the use of the common cyclic deformation curve $\tilde{\sigma}_i = f(\tilde{\varepsilon}_i, T)$. At that time, plastic strains with reversed sign can occur. The computation of the first half-cycle defines the intensity ranges of all loop parameters $\Delta\sigma_i$, $\Delta\varepsilon_i$, and $\Delta\varepsilon_{pi}$ for the constant cycles. It is not required carrying out the computations of the subsequent half-cycles because the results will be the same in the set statement.

4.3 Thermal Fatigue

4.3.1 Non-isothermal Cyclic Processes

4.3.1.1 Role of Temperature Ranges

All cycle parameters depend on temperature. Experiments show that yet at stable cycles with a great temperatures range, the heating and cooling half-cycles can essentially differ, despite closure of a cycle, because of differing relations between their elastic and plastic sites.

When the change of temperature field during a cycle may be neglected, the cycle process can be considered as quasi-isothermal that allows defining its parameters by average temperature for a cycle. For essentially non-stationary cycles, the approximation of each half-cycle should be fulfilled separately under its average temperature.

The role of temperature ranges ΔT appears in three ways. Temperature stresses are summarized with stresses from external loads and join in the common range of stresses $\Delta\sigma_i$. Temperature dependence of cyclic deformation diagrams changes the calculated stress and strain ranges. The LCF characteristics of a material deteriorate as temperature rises.

4.3.1.2 Thermocyclic Tests

Tests for thermal fatigue, i.e., at cyclic heating and cooling, are usually carried out in tension–compression under hard loading (with constant total strain range) when the mechanical strain part $\Delta\varepsilon_\sigma$ is completely defined by thermal expansion $\Delta\varepsilon_T = \Delta(\alpha_T T)$

$$\Delta\varepsilon = \Delta\varepsilon_\sigma + \Delta\varepsilon_T = \text{const.} \tag{4.10a}$$

The strain range $\Delta\varepsilon_\sigma$ is summarized by elastic, plastic, and creep parts

$$\Delta\varepsilon_\sigma = \Delta\varepsilon_e + \Delta\varepsilon_p + \Delta\varepsilon_c. \tag{4.10b}$$

Such tests at $\Delta\varepsilon = 0$ have originally been offered by Coffin (1954) and then repeated by many other researchers.

Heating is usually provided by the transmission of electric current through the tubular specimen, and cooling—by air blowing through it. At a "saw-shape" cycle of temperature change $T_{\min} \Leftrightarrow T_{\max}$, trials go without retention at the maximum temperature that allows eliminating the influence of creep. Reviews of thermal fatigue test methods are presented in a number of works, for example, Shorr and Dul'nev (1964), Manson (1966), and others.

At full fastening of the specimen's ends (with $\Delta\varepsilon = 0$), compressing stresses from the temperature expansion change synchronously with temperature: The greatest compressing stress and plastic strain turn out at the maximum temperature, which leads to residual tensile stress at cooling.

In the course of tests at the set range of temperature ΔT, the arising stresses are measured up to destruction at N_d cycles. At a stable state of a material and absence of creep, the ranges of stress $\Delta\sigma$ remain constant, which allows calculating a function $N_d(\Delta\sigma)$ along with a function $N_d(\Delta T)$. This dependence is true only for that range of temperature at which the tests were operated. Strain characteristics are calculated using Eq. 4.3a. Results of such tests can be used for the estimation of cycle durability for given test conditions or for comparison of fatigue strength for various materials.

In order to establish the function $N_d(\Delta T)$ at constant strain range $\Delta\varepsilon$ but different stress $\Delta\sigma$ ranges, adjustable elastic shims are introduced into devices of Coffin's type.

4.3.1.3 Thermal LCF

A temperature range at the fixed minimum temperature of a cycle (usually normal) is defined by the maximum temperature $T_{\max}$.

Due to the results of thermocyclic tests, the number of cycles to failure can be approximately defined by the equation (Dul'nev in Birger and Shorr 1975)

$$N_d \approx 10^{\frac{A - T_{\max}}{B}}, \tag{4.11}$$

where $T_{\max}$ is taken in degree centigrade.

The factors A and B in Eq. (4.11) due to tests of a number of heat-resisting alloys may be estimated as A = (1000–1400) °C, B = (100–250) °C.

Results of thermal fatigue tests of different materials can be also approximately generalized by the formula

$$\Delta\varepsilon = \left(C_p + \frac{\sigma_B}{E}\right)\Delta\bar{\varepsilon}, \tag{4.12}$$

where the constant C_p is defined by Eq. (4.8) at $m_C \approx 0.6$.

Dependence of non-dimensional strain $\Delta\bar{\varepsilon}$ on cycle number N_d for five heat-resistant alloys tested at the temperatures $T_{\max} = 700 - 900\,°\mathrm{C}$ and $T_{\min} = 100\,°\mathrm{C}$ is shown in Fig. 4.8 (Dul'nev in Birger and Shorr 1975) where the area of experimental points is blacked out. Material characteristics σ_B, E, and ψ in Eqs. (4.8) and (4.12) correspond to temperature $T_{\max}$.

Results of thermocyclic tests can be also presented in the form of dependence $\Delta\varepsilon(N_d)$.

In logarithmic coordinates, within a range of cycle's number $10^2 - 10^4$, this averaged dependence for a constant temperature $T_{\max}$ is close to a straight line, i.e., can be displayed, e.g., by the equation

$$\Delta\varepsilon N^m = C, \tag{4.13}$$

where m, C are material characteristics depending on temperature $T_{\max}$.

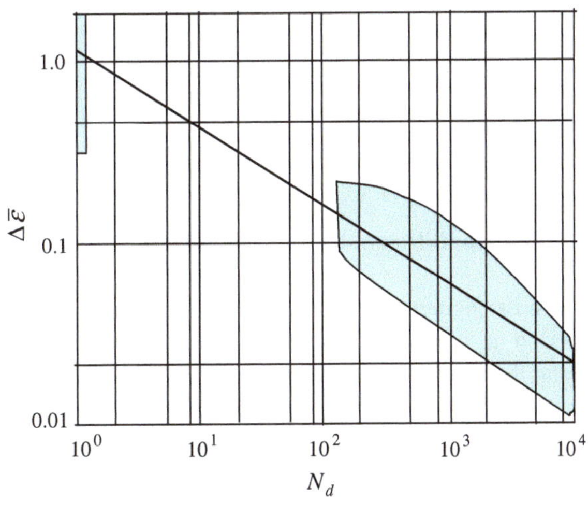

Fig. 4.8 Thermal fatigue diagram for five heat-resistant alloys

4.3.2 Factors Influencing Thermal Fatigue

The structures' resistance to thermal fatigue depends on a variety of factors (see, e.g., Manson 1966; Dul'nev in Birger and Shorr 1975). Influence of cycle duration will be considered in Chap. 7 in connection with a creep role. The main factors affecting thermal are listed below.

Static load. Thermocyclic loading of machine parts is usually combined with action of constant forces. As a result, the cycle of stress alteration becomes asymmetrical. As well as in case of high cycle fatigue, static loads can considerably reduce the number of cycles prior to failure, especially if the maximum temperature of the cycle increases.

Surface hardening leads to reduction of a material plasticity and usually reduces the resistance to thermal fatigue, especially at the large strain amplitudes.

Material structure and composition. Many observations show that increase of grain sizes reduces the resistance to thermal fatigue.

Resistance to thermal fatigue at the elevated temperature depends not only on high-temperature strength, but also on the coefficient of thermal expansion and heat conductivity.

In case of large amplitudes of plastic strains appearing in each cycle, the high plasticity of a material for all ranges of working temperatures is necessary. For small plastic strains, the material plasticity is of less importance in comparison with the strength properties.

Heat-resistant coatings result in an increase of heat stability if they possess sufficient plasticity providing non-separable adjacency of coating to the protected surface.

Stress concentration in the area of holes, fillets, fluting planes, etc., can considerably reduce (in several times) the number of thermal cycles prior to failure. The presence of stress concentrators leads to the local changes of temperature field and redistribution of temperature stresses. If plastic deformation arises in the mentioned areas, the maximum stresses decrease comparing with that for the material elastic behavior, while maximum strains increase.

4.3.3 Synchronous Non-isothermal Cycles

4.3.3.1 Conditions of Cycle Stabilization

In most engineering situations, when the temperature field changes with the same period as the loads, stable cycles with constant stress/strain relations can be established in all points of the body. In this case, the deformation diagrams for even and odd half-cycles might be different due to the influence of temperature change.

It has been demonstrated in Sect. 4.1.2.2 that cycles do not become stable at once.

If temperature in all points of a body changes so that extreme temperatures coincide with extreme loads, and the temperature's drop at cooling coincides with the end of unloading, then all cycles *of loading/heating* become identical. In the absence of residual stresses at moderate temperatures, the full *stress/strain* cycle (see half-cycle $k = 2$ in Fig. 4.4a, b) would also be closed from the beginning.

In a common case, due to arising of thermal residual stresses, the cycle terminates only after the stresses and strains at the beginning and at the end of a cycle become identical. This happens at the end of the second full cycle (i.e., at $k = 4$).

In some cases, loads and temperature alteration can occur simultaneously with the same period, but in different phases. In disks of turbomachines, stresses from centrifugal forces quickly rise at increase of rotational speed, while the growth of temperature and thermal stresses occur much slower as far as the disk is heated up; at the machine shutdown, stress from centrifugal forces quickly falls, while the disk is cooled more slowly. Such non-isothermal cycle repeats at each start of the machine. In these conditions, various loads change with an identical period from zero-to-maximum value which they do not reach simultaneously; but in the end of each cycle (including the first one), all loads come back to the initial zero value.

Under simultaneous change of loadings and temperatures, calculation of each half-cycle should be carried out by the step-by-step method described to Chap. 3 for unidirectional non-isothermal processes. Instead of this, it is possible to assume that within each half-cycle, change of loads at constant temperature and temperature change at constant loads occur consecutively.

Next example shows that non-isothermal cyclic processes are affected to a large extent by temperature dependence of the elasticity modulus (Shorr 2013).

4.3.3.2 Numerical Example

As an example, we consider formation of a closed cycle for the following non-isothermal problem. A rod is periodically loaded in the axial direction and heated up so that its mechanical strain $\varepsilon_\sigma = \varepsilon - \alpha_T \Delta T$ changes on the symmetric sawtooth law of constant range $\Delta\varepsilon_\sigma$ (Fig. 4.9a). All cycle parameters are displayed versus process mark t. The temperature changes step wisely from T_1 to T_2 and back (Fig. 4.9b).

At the beginning of the process, there are $\varepsilon_\sigma = 0$ and $T = T_2$ (see points O in Fig. 4.9).

In Fig. 4.10a, a process of stress and mechanical strain variation for three half-cycles for the example of Fig. 4.9 is displayed by bold arrows. Elastoplastic diagrams for temperatures T_1 and T_2 are shown by double lines and the borders of the elastic area for these temperatures—by lines I and II. The bilinear approximation of diagrams is used. The ideal Bauschinger's effect is assumed.

The corresponding graphics of plastic strain and stress dependences on process mark t are shown above in Fig. 4.9c, d. Compatible points in all graphics are noted by identical numerals, and the termination moments for three half-cycles are allocated by asterisks. Jumps in stresses are connected with transitions to other cycle

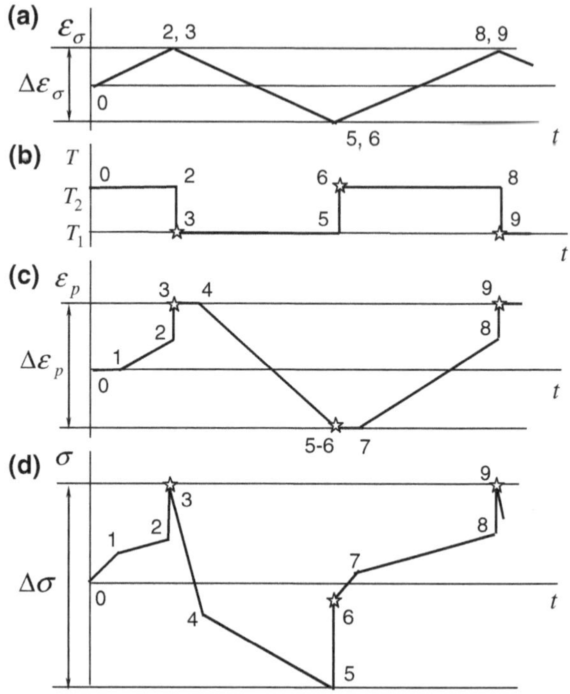

Fig. 4.9 Example of transition to closed deformation cycle at non-isothermal loading

deformation diagrams induced by temperature change under the current plastic strain. Computation reveals the values of stress $\Delta\sigma$ and plastic strain $\Delta\varepsilon_p$ ranges.

From Figs. 4.9 and 4.10, it is visible that after the third half-cycle, i.e., before the following second unloading (point 9), initial conditions are just the same as before the antecedent first unloading (compare points 3 and 9). After this moment, the subsequent complex cycles become identical.

The process of the previous problem has begun from the point in which the strain has the average value for a cycle, and the temperature was maximal. Generally, the average values of loads are not defined in advance. Moreover, all loads usually grow from zero, and a temperature—from normal conditions, which correlates with point 5 in Fig. 4.9a, b. Assuming that the stable cycle will start from a point lying on a border of the coldest area (in the given example in a point 5), the similar analysis shows that the cycle becomes closed not after the third half-cycle, but after the second one, as shown in Fig. 4.10b. The range values of stresses and plastic strains remain the same, as in the previous case.

If a constant deformation cycle is established after the several non-stationary periods, then the diagram corresponding to the first established half-cycle $k = 1$ should be accepted for definition of the stable loop variables.

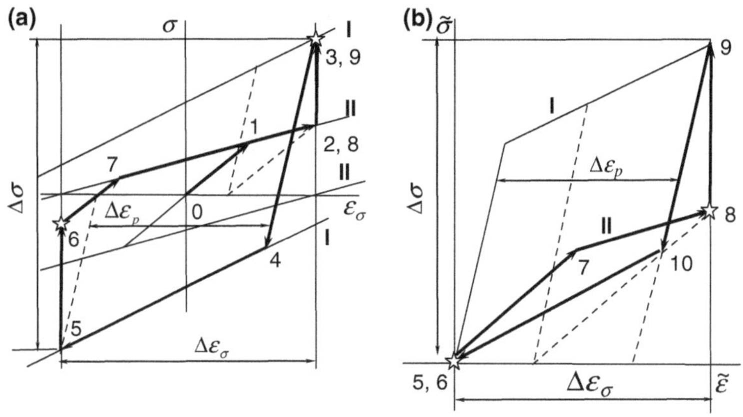

Fig. 4.10 Process of stress/strain variation at non-isothermal loading (see Fig. 4.9)

4.4 LCF Under Variable Cycles

Detailed step-by-step 3D computations of cyclic stress/strain state at varying loads and temperatures are possible with the use of universal software codes (see Chap. 5).

But in case of experimental data deficiency, a computation of consecutive cycles becomes a rather mathematical procedure, and not modeling of a real physical process. With reference to a great number of cycles, it is more rational to use approaches based on specific properties of cyclic processes and special computational methods.

The stress/strain cycles are variable because of unsteady loading and heating or unstable material state. The both cases are considered below.

4.4.1 Unsteady Loading and Heating

4.4.1.1 Cyclic Damage

For multiple deformation cycles with constant value of plastic strain range $\Delta\varepsilon_p$ in each half-cycle, the number of cycles prior to local LCF failure can be directly defined using experimental diagrams of cyclic deformation.

In order to use these diagrams for LCF calculation at varying cycles, different methods of material damage summing have been proposed.

The simplest measure of *cyclic damage* during N half-cycles at constant plastic strain range $\Delta\varepsilon_p$ is a ratio of the number N to half-cycle number N_d prior to failure

$$D(N) = \frac{N}{N_d}, \tag{4.14}$$

where $N_d(\Delta\varepsilon_p, T)$ depends on $\Delta\varepsilon_p$ and temperature T, e.g., due to Eq. (4.4).

Assuming uniform damage accumulation during entire cyclic process, we find a constant damage increment for each half-cycle

$$\Delta D = \frac{1}{N_d}. \tag{4.15}$$

At the beginning of loading, there is $D = 0$, at a failure point

$$D_d = D(N_d) = 1. \tag{4.16}$$

As well as in Chap. 3, the term "element damage" describes any irreversible structural changes leading eventually to destruction of the material element.

Multiplying numerator and denominator of Eq. (4.14) by a constant plastic strain range $\Delta\varepsilon_p$, we obtain

$$D(N) = \frac{\varepsilon_\Sigma^p(N)}{(\varepsilon_\Sigma^p)_d}, \tag{4.17}$$

where

$$\varepsilon_\Sigma^p(N) = N\Delta\varepsilon_i^p \tag{4.18}$$

is named *accumulated plastic strain* (APS) over N identical half-cycles $\Delta\varepsilon_i^p$ and

$$(\varepsilon_\Sigma^p)_d = \varepsilon_\Sigma^p(N_d) = N_d\Delta\varepsilon_i^p \tag{4.19}$$

is the APS magnitude reached at the failure point (Temis 1989).

A ratio

$$\bar{\varepsilon}_\Sigma^p(N) = \frac{\varepsilon_\Sigma^p(N)}{\varepsilon_\Sigma^p(N_d)} = D(N) \tag{4.20}$$

presents a *relative accumulated plastic strain* (RAPS) during N half-cycles. This factor coincides with material cyclic damage D during the same half-cycle number.

A method of linear summation of damages, expressed in terms of ratio N/N_d and apparently offered by Palmgren, is sometimes named "Miner's rule."

4.4.1.2 LCF Charts for the Stable State of a Material

Equation (4.4), bonding destroying cycle number N_d of LCF with constant range of plastic strain, can be transformed to

$$(\varepsilon_{\Sigma}^{p})_d N_d^{m_p - 1} = C_p. \tag{4.21}$$

Equations (4.4) and (4.16) may be written in logarithmic coordinates

$$\begin{aligned} \log \Delta\varepsilon_p &= \log C_p - m_p \log N_d, \\ \log(\varepsilon_{\Sigma}^{p})_d &= \log C_p + (1 - m_p) \log N_d. \end{aligned} \tag{4.22}$$

Corresponding plots for a real case are displayed in Fig. 4.11a where the first Eq. (4.22) is shown by the bold line I, and the second one (log of the APS)—by the line II. Strains are expressed in percentage, value of relative residual elongation at rupture is assumed to be equal to $\delta = 9\,\%$, $m_p = 0.5$, and the APS is marked $\varepsilon_{p\Sigma}$.

Equations (4.22) has the common beginning at the point $N_d = 1$ where $(\varepsilon_{\Sigma}^{p})_d = \Delta\varepsilon_p = C_p = \delta$. Then, they disperse into two straight lines, so as the line I demonstrates a usual experimental curve of a LCF. At any constant range of the plastic strain denoted by point A, the cyclic failure occurs in point B with corresponding APS shown in point C of the line II.

Equations (4.22) yields

$$\log(\varepsilon_{\Sigma}^{p})_d = \log\, \Delta\varepsilon_p + \log N_d, \tag{4.23}$$

so the dependence $\log(\varepsilon_{\Sigma}^{p})_d = f(\log N_d)$ is easily constructed on the base of the experimental line $\log \Delta\varepsilon_p = f(\log N_d)$.

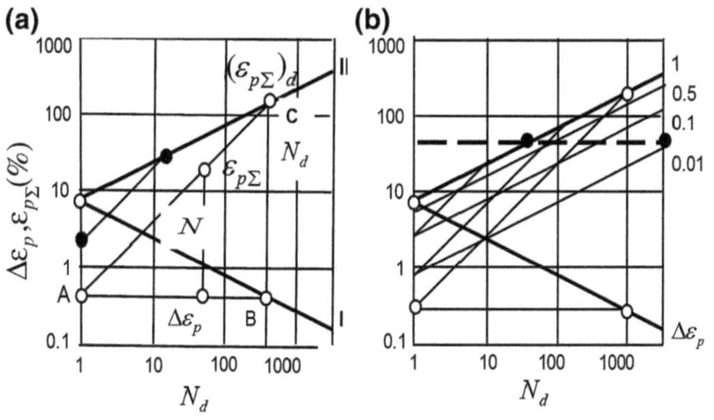

Fig. 4.11 Dependence between destroying half-cycle number N_d and plastic strain parameters $\Delta\varepsilon_p$ and $\varepsilon_{p\Sigma}$

With variation of the exponent m_p, slopes of the lines I and II change, but the distance between them all of the time is defined by the varying value of $\log N_d$. For usual values $m_p = 0.5 - 0.6$, the lines are almost equally (but inversely) inclined to the axis of $\log N_d$.

The number N_d of destroying cycles can be found as a function of $\Delta\varepsilon_p$ or $(\varepsilon_\Sigma^p)_d$ by any of the following equations

$$N_d = \left(\frac{C_p}{\Delta\varepsilon_p}\right)^{\frac{1}{m_p}} = \left(\frac{(\varepsilon_\Sigma^p)_d}{C_p}\right)^{\frac{1}{1-m_p}}. \tag{4.24}$$

Thin line AC shows the increase of the strain ε_Σ^p as half-cycle number N rises. From Fig. 4.11a, it is visible that destruction occurs not at certain size of APS, but in very wide range of its values depending on at what range of plastic strain their summation is carried out (cp. inclined lines with light and dark points).

In Fig. 4.11b, the inclined lines of APS corresponding to constant damage values $D = 0.01-1$ where $D = (\bar{\varepsilon}_\Sigma^p)_d$ are displayed. The values of damage D are indicated in the right scale of the graph. A dashed line referring to one of the constant quantity $\varepsilon_{p\Sigma} \approx 50$ is also shown. It is seen that the identical quantities APS answer values of damage $D = (\bar{\varepsilon}_\Sigma^p)_d$ which differ by several orders of amplitude.

Thus, under elastoplastic cyclic loading, the APS quantity is not a material structural parameter reflecting the damage accumulation and must be considered only as a possible form of representation of LCF test results for various ranges of plastic strains. There are more grounds to present the current value of the governing damage parameter χ through the relative accumulated plastic strain (RAPS)

$$\chi(N) = \bar{\varepsilon}_\Sigma^p(N), \tag{4.25}$$

as it was proposed by Temis and Shorr (2011).

In the theory of plasticity, the concept of "accumulated plastic strain" ε_Σ^p (APS), often named as Odqvist's (1933) parameter, is widely used for the estimation of material damage increasing during any deformation process, see Chap. 3.

In case of uniaxial processes, APS is defined by summation of absolute values of plastic strain increments $\varepsilon_{p\Sigma} = \int |d\varepsilon_p|$. The value of APS is assumed to be $\varepsilon_{p\Sigma} = \delta$ at the moment of material's failure. It makes sense to distinguish the APS from the current plastic strain $\varepsilon_p = \int d\varepsilon_p$ only for sign-variable processes when during some moments, the inequality $\varepsilon_p d\varepsilon_p < 0$ holds true. Always, the value of APS is $\varepsilon_{p\Sigma} \geq \varepsilon_p$.

For multi-axial process, the APS is usually defined as $\varepsilon_{p\Sigma} = \int (d\varepsilon_p)_i$ where $(d\varepsilon_p)_i = (2d\varepsilon_{ij}^p d\varepsilon_{ij}^p/3)^{0,5}$ is the intensity of plastic strain increments.

Under unidirectional proportional loading (even with unloading into elastic area and repeated loading in the previous direction), the APS coincides with the current value of plastic strain intensity ε_i^p described by the generalized deformation diagram. In this case, the maximum value of ε_Σ^p for structural materials usually lies within $\delta \approx (10 - 50)\%$.

For cyclic processes, a quantity of APS—ε_{Σ}^{p} increases in each half-cycle for the width of a hysteretic loop $\Delta\varepsilon_{i}^{p}$ so after N half-cycles, it becomes

$$\varepsilon_{\Sigma}^{p}(N) = \sum_{k=0}^{k=N} (\Delta\varepsilon_{i}^{p})^{(k)}, \tag{4.26}$$

and for great number of cycles can reach the values $\sim$500 % and more (see, Fleming and Temis 2002; Putchkov et al. 1995).

Therefore, the influence of APS on cyclic processes cannot be in any way described by the results of sign-constant loading tests.

4.4.1.3 Summation of Cyclic Damage

For the approximate computation of cyclic damage D received during many varying half-cycles, a method of *damage linear summation* can be used. According to this method, the damage acquired by a material during N half-cycles is

$$D(N) = \sum_{k=0}^{k=N} \Delta D^{(k)}, \tag{4.27}$$

where $\Delta D^{(k)}$ is the damage received during the kth half-cycle.

Summation of damages due to Eq. (4.26) encloses all half-cycles, including the initial one ($k = 0$), and is valid both for identical and for various cycles.

Assuming that Eq. (4.15) is correct for any half-cycle and the damage process does not depend on an order of alternation of cycles, Eq. (4.27) yields

$$D(N) \approx \sum_{k=0}^{k=N} \frac{1}{N_d\left[\Delta\varepsilon_p^{(k)}, T^{(k)}\right]}, \tag{4.28}$$

where $N_d(\Delta\varepsilon_p^{(k)}, T^{(k)})$ depends on $\Delta\varepsilon_p^{(k)}$ and temperature $T^{(k)}$ due to Eq. (4.4).

The destruction would occur at the total number of half-cycles $N_{d\Sigma}$ when

$$D(N_{d\Sigma}) = \sum_{k=0}^{k=N_{d\Sigma}} \frac{1}{N_d\left[\Delta\varepsilon_p^{(k)}, T^{(k)}\right]} = 1. \tag{4.29}$$

Tests show that in various situations, including varying temperature, destructions from LCF may be described better using *nonlinear summation*. Calculation accuracy can be also improved if some experimental factor a_{Σ} is introduced in Eqs. (4.15) and (4.29) instead of unity. A quantity of this factor is usually $a_{\Sigma} = 0.6{-}0.8$, but in some cases, the values up to $a_{\Sigma} = 1.2$ were observed (Dul'nev in Birger and Shorr 1975).

A value of APS accumulated during the N cycles is calculated by Eq. (4.26), and an average size of plastic increments referring to N half-cycles is

$$\Delta\varepsilon_m^p = \frac{1}{N}\varepsilon_\Sigma^p(N). \tag{4.30}$$

At destruction point there is

$$(\Delta\varepsilon_m^p)_d = \frac{1}{N_{d\Sigma}}\varepsilon_\Sigma^p(N_d). \tag{4.31}$$

During varying loading conditions, the values of plastic range $\Delta\varepsilon_p^{(k)} = f(N)$ can alternate under any law, for example, as shown in Fig. 4.12 by a curve AD. Accumulated strain ε_Σ^p increases along the corresponding curve AE. The function $\varepsilon_\Sigma^p(N)$ may be defined by consecutive summation of ranges $\Delta\varepsilon_p^{(k)}$ due to Eq. (4.26) that can be resolved into the average range $\Delta\varepsilon_{pm} = \varepsilon_\Sigma^p(N)/N$. (In Fig. 4.12, it is accepted that $\Delta\varepsilon_{pm}$ coincides with the initial range in point A).

The LCF failure occurs at the number of cycles N_d corresponding to the value $\left(\varepsilon_{p\Sigma}\right)_d$ in point C. The relevant average value of strain range $\Delta\varepsilon_{pm} = (\varepsilon_{p\Sigma})_d/N_d$ is shown by point B. It is possible to expect that positions of points B and C at a stable state of a material are near to corresponding points on lines I and II, received at constant strain cycles with the average range $\Delta\varepsilon_{pm}$.

The value N_d can be defined by Eq. (4.4) in which it is necessary to take $\Delta\varepsilon_p = \Delta\varepsilon_{pm}$ or by Eq. (4.20) using APS value $(\varepsilon_\Sigma^p)_d = \Delta\varepsilon_{pm}N_d$.

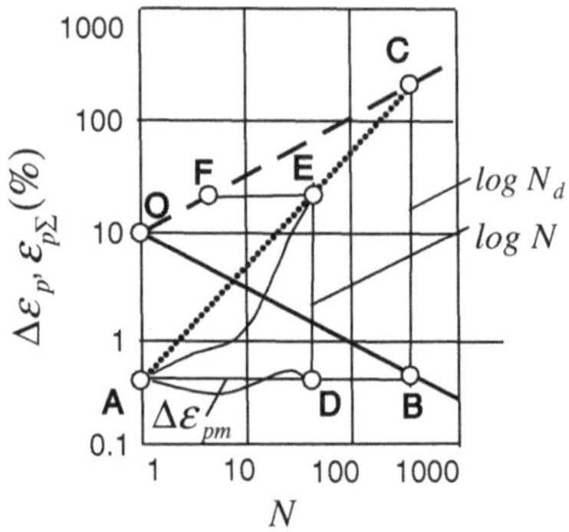

Fig. 4.12 Dependence between destruction half-cycle number N_d and plastic strain parameters at variable deformation cycles

4.4.1.4 Multi-regime Cycles

In practice, some sequence of different cycles often forms a typical *multi-regime (complex) operational cycle.* If the complex cycles are identical, they generate a stable elastoplastic cyclic process.

Thus, many machines are maintained in various service conditions which can be presented as a periodic set n of several most typical regimes q: operation at peak, nominal, intermediate, and lowered loads. Repeated transitions from one regime to another are defined as deformation cycles with different stress and temperature ranges. Under such conditions, deformation cycles become variable even for a stable state of the material.

An order of the regimes alteration usually may be neglected. Then, the homogeneous regimes within each set may be combined into groups, with cycle numbers being summed up. The cycle damage accumulation D_{cc} is computed for the complex operation cycle, and then for all such cycles, $D_{\Sigma} = nD_{cc}$.

Damage D_{cc} accumulated during a complex cycle is determined by Eq. (4.28)

$$D_{cc} = \sum_{k=1}^{k=q} \frac{N^{(k)}}{N_d^{(k)}\left[\Delta\varepsilon_p^{(k)},\ T^{(k)}\right]} \tag{4.32}$$

where $k = 1, 2, \ldots, q$, $N^{(k)}$ is a number of cycles corresponding to the regime k, and $N_d^{(k)}$ is a number of cycle prior to failure under this operating regime.

The number N_d of complex operating cycles prior to failure corresponding to the magnitude $D_{\Sigma} = 1$ can be found from the equality

$$N_d = D_{cc}^{-1}. \tag{4.33}$$

For calculation of cyclic damage accumulation D_{cc} over a complex cycle, it is necessary to present a process $\varepsilon_i^p = f(t)$ as a set of the closed cycles of various ranges $(\Delta\varepsilon_i^p)^{(k)}$, see Eq. (4.26). The greatest damage, as a rule, brings the total operating range of plastic strains $(\Delta\varepsilon_i^p)_{\max} = \varepsilon_{i,\max}^p - \varepsilon_{i,\min}^p$, which always must be included in the analysis. In order to calculate contributions of other regimes into damage accumulation, various modifications of a "rainflow" method (see, e.g., Downing and Socie 1982) are used.

Reliability of LCF estimations at complex cycles transformed into combination of the simple cycles must be further investigated.

4.4.1.5 Quasi-stable Cycles

There are some complex process whose separate simple cycles can be considerably different, but the *average* plastic increment $\Delta\varepsilon_{pm}$ for the complex cycles is stable from the process beginning or is quickly stabilized in the course of increase of the cyclic number. Such quasi-stable cycles have important engineering value.

Assuming the increment $\Delta\varepsilon_{pm}$ to be constant up to destruction, we can find the value $N_d = f(\Delta\varepsilon_{pm})$ and calculate the relative accumulated plastic strain (RAPS) corresponding to the current number N of complex cycles

$$\bar{\varepsilon}^p_\Sigma(N) = \frac{N}{N_d} = D(N). \tag{4.34}$$

The factor RAPS varies within the same borders as a damage factor D, i.e., in the interval $0 \le \bar{\varepsilon}^p_\Sigma(N) \le 1$.

Sometimes, cyclic loading under casual alternation of stress and strain ranges can also be considered as quasi-stable with constant value of average increment $\Delta\varepsilon_{pm}$. The value $\Delta\varepsilon_{pm}$ can be found by calculations or measures for a limited part N of the casual process. Then, the number of the cycles up to failure $N_d = f(\Delta\varepsilon_{pm})$ may be calculated by Eq. (4.4).

4.4.2 Unstable Material Characteristics

4.4.2.1 Unstable Cycles and Their Stabilization

Variability of a deformation cycle under constant conditions of loading and heating can be provoked by material cyclic hardening or softening.

Rates of process stabilization (see Sect. 4.1.2.2) depend on the level of stresses and temperatures. Sometimes, transition to an almost stable state comes to the end in two–three cycles, while for some materials, it occurs during many hundreds of cycles or continues until structure failure. Non-monotonic change of characteristics accompanied by transition from hardening to softening and visa versa is also possible.

Analysis of deformation diagrams for cyclic processes in dependence of non-stationary development of the plastic strain ε_p requires instrumental measurement of process parameters $\sigma(t)$, $\varepsilon(t)$ in the course of specimens testing. Using this information, the unclosed hysteretic loops of $\sigma(\varepsilon)$ can be consecutively constructed and increments of elastic $\Delta\varepsilon_e$ and plastic $\Delta\varepsilon_p$ strains for each half-cycle can be established.

Cycle evolution at non-stable material state proceeds, as a rule, rather smoothly that allows considering each cycle as quasi-closed and overall deformation process as quasi-cyclic.

4.4.2.2 Approximation of Cyclic Deformation Diagrams

Deformation cycles for unstable state of a material become open. Therefore, computations for each half-cycle must be performed separately, using the final conditions of the previous k half-cycle as the initial conditions of subsequent $(k+1)$ one. If stabilization occurs after several cycles, then the cyclic deformation diagrams $\tilde{\sigma}_i = f^{(k)}(\tilde{\varepsilon}_i)$ can be represented in the graphic or tabular form at different temperatures. At a great number of unstable cycles, it is necessary using the approximate analytical representation of cyclic diagrams.

All half-cycles of elastoplastic deformation have qualitatively similar character in the form of two or three sites—linearly elastic, nonlinearly plastic, and in many cases (at the large width of a hysteretic loop) almost linearly plastic site with constant tangential modulus. This allows a principle of similarity of half-cycles or their linear and nonlinear parts (Temis 2005) to be used for approximation.

Starting from Masing (1926), it is accepted to take the monotonous static loading curve $\sigma = f^{(0)}(\varepsilon)$ of initial (zero) half-cycle as a basis for the curve $\tilde{\sigma} = f^{(k)}(\tilde{\varepsilon})$ in tracking coordinates of any half-cycle $k = 1, 2, \ldots$ description. The various formulas using such approach are described in many reviews on material cyclic deformation, i.e., Moskvitin (1965), Shneyderovich (1968), Putchkov and Temis (1988), and others.

However, it is more rational to accept a curve of the first full half-cycle or the first unloading $\tilde{\sigma} = f^{(1)}(\tilde{\varepsilon})$ as the such basis diagram (see for example, Gusenkov 1979). As shown in many experiments, the curves of subsequent half-cycles differ from this curve much less than from the initial one. The function $\tilde{\sigma} = f^{(1)}(\tilde{\varepsilon})$ is always detected at cyclic tests, and its experimental definition does not meet any difficulties.

In the plastic area, it is convenient to transform separately the yield point $\sigma_y^{(k)}$ and a nonlinear curve at a site $\tilde{\sigma}^{(k)} - \sigma_y^{(k)}$. The latter can be represented by the function $\tilde{f}^{(k)} = f^{(k)} - \sigma_y^{(k)}$ describing the stress increase within the plastic area. Yield point in strains $\varepsilon_y^{(k)}(T) = \sigma_y^{(k)}(T)/E(T)$ and a function $\tilde{f}^{(k)}$ should be defined experimentally.

In Fig. 4.13, a cyclic diagram of the first full half-cycle is shown by line 1, where designations $\sigma_y^{(1)} = \sigma_{y1}$ and $\tilde{\sigma}^{(1)} - \sigma_{y1} = \tilde{f}^{(1)}(\tilde{\varepsilon})$ are taken. Note, from Masing's approach, $\sigma_{y1} = 2\sigma_{y0}$.

The diagram of cyclic deformation for unstable state of a material may be described for a half-cycle (k) by various means, e.g., by the equations

$$\tilde{\sigma}^{(k)}(T) = \begin{cases} E(T)\tilde{\varepsilon}^{(k)} & \text{at } \tilde{\varepsilon}^{(k)} \leq \varepsilon_y^{(k)}(T), \\ E(T)\varepsilon_y^{(k)} + \tilde{f}^{(k)}\left\{ \left[\tilde{\varepsilon}^{(k)} - \varepsilon_y^{(k)}(T)\right], T \right\} & \text{at } \tilde{\varepsilon}^{(k)} > \varepsilon_y^{(k)}(T) \end{cases}. \tag{4.35}$$

The functions $\varepsilon_y^{(1)}(T)$ and $\tilde{f}^{(1)}\{ [\tilde{\varepsilon} - \varepsilon_y^{(1)}(T)], T\}$ can be defined by the test for the first half-cycle, $k = 1$.

Fig. 4.13 Cycle curve transition from the 1st full half-cycle to half-cycle k

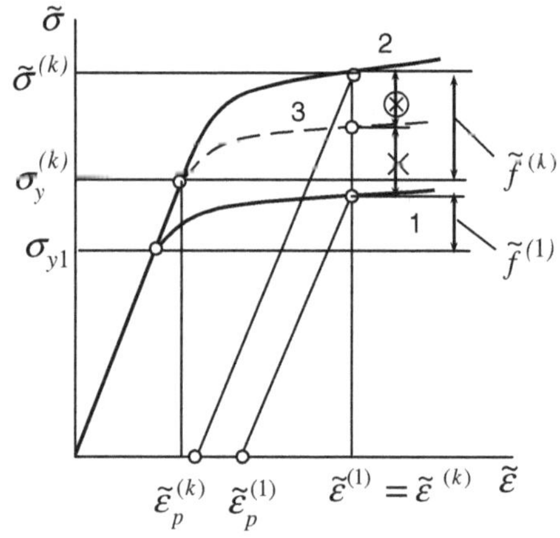

Yield point in strains $\varepsilon_y^{(k)}$ and curve $\tilde{f}^{(k)}$ for half-cycle k can be expressed by corresponding parameters of the first half-cycle as

$$\varepsilon_y^{(k)} = \frac{\sigma_y^{(k)}}{E} = \frac{A(k)\sigma_{y1}}{E}, \tag{4.36}$$

$$f^{(k)} = B(k)f^{(1)}. \tag{4.37}$$

At a stable state of a material, there are $A(k) = B(k) = 1$, and for non-isothermal cycles, they depend on temperature.

At combined stress states, Eq. (4.35) is represented in the form of relations between stress and strain intensities.

In Fig. 4.13, a specification of half-cycle k curve is schematically shown. The basis dependence $\tilde{\sigma} = f^{(1)}(\tilde{\varepsilon})$ is presented by curve 1 assuming $A(k) = 1.3$ and $B(k) = 2$. The dotted curve 3 shows a curve 1 transferred to a new position corresponding to new value of the yield point $\sigma_y^{(k)} = 1.3\sigma_{y1}$. Increments of the yield point $\sigma_y^{(k)}$ and the nonlinear function part $f^{(k)} = 2f^{(1)}$ are shown by badges $\times$ and $\otimes$, respectively. The effect of hardening that leads to appreciable reduction of a plastic deformation part from $\tilde{\varepsilon}_p^{(1)}$ to $\tilde{\varepsilon}_p^{(k)}$ at the same value $\tilde{\varepsilon}^{(1)} = \tilde{\varepsilon}^{(k)}$ is clearly visible.

In the literature, various forms of the cycle description and of a choice of the governing parameter χ are offered (see, e.g., Temis 1989, 1994, 2005; Chaboche 1989). Sometimes, such dependence is described by the aid of one function that complicates approximation. In some works, a finer description of changing hysteretic loops is reached by the increase of the quantity of experimental factors. However, a more precise approximation of dependence $\tilde{\sigma}^{(k)} = f^{(k)}(\tilde{\varepsilon}^{(k)}, \chi)$, obtained for a specific material under certain test conditions, does not raise reliability of LCF computation for other materials. It is necessary to keep in mind that different approximate transformations of curves $\tilde{\sigma} = f(\tilde{\varepsilon})$ can bear a formal character and do not show the real physics of cyclic processes.

Computations show that as far as the cycle number N increases, the RAPS usually increases as well as absolute APS. However, sometimes a small temporary reduction of value $\bar{\varepsilon}_{\Sigma}^{p}$ is possible if current values $\Delta\varepsilon_p$ decrease (in case when the destroying number of half-cycles $N_d(N)$ increases faster than the average value $\Delta\varepsilon_{pm}(N)$ decreases). In these cases, it is necessary to leave the parameters of a half-cycle without change.

4.4.2.3 Failure at a Different Loading Program

Experimental dependences $\Delta\varepsilon_i^p = f(N_d)$ and $\varepsilon_{\Sigma d}^p = f(N_d)$ for unstable state of a material become curvilinear in logarithmic coordinates. Assuming valid a principle of damage linear summation, we can calculate the functions $\varepsilon_{\Sigma}^p(N)$ and $\Delta\varepsilon_m^p(N)$ for any loading process using Eqs. (4.30) and (4.31) and find then points of their crossing with corresponding LCF curves. These points can roughly define a number N_d of cycles prior to failure for the given process.

In Putchkov et al. (1995), a diagram $\left(\varepsilon_{\Sigma}^p\right)_d = f(N_d)$ for one of the titanic alloy obtained by data processing of test results by both authors and other authors referred to different loading programs has been presented. In fact, specific dependences $\varepsilon_{\Sigma d}^p = f(N_d)$ correspond to each loading program. Some difference in test results at soft and rigid loadings also specifies small instability of materials properties. In spite of this, a general scatter of experimental points referred to various test programs allows, with comprehensible accuracy, the averaging of test results within cycle number $N_d \leq 1000$ (corresponding to $(\varepsilon_{\Sigma}^p)_d \leq 300\,\%$) using a *single straight line* $\log N_d = f[\log(\varepsilon_{\Sigma}^p)_d]$ analogous to the line II in Fig. 4.11a. An averaged value of the function $\log N_d = f(\log \Delta\varepsilon_{pm})$ has also a view of a straight line analogous to the line I in this figure.

A function $\Delta\varepsilon_{pm}(N_d)$ is, in essence, a curve constructed due to Eq. (4.4), but extended to a case of variable cycles, including a material unstable state. The possibility of such approximation of test results, referred to different loading programs, by a common function $\varepsilon_{\Sigma d}^p(N_d)$ demands further research.

In this way, the experimental diagram obtained at a stable state of a material can be used for approximate computation of cycle number N_d at any change of half-cycles ranges $\Delta\varepsilon_p = f(N)$. Let us assume the ranges $\Delta\varepsilon_p^{(k)}$ to be stepwise alternated, taking values of $\Delta\varepsilon_p^{(k)} = 0.5\,\%$ at $k = 1, 2, 5, 6 \ldots$ and $\Delta\varepsilon_p^{(k)} = 1.5\,\%$ at $k = 3, 4, 7, 8\ldots$, etc. In Fig. 4.14a, this area is blacked out.

The average value $\Delta\varepsilon_{pm} = f(N)$ marked by curve AEC quickly (beginning from a point E) takes the value of $\Delta\varepsilon_{pm} = 1\%$. Point D on the line II correlates with point E. The value of APS increases by shaped curve AB. Due to Eq. (4.29) for value $\Delta\varepsilon_{pm} = 1\%$ at $C_p = \delta = 10\%$, and $m_p \approx 0.5$, we find $N_d = 100$ half-cycles (point C). At that time, Eq. (4.30) gives the value $(\varepsilon_{\Sigma}^p)_d = 100\,\%$ (point B).

4.4.2.4 Non-isothermal Two-Level Process

In Fig. 4.14b, two pairs of curves referred to cyclic damage at two temperatures $T_2 > T_1$ are shown. Material characteristics δ and m_p are assumed to be temperature dependent. Pair lines I refer to function $\bar{\varepsilon}_{\Sigma}^{p} = f(N_d, T)$, the second pair II—to function $\Delta\varepsilon_p = f(N_d, T)$.

Let us consider a two-level process of loading and heating shown in Fig. 4.14a (with the same designations of characteristic points).

For the set program of cyclic loading $\Delta\varepsilon_p = f(N)$, a current quantity of accumulated plastic strain $\varepsilon_{\Sigma}^{p} = f(N)$ does not depend on temperature (points A and C). But a relative value $\bar{\varepsilon}_{\Sigma}^{p}$ characterizing damage accumulated during the cycles N will be different. Failure under higher temperature T_2 will occur earlier than under the lower one (cp. points D and B).

After temperature increases from T_1 to T_2 at intermediate point E, growth of $\bar{\varepsilon}_{\Sigma}^{p}$ accelerates and destruction will occur at some point F allocated between points D and B as noted by asterisk.

In Fig. 4.15, an example of definition of the damage number N_{eqv}, d for an operation at two regimes is shown. The pattern meets the following data: $N_{1d}(T_1) = 2000$, $N_{2d}(T_2) = 1000$, $N_1 = N_2 = 500$, and $N_{\Sigma} = 1000$.

Under constant temperatures T_1 and T_2, the failure occurs accordingly in points B and D. The damage under the loading program $T_1 \rightarrow T_2$ will progress along the trajectory OAE, and under the program $T_2 \rightarrow T_1$—along the trajectory OCE.

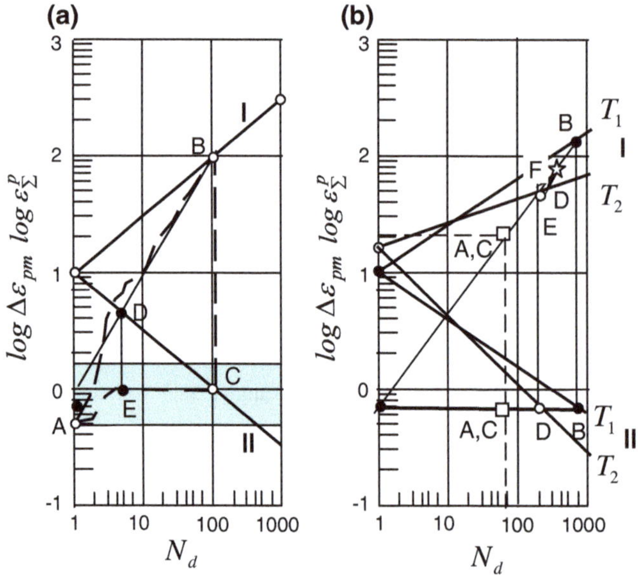

Fig. 4.14 Calculation of half-cycle number N_d prior to failure: **a** at double-level loading, **b** at different temperatures

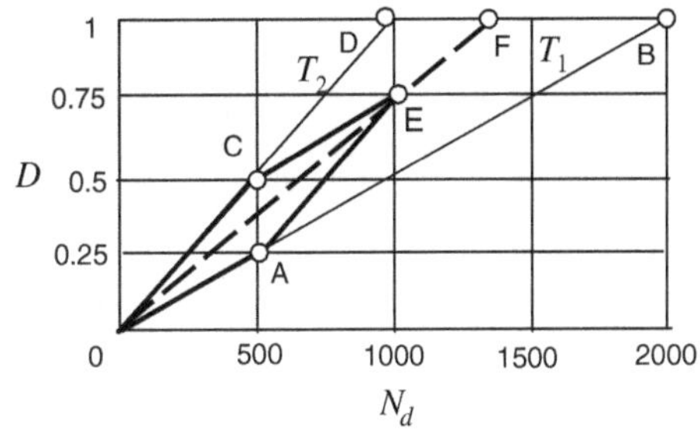

Fig. 4.15 Calculation of damage cycles N_d at non-isothermal loading

A total damage $D = 0.75$ for $N_\Sigma = 1000$ is identical in both trajectories. The line OE corresponds to an "equivalent" regime with $N_{eqv} = N_\Sigma$. If we assume the further loading to proceed under the same equivalent regime (along the dashed line EF), we find $N_{eqv}, d = 1333$ (point F), which corresponds to computation by Eqs. (4.32)–(4.33).

References

Arutjunjan RA (1964) About cyclic loading of elastoplastic medium (in Russian). Izv.AN SSSR, Mekh. i Matem 4:89–92

Bauschinger J (1881) Uber die Veranderung der Elastizitatsgrenze und des Elastizitats moduls Verschiedener Metall. - Zivilingenier: 284–348

Birger IA, Balashov BF (eds) (1981) Constructive strength of materials and parts of gas-turbine engines (in Rusian). Mashinostroyenie, Moscow

Birger IA, Mavljutov RR (1986) Strength of materials (in Russian). Nauka, Moscow

Birger IA, Shorr BF (eds) (1975) Thermal strength of machine parts (in Russian). Mashinostroenie, Moscow

Chaboche JL (1989) Constitutive equations for cyclic plasticity and cyclic viscoplasticity. Int J Plast 5:247–302

Chaboche JL, Nouailhas D (1989) Constitutive modeling of ratchetting effects Part I: experimental facts and properties of the classical models. J Eng Mater Technol 111(4):384–392

Coffin LF (1954) A study of the effects of cyclic thermal stresses on a ductile metal. Trans ASME Ser A 76(6):931–950

Dowing SD, Socie DF (1982) Simple rainflow counting algorithm. Int J Fatigue 4:31–40

Fleming WJ, Temis JuM (2002) Numerical simulation of cyclic plasticity and damage of an aluminium metal matrix composite with particulate SiC inclusions. Int J Fatigue 24:1079–1088

Gusenkov AP (1979) Strength under isothermal and non-isothermal low-cycle loading (in Russian). Nauka, Moscow

Kachanov LM (1969) Foundation on theory of plasticity (in Russian), 2nd edn. Nauka, Moscow

Koiter WT (1960) General theorems for elastic-plastic solids. In: Sheddon I, Hill R (eds) Progress in solid mechanics, vol 1. North-Holland Publishing Company, Amsterdam

Manson SS (1966) Thermal stress and low-cycle fatigue. McGraw-Hill Company, New York

Masing G (1926) Eigenspannungen und Verfestigung beim Messing. Proceedings of 2nd international congress on applied mechanics, Zurich, pp 332–335

Melan E (1938) Zur Plastizität des raumlichen Kontinuums. Ing Arch, No.9: 116

Moskvitin VV (1965) Plasticity under variable loadings (in Russia). MGU, Izd, Moscow

Odqvist FKG (1933) Verfestigung von flussisenahnlichen Körpern. ZAMM 13(H.5):360–363

Putchkov IV, Temis YuM (1988) Analytical formulas for elastoplastic response of structures materials under cyclic loading (in Russian). Problemy Prochnosty 9:18–22

Putchkov IV, Temis YuM, Dowson AL, Damri D (1995) Development of a finite element based strain accumulation model for the prediction of fatigue lives in highly stressed Ti components. Int J Fatigue 17(6):385–398

Roytman IM, YaB Fridman (1947) About metals tests under repeatedly change loadings at plastic domain (in Russian). Zavodskaya Laboraturita 4:452–463

Shneyderovich RM (1968) Strength under static and repeatedly static loadings (in Russian). Mashinostroenie, Moscow

Shorr BF (2013) On some problems and limitations in the theory of thermoplasticity (in Russian). In: Contemporary problems of thermo-visco-plasticity in applied analysis tasks of advanced structures and technologies. Trudy VI Shkoly-Seminara, MAMI, Moscow, pp 125–143

Shorr BF, Dul'nev RA (1964) Investigations of thermal stresses and creep at varying temperatures (Review in Russian). J Zavodskaya Laboratoria, No. 3, Moscow, pp 340–347

Smith RW, Hirschber MH, Manson SS (1963) Fatigue behaviour of materials under strain cycling in low and intermediate life range. NASA Tech. Note, D-1574

Temis JuM (1994) Applied solution methods of thermal plasticity problems (in Russian). In Frolov KV et al (eds) Machinary, Encyclopedia. Machines dynamics and strength, vol 1–3, B.1. Kolesnikov KS (ed). Mashinostroenie, Moscow, pp 231–236

Temis YuM (1989) Plasticity and creep of gas turbine engine components under cyclic loading (in Russian). In: Problems of strength and dynamics in air-engine engineering, CIAM Trudy 1237, No. 4, Moscow, pp 32–50

Temis YuM (2005) Plasticity and creep modeling of turbo engines constructive materials (in Russian). In: Proceedings of 49th international conference AAI, Part 2, MAMI, Moscow, pp 25–76

Temis YuM, Shorr BF (2011) About a role of accumulated plastic strain in the theory of plasticity (in Russian). Bulletin of Nijegorogsky University, NNGU, No. 4, part 4, pp 1879–1881

Troshchenko VT, Lebedev AA, Strijalo VA, Stepanov GV, Krivenjuk VV (2000) Mechanical behavior of materials at different load terms (in Russian). Logos, Kiev

Chapter 5
Unsteady Thermal Plasticity

Abstract A topic of this chapter is the general approach to computation analysis of unsteady elastoplastic deformations in machine parts and mechanical systems under non-isothermal loading, based on modern experimental data and theoretical developments. It is shown that widespread variants of theories of plasticity, including strain hardening model, enable satisfactory evaluation of the plastic strain rate, associated with transient processes, only provided that strains increase along an invariable direction. The theory of stabilized anisotropic hardening (SAH theory) developed over the last few decades is outlined. This theory takes into consideration loading history and material adaptability along with separation of stresses into so-called active and residual micro-stress parts. A comparison with numerous experimental results demonstrates efficiency of the SAH theory for calculating the plastic strain rate under sign-varying multi-axial proportional loading. An experimentally justified extension of the SAH theory proposed by the author for analyzing more elaborate case of non-proportional (out-of-phase) non-isothermal loading and conditionally named "the model of stabilized anisotropic-ray hardening" (the SARH theory) is described in detail. The theory starts from the concept of the evolvement of anisotropy deformation along a "hardening axis," which is collinear to a varying vector of the aforementioned active stresses, and also relies on separation of residual micro-stresses into steady (isotropic) and unsteady (anisotropic) parts; in doing so, unsteady micro-stresses vary under rotation of the hardening axis. The constitutive relations written in the form of the projections of micro-stresses, plastic strains, and their increments onto this axis govern arbitrary non-isothermal deformations. A step-by-step procedure of numerical solution of a thermal plasticity problem using the SARH model is also described. Numerical examples illustrate the applicability of the proposed fairly simple model for restoring the plastic strain rate and direction of plastic strain observed for specimens loaded over a wide range of magnitudes and orientations of the stress vector. The possibility of estimating damage accumulation and failure conditions is addressed for general elastoplastic non-isothermal process. A brief survey of the publications (both in Russian and English) on the SAH theory of elastic plastic deformations of solids subject to sign-varying loading is presented.

B.F. Shorr, *Thermal Integrity in Mechanics and Engineering*,
Foundations of Engineering Mechanics, DOI 10.1007/978-3-662-46968-2_5

Under the theory of unsteady thermal plasticity, we shall understand the collection of the scientific approaches to the analysis of the specific problem. These approaches use the results of the experimental investigations of thermomechanical materials behavior in conjunction with the basic principles of the theory of solids.

For useful reviews of specific problems, the reader is referred to Il'ushin (1948), Temis (1994b), Lemaitre and Chaboche (1990, 2004), and others.

In a large number of engineering applications, unsteady deformation processes can be considered as being close to proportional when direction of the stress tensor changes inconsiderably under repeated loading/unloading. Significant number of investigations, beginning from Hohenemser and Prager (1932), Odquist (1933), and others, was devoted to the problems of plastic yielding under combined (multi-axial) stress state in case of proportional loadings.

Engineering problems, where plastic strain develops under any essentially varying non-proportional combined stress, are rather seldom in practice. They may arise, for example, when the boundary conditions of a structure's part have been changed due to the thermal expansion. However, these problems are of great scientific interest, and considerable emphasis on theoretical and partly experimental analysis of these problems has been placed over the last years.

Some points, concerning theoretical and engineering aspects of the material unsteady thermal plasticity are stated below. In this monograph, we focus our attention on the approaches which can serve on the one hand as a basis for a practical engineering design, and on the other hand would specify perspective directions of further scientific researches in the field of thermal plasticity.

5.1 Experimental Backgrounds

For *monotonous increase of stresses without unloading* considered in Chap. 3 allocation of the elastic area and its borders are not important, which allows the representation of deformation diagrams in the form of the generalized dependences between stress and strain intensities $\sigma_i = f(\varepsilon_i)$. For *cyclic loading* with a large number of identical or almost identical cycles considered in Chap. 4, it is also possible to use a generalized cyclic dependence between intensities of stress and strain ranges $\Delta\sigma_i = f(\Delta\varepsilon_i)$. For *other types of loading*, considered below, experiments demonstrate the presence of two deformation areas—elastic and plastic ones—with essentially different properties, though the threshold between these areas cannot be defined precisely. For analysis of arbitrary elastoplastic sign-changing processes, it is essential to examine some experimental data in addition to that presented in the previous chapters.

5.1.1 Diagrams of Arbitrary Alternating Deformation

5.1.1.1 Common Features of the Deformation Diagrams

To build theoretical "models of material behavior" being in agreement with experimental data in spite of the large variety of deformation diagrams, it is necessary to diverge from some individual features and reveal the most common ones.

In diagrams obtained by arbitrary sign-changing loading of structural materials and shown in Fig. 5.1, four types of curves can be plotted: 1—a curve of the initial loading $\tilde{\sigma}_0(\tilde{\varepsilon}_0)$, 2—lines of primary unloading from this curve, 3—a line of repeated loading in the original direction, and 4—lines of further loadings in any directions, including the one opposite to the initial. In Chap. 4, it has been assumed that each line except of the first one can be presented as the dependence $\tilde{\sigma}(\tilde{\varepsilon}, \chi)$, where the origin of the current (local, running) coordinate system $\tilde{\sigma}, \tilde{\varepsilon}$ is allocated in the point of the last change of a loading direction. This point may be conventionally named "the point of unloading." The definition of the structural function χ, which affects a current curve $\tilde{\sigma}(\tilde{\varepsilon}, \chi)$, with reference to cyclic processes has been also given in Chap. 4. Global coordinates σ, ε referred to the points of the curve $\tilde{\sigma}(\tilde{\varepsilon}, \chi)$ are defined as

$$\sigma = \breve{\sigma} \pm \tilde{\sigma}, \quad \varepsilon = \breve{\varepsilon} \pm \tilde{\varepsilon}, \tag{5.1}$$

where $\breve{\sigma}, \breve{\varepsilon}$ are the global coordinates of the point of unloading, and the signs ($\pm$) specify the direction of the local coordinates that start from this point.

The first unloading can start at any point of the initial loading curve, and the subsequent alternating loadings—at any point of the deformation curves prior to failure of a body.

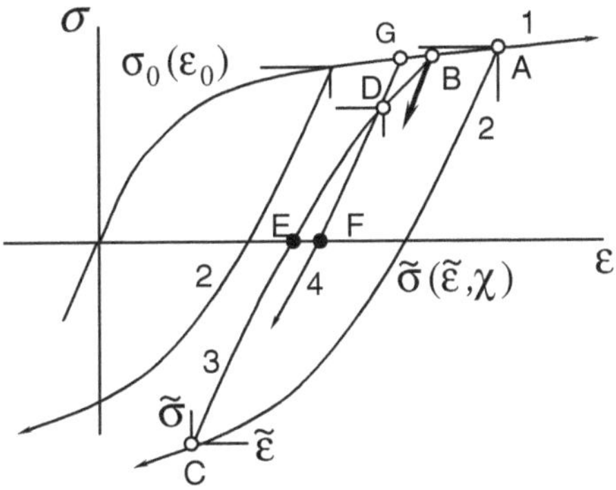

Fig. 5.1 Different elastoplastic deformation diagrams in case of alternating loading

Common features of the experimental deformation curves are as follows:

- In any point of the curves, irrespective of the loading sign, material factors of rigidity E_σ (moduli) or compliance $a_\sigma = E_\sigma^{-1}$ defined by the derivatives

$$E_\sigma = \frac{d\sigma}{d\varepsilon} \quad \text{and} \quad a_\sigma = \frac{d\varepsilon}{d\sigma}, \tag{5.2}$$

are always positive.
- Character of the initial curve $\sigma_0(\varepsilon_0)$, as a rule, considerably differs from other curves $\tilde{\sigma}(\tilde{\varepsilon}, \chi)$, which may depend on the structural function χ.
- Under arbitrary alternating deformation, concepts of "loading" and "unloading" become conditional in a sense that they may be defined only relatively to the direction of the original loading.
- Any point D on the curve 3 can be reached in a number of ways: after elastic unloading from point G or after repeated loading from point C. For periodical (cyclic) processes, further unloading happens along the curve DF. In the common case, two other ways with different rates are also possible: reversion to point G without altering the plastic strain, and further motion along curve DB which leads to growth of the plastic strain.

Aspects of material behavior specified in this chapter may be established representing total strain as the combination of elastic and plastic parts. After altering the loading direction, the local coordinate system $\tilde{\sigma}(\tilde{\varepsilon}, \chi)$ changes its origin and orientation if a new plastic strain appears prior to this moment. When the change of the loading occurs within the elastic area, the coordinate system does not change its origin and orientation. Thus, the use of the coordinates $\tilde{\sigma}(\tilde{\varepsilon}, \chi)$ becomes intricate.

For multi-axial loading, the concept of total strain, considered as a sum of elastic and plastic parts, leads to the representation of strain intensities in the form $\varepsilon_i = \varepsilon_i^e + \varepsilon_i^p$.

Under certain conditions, when elastoplastic loading continues along the line DB in Fig. 5.1, point B coincides with point A, and the loop A–C–E–B becomes closed, but in general, the positions of points B and A differ. In case of further loading from any point D in the same direction, shown by the arrows $D \to B \to A$, plastic strains grow, while by unloading along the opposite way $D \to F$, they would remain constant. If the unloading starts in the point B, it will go as shown in Fig. 5.1 with the bold arrow.

Diagrams for repeated torsion of the thin-walled tubes $\tau_0(\gamma_0)$ in terms of tangential stresses τ and shears γ have the same features as for tension/compression. These diagrams, plotted in terms of stress and strain intensities as functions $\sigma_i(\varepsilon_i)$, become rather identical. Some distinctions might be caused by the mean stress, which is zero in case of torsion and nonzero for tension/compression, and also by the inhering anisotropy.

Experiments reveal that the elasticity modulus might be regarded as a function of the temperature $E(T)$; it depends on the parameter χ very little.

5.1.1.2 Isotropic Hardening and Softening

Some information on isotropic and anisotropic material characteristics has been presented in Chap. 4. Below, additional notes with respect to sign-changing processes are given.

In inherent state and under elastic deformation, the majority of structural materials may be usually considered as isotropic (inherently anisotropic materials will be considered in Chap. 10). During the deformation process, the change of some characteristics, for example a width $h_e = 2\sigma_y^*$ of the elastic area (shown in Fig. 5.2) is isotropic. An increase of the width h_e reflects the strain hardening (strengthening), and its decrease—the strain softening (loss of strength) of the material.

In order to find the material isotropic characteristics from a deformation diagram $\sigma(\varepsilon, T)$, it is necessary to repeat unloading and obtain a dependence of the elastic area width h_y versus a structural function χ. The change of isotropic characteristic is mostly irreversible but sometimes can be partially restored.

In case of multi-axial deformation, a number of material properties remain isotropic, in particular, elasticity moduli, a radius of the elastic area, and some other which will be considered below in Sect. 5.3.

5.1.1.3 Deformation Anisotropy

Deformation anisotropy affecting a location of the elastic area may be defined by the motion of the center of this area. In Fig. 5.2, as above in Figs. 4.1–4.3 of Chap. 4, locations of central points of the elastic area during the deformation process are shown with dashed lines. Unlike isotropic hardening, displacement of the elastic area is always reversible and a sign of its displacement increment coincides with the sign of stress.

In Fig. 5.2, possible allocations of the elastic area with constant width h_e for some moments of loading/unloading are shown, and the trajectories of motion of the elastic area central points are displayed by arrows. White rhombic points show the allocation of boundaries of the elastic area under development of significant plastic strains in the "positive" direction, and black rhombic points—for deformation in the opposite direction. At the identical value of the plastic strain ε_p, the

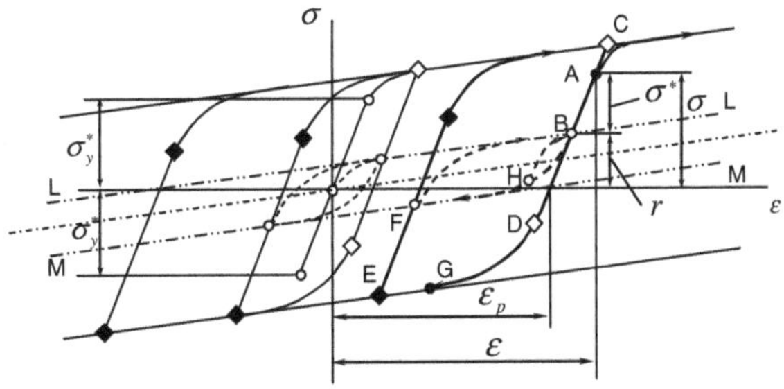

Fig. 5.2 Deformation diagram for any loading trajectories

elastic area can occupy various positions within the boundaries specified, depending on the previous loading process.

Experiments show that the possible positions of the area center curve are limited for a material stable state by two straight lines LL and MM, which are nearly parallel each other. Transition from one border to another takes place after the stress alters direction and plastic strain recommences in the opposite direction, for example along the curve BHF.

An increment of plastic strain on stress is $d\varepsilon_p = a_p d\sigma$. The rate of its development is determined by the factor of plasticity compliance a_p which is the inverse of modulus of plasticity $a_p = E_p^{-1}$, where

$$E_p = \frac{d\sigma}{d\varepsilon_p} = \frac{E_t}{1 - E_t/E}. \tag{5.3}$$

Figure 5.2 demonstrates that the slopes of the curves $\sigma(\varepsilon)$, equal to $E_t \approx E_p$, differ at the points located on the opposite borders (see for example, points C and D). The slopes of both borders straight lines LL and MM are equal to the *minimum* modulus of plasticity $E_{p,\min}$ or tangential modulus

$$E_{t,\min} = \frac{d\sigma_y(\varepsilon)}{d\varepsilon}. \tag{5.4}$$

In engineering materials, visible deformation anisotropy always appears, while the effects from isotropic hardening may be less significant.

5.1.2 *Prehistory of Loading—Adaptability*

5.1.2.1 Prehistory of Loading

All facts that have been presented above demonstrate that a deformation process depends on the prehistory of loading and deformation.

In case of a sign-changing deformation, the principal effect of prehistory means that after each change of loading directions, including the stable state of a material, the rate of plastic strain varies (see also Chap. 4). In order to reflect this effect in calculations, it is necessary to store in memory the data referring to the points of deformation curves, where all of the previous changes of stresses sign have occurred. Such requirements for calculations are too difficult.

After a number of forgoing changes of the stress sign, the role of each of them should be gradually weakened, and the influence of prehistory becomes limited. Il'ushin (1963) has drawn attention to the limited area of prehistory and named it "length of a delay trace." But he did not connect this effect directly with sign alterations of strains.

An approach providing the account of the fading effect of loading prehistory without detailed analysis of preceded strain process will be stated below.

5.1.2.2 Adaptability Under Constant and Variable Loadings

Concept of Adaptability

This concept can be applied to both material and mechanical system.

In the theory of plasticity, adaptability is usually referred to the conditions, which allow elastoplastic deformations of a system to be transferred into purely elastic ones (Grüning 1926; Prager 1957; Koiter 1960 and others). We shall consider the adaptability in a more broad sense.

Under varying loads and at elevated temperatures high-loaded machine parts can operate for a long time only under conditions when their reaction to repeated mechanical and thermal influences remains at least approximately stable. Stability is shown in preservation of the parts dimensions and in accumulation of internal "micro-damages" with almost constant (and moderate) rate. Many structural materials possess similar property within certain limits. For cyclic loading, the adaptability occurs in stabilization of deformation cycles with constant moderate rate of increase of accumulated damage (see Chap. 4).

In many cases, a stable state is established after some transient stage which can sometimes be neglected, but in general, this stage must be taken into account in design methods. Under isotropic hardening, the elastic area extends with reduction of plastic zones, and the transient stage cannot include many cycles. Long-term softening of material usually makes a machine part unserviceable.

Adaptability in generalized sense is the stable response of a material to the identical thermomechanical effects during some time period.

It is obvious that such adaptability is possible only under stable (or stabilized) state of a material, while reversible anisotropy deformation can take place. Effect of fading influence of loading prehistory promotes adaptability.

For stable cyclic processes, all local diagrams $\tilde{\sigma}(\tilde{\varepsilon})$ after some transient period become practically identical and the width of elastic area becomes constant.

After the rate of plastic strain changes the direction, tangential modulus E_t quickly decreases from elastic value $E_{t,\max} = E$ to a minimum constant value $E_{t,\min} \approx E_{p,\min}$ as was shown in Fig. 5.2 for any trajectory of loading. Adaptability at elevated temperatures has a specific character due to creep (see Chap. 7).

Non-stationary and Non-isothermal Processes

At unstable state of a material, the width of elastic area, distance between border lines LL and MM, their inclination (i.e., the modulus $E_{p,\min}$), and the form of a nonlinear part of strain curves vary during loading.

An approach for analysis of complex non-stationary processes that takes into account influence of a loading prehistory will be discussed below.

Under non-stationary temperature fields, all specified mechanical characteristics of a material are functions of space coordinates and time.

5.2 Stabilized Anisotropic Hardening—Uniaxial Stressed State

5.2.1 Isothermal Processes

5.2.1.1 Active and Residual Micro-stresses

Kadashevich and Novojilov (1958) represented the total stress σ for elastoplastic process as a sum of "*active*" stress σ^* and "*internal residual micro-stresses*," which we shall denote r

$$\sigma = \sigma^* + r. \tag{5.5}$$

This concept was apparently first introduced by Ishlinsky (1954) in application to linear hardening. Physical interpretation of this decomposition can be found in Novojilov (1964).

As early, Reuss (1930) had paid attention on possibility of micro-stresses origin linked with plastic deformations.

Active stresses vary within the elastic area $-\sigma_y^* \leq \sigma^* \leq \sigma_y^*$ (see Fig. 5.2 where micro-stresses are shown in their limiting position referred to plastic strain ε_p). The quantity σ_y^* represents the "active" yield point equal to the half of the elastic area width $h_e = 2\sigma_y^*$. For a stable material state and constant temperature, the quantity $\sigma_y^* =$ const; generally, it is a function of plastic strain and temperature $\sigma_y^* = f(\varepsilon_p, T)$, i.e., depends on isotropic hardening or softening (see below Eqs. (5.25a, b)–(5.26)).

At full unloading (to $\sigma = 0$) from stresses exceeding a yield point, residual stresses equal to $\sigma^* = -r$ are retained in a body. At repeated loadings from the stress $\sigma^* = 0$ (points *B* or *F* in Fig. 5.2), plastic yielding is renewed at different full stresses $\sigma = r + (\mathrm{sign}\sigma)\sigma_y^*$ depending on the loading direction.

Residual micro-stresses r are determined by current coordinates of the middle line of elastic area. They are a function (commonly nonlinear) of plastic strains. The derivative

$$E_p = \frac{dr(\varepsilon_p)}{d\varepsilon_p}, \tag{5.6}$$

represents the usual modulus of plasticity connected at $\sigma_y^* = \text{const}$ with tangential modulus E_t by Eq. (5.3).

The stresses r can be treated as additional stresses arising due to anisotropic hardening. They are absent in an ideally plastic body. Further, we will use the term "micro-stresses" although it has a conditional sense. Definitions "residual" and "internal" will be dropped.

Until inequality $|\sigma^*| < \sigma_y^*$ takes place, plastic strains and micro-stresses keep their previously attained values and boundary positions of the elastic area do not vary, i.e., $d\varepsilon_p = 0$, $dr = 0$, $d\sigma_y^* = 0$, $d\sigma^* = d\sigma$, and $d\varepsilon_e = d\sigma/E$.

At $|\sigma^*| = \sigma_y^*$, the active stresses reach one of the borders of the elastic area. If the total stress continues changing in the direction of the active stress, elastic and plastic strains together with micro-stresses increase in the same direction. Conditions of the beginning or renewal of yielding can be presented in the form

$$|\sigma^*| = \sigma_y^*, \tag{5.7}$$

$$\sigma^* d\sigma > 0. \tag{5.8}$$

In this case, active stresses at the boundaries of the elastic area reach their limiting value σ_y^* and do not change at $\sigma_y^* = \text{const}$, while the increments of micro-stresses and total ones are identical

$$dr = d\sigma = E_p d\varepsilon_p. \tag{5.9}$$

5.2.1.2 Stabilized Anisotropic Hardening

Material Model

The concept of the SAH theory (stabilized anisotropic hardening) is based on the phenomenological model of elastoplastic material, providing description of adaptability (in the generalized sense specified above) in dependence of loading history and direction (Shorr 1975; Birger and Shorr 1975; some other works are described below in Sect. 5.4.3.2).

In Fig. 5.3, position of any point A in coordinates r, ε_p is shown. A micro-stress r can be presented by a sum of steady (linear) part $r' = E_{p0}\varepsilon_p$ and unsteady (nonlinear) part r''

$$r = r' + r''. \tag{5.10}$$

The latter varies within the limits of $-r_{\lim} \le r'' \le r_{\lim}$.

For the given value of ε_p, micro-stresses can vary within the area $E_{p0}\varepsilon_p - r_{\lim} \le r \le E_{p0}\varepsilon_p + r_{\lim}$, where E_{p0} is the minimum quantity of the plasticity

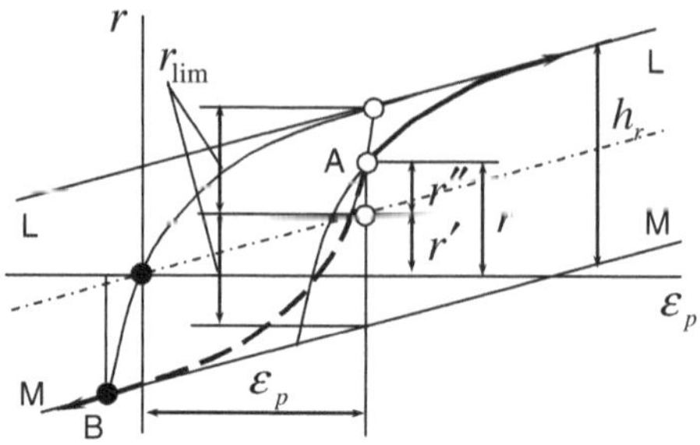

Fig. 5.3 Design model of stabilized anisotropic hardening

modulus referred to the stabilized portion of the deformation diagram, $r_{\lim}$ is a half of the distance between limiting lines LL and MM. Equations of these lines are

$$r_{\lim}^{\pm} = E_{p0}\varepsilon_p + \mathrm{sign}\sigma^* r_{\lim}. \tag{5.11}$$

At $\sigma^* > 0$, micro-stresses tend to the line LL, and at $\sigma^* < 0$—to the line MM.

The different rate of the micro-stress increment dr can be described by assumption that the micro-stress rate at a point A is proportional to the distances $l^{\pm}$ from this point to that limiting line, in which direction a loading takes place.

An absolute quantity of such distance is

$$l^{\pm} = (r_{\lim} - r)\mathrm{sign}\sigma^* = r_{\lim} + \mathrm{sign}\sigma^*(E_{p0}\varepsilon_p - r). \tag{5.12}$$

At the active stress σ^* of any sign, we obtain

$$dr = E_p(r, \varepsilon_p)d\varepsilon_p, \tag{5.13}$$

where the variable modulus of plasticity E_p is

$$E_p(\varepsilon_p, r) = E_{p0} + \beta_p[r_{\lim} + \mathrm{sign}\sigma^*(E_{p0}\varepsilon_p - r)]. \tag{5.14}$$

Here, $\beta_p \geq 0$ is a parameter of hardening stabilization.

A renewal of plastic current with increase ε_p from any point A in a former direction, in which it has occurred till the moment of the last unloading, is possible only at preservation of a $\mathrm{sign}\sigma^*$ (in given case at $\mathrm{sign}\sigma^* = +1$, see continuous curve in Fig. 5.3). Plastic strain of the oppose direction can arise if the active stresses arrive at another boundary of the elastic area and a $\mathrm{sign}\sigma^*$ changes on the opposite, from (+) to (−). At $\mathrm{sign}\sigma^* = -1$, there is an algebraic reduction of plastic strain (dash curve in the same figure).

Current value of the plasticity modulus E_p depends on the process character.

Equations (5.13)–(5.14) satisfy the following conditions:

- At the center line $r = E_{p0}\varepsilon_p$ (with $l^{\pm} = r_{\lim}$) the rate of stresses increase is

$$\left.\frac{dr}{d\varepsilon_p}\right|_{\varepsilon_p=0} = E_{p0} + \beta_p r_{\lim}. \tag{5.15}$$

- If plastic strain increases in certain direction and stresses r approach limiting values $r_{\lim}$ with $l^{\pm} = 0$, the rate tends to minimum quantity

$$\left.\frac{dr}{d\varepsilon_p}\right|_{l^{\pm}\to 0} = E_{p0}. \tag{5.16}$$

In this case, the rate of plastic strain $d\varepsilon_p/dr = E_{p0}^{-1}$ becomes the greatest.

- If under $r = r_{\lim}^{\pm}$ the active stress changes its sign the "distance" from r to the opposite limiting line becomes maximum value $l^{\pm} \to 2r_{\lim}$, the rate of stresses increase is the greatest

$$\left.\frac{dr}{d\varepsilon_p}\right|_{l^{\pm}\to 2\sigma_{a0}} = E_{p0} + 2\beta_p r_{\lim}, \tag{5.17}$$

and the rate of plastic strain is minimum.

Equations (5.13)–(5.14) describe various rate of plastic strains depending on material characteristics $r_{\lim}, E_{p0}, \beta_p, \sigma_y^*$, current values r, ε_p and the signσ^*. These dependencies reflect the most important effect of prehistory of loading that answers experimental data.

Equations (5.10), (5.13), and (5.14) yield

$$dr = dr' + dr'' = [E_{p0} + E_p''(r, \varepsilon_p)]d\varepsilon_p, \tag{5.18}$$

where the variable modulus of plasticity $E_p = E_{p0} + E_p''$ consists of E_{p0} and

$$E_p'' = \beta(r_{\lim} - \operatorname{sign}\sigma^* r'') \tag{5.19}$$

and also

$$r'' = r - E_{p0}\varepsilon_p. \tag{5.20}$$

If the yielding conditions (5.7)–(5.8) are valid, Eq. (5.13) can be integrated in a variable ε_p.

Sign-Constant Loading

For a loading stage within which the sign of active stress remains constants and initial conditions are $r = r_{in}$ and $r'' = r_{in}''$ at $\varepsilon_p = \varepsilon_{p,in}$, we obtain

$$r = E_{p0}\varepsilon_p + \text{sign}\sigma^* r_{\lim} - \left(E_{p0}\varepsilon_{p,in} + \text{sign}\sigma^* r_{\lim} - r_{in}\right)\exp\left[-\beta_p \text{sign}\sigma^*\left(\varepsilon_p - \varepsilon_{p,in}\right)\right] \tag{5.21}$$

or for unsteady plastic strain

$$r'' = \text{sign}\sigma^* r_{\lim} - (\text{sign}\sigma^* r_{\lim} - r''_{in})\exp[-\beta_p \text{sign}\sigma^*(\varepsilon_p - \varepsilon_{p,in})] \tag{5.22}$$

Equations (5.21)–(5.22) describe a family of the curves $\sigma_a = f(\varepsilon_p)$ shown in Fig. 5.2, allocating from them the curve which is passing through a point r_{in}, $\varepsilon_{p,in}$. In particular, for the first loading at $r_{in} = \varepsilon_{p,in} = 0$ with $\text{sign}\hat{\sigma}^* = +1$, it yields

$$r = E_{p0}\varepsilon_p + r_{\lim}[1 - \exp(-\beta_p\varepsilon_p)]. \tag{5.23}$$

Description of micro-stresses in the form of the sum of the linear unfading part $r' = E_{p0}\varepsilon_p$, defining by current value of plastic strain, and a nonlinear fading part $r'' = r - r' = r_{\lim}[1 - \exp(-\beta_p\varepsilon_p)]$ depending on loading prehistory, answers the Novojilov's (1964) theoretical representations and experimental data. Below, it is shown that such partition is necessary for analyzing arbitrary multi-axial processes.

Equation (5.22) allows finding material characteristics by the use of experimental diagrams of the stabilized deformation. A value of the limit modulus of plasticity E_{p0} is represented through tangential modulus E_t at linear diagram section by Eq. (5.3) and limit value of micro-stresses—as half of distance between lines LL and MM, $r_{\lim} = 0.5\,h_a$.

It follows from Eq. (5.22) for $\sigma^* > 0$

$$\bar{r}'' = \frac{r''}{r_{\lim}} = 1 - (1 - \bar{r}''_{in})\exp[-\beta_p(\varepsilon_p - \varepsilon_{p,in})], \tag{5.24a}$$

where from

$$\beta_p = \frac{1}{\varepsilon_p - \varepsilon_{p,in}} \ln\frac{1 - \bar{r}''_{in}}{1 - \bar{r}''}. \tag{5.24b}$$

The ratio $\bar{r}''$ can be taken for values of strain ε_p and micro-stress r in any point of the experimental curves $r(\varepsilon_p)$. The curve of the first loading with initial conditions $r''_{in} = \varepsilon_{p,in} = 0$ may be used for a material stable state.

Hardening Stabilization Parameter

Since Eqs. (5.13)–(5.14) and Eq. (5.23) only approximately display real dependence $r(\varepsilon_p)$, the values of β_p calculated for points with different ratio $\bar{r}''$ might be various. In dependence on an expected area of the stress/strain state, different values of $\bar{r}''$ can be used. For a strain area commensurable with strain yield point

$\varepsilon_y = \sigma_y/E$, the size β_p can be averaged for $0.3 < \bar{r}'' > 0.9$. For the analysis of problems with $\varepsilon_p \gg \varepsilon_y$, it is useful to take $\beta_p \approx 4.5/\breve{\varepsilon}_p$ where the strain $\breve{\varepsilon}_p$ refers to the initial point of a linear section of deformation diagram $\sigma(\varepsilon)$ and $r'' \approx r_{\lim}$.

In Fig. 5.4, curve 1 shows the experimental dependence $r(\varepsilon_p)$ corresponding to the curve 1 of the diagram in Fig. 3.7 of Chap. 3, and a straight line 2—the limit tangent to this curve with slope equal to $E_{p0} \approx 9600$ MPa.

Having elongated the line 2 before crossing with the axis of ordinate, we find the value $r_{\lim} \approx 365$ MPa. The experimental dependence $r'' = f(\varepsilon_p)$ is shown by white dots. Averaged value β_p obtained by six points is $\beta_p \approx 335$ and leads to design dependence $r'' = r_{\lim}[1 - \exp(-\beta_p \varepsilon_p)]$ shown by curve 3.

In Bondar' and Danshin (2008), some computer-based determinations of material characteristics, including $r_{\lim}$ (denoted in this work σ_{a0}), E_{p0}, and β_p are described; e.g., parameter β_p is defined as a mean-square quantity on several points of an experimental curve $r(\varepsilon_p)$. Material characteristics E_{p0}, $r_{\lim}$, and β_p for two steel grades are presented in Table 5.1.

Computation results for a number of problems are also presented in the reference work.

Sign-Changing Loading with Isotropic Hardening

Equation (5.5) according to conditions (5.8) yields

$$d\sigma = \mathrm{sign}\sigma^* d\sigma_y^* + dr = (E^* + E_p)d\varepsilon_p, \tag{5.25a}$$

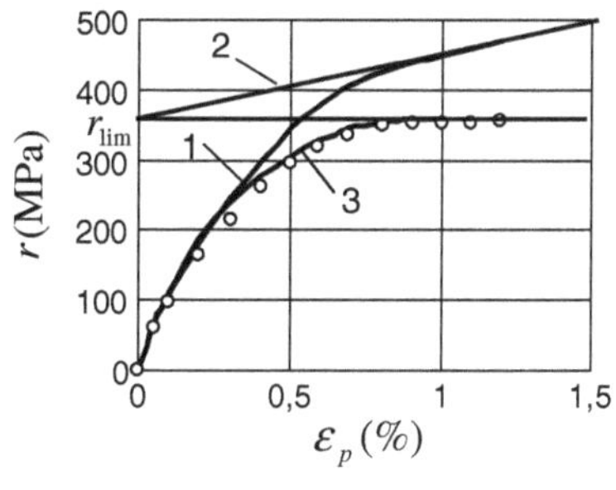

Fig. 5.4 Experimental and design dependences $r(\varepsilon_p)$ and $r'' = f(\varepsilon_p)$

Table 5.1 Material characteristics E_{p0}, $r_{\lim}$, and β_p of the SAH model

Material characteristic		Steel	
		Structural	Stainless
Plasticity modulus E_{p0}	MPa	9220	7400
Limiting unsteady micro-stresses $r_{\lim}$	MPa	112	130
Parameter of stabilization β_p	–	760	280

Bondar' and Danshin (2008)

where the modulus

$$E^* = \text{sign}\sigma^* \frac{d\sigma_y^*}{d\varepsilon_p} \tag{5.25b}$$

describes the rate of the active yield point alteration due to material isotropic hardening or softening. For stable material state $E^* = 0$, in the case of expansion of elastic area $E^* > 0$, and at its narrowing $E^* < 0$. As a rule, $E^* + E_p > 0$ and sign $d\varepsilon_p =$ sign $d\sigma$.

The basic equation of elastoplastic deformation at sign-changing direction of isothermal loading in the 1D statement becomes

$$d\varepsilon = d\varepsilon_e + d\varepsilon_p = \left[\frac{1}{E} + \frac{1}{E^* + E_p}\right] d\sigma, \tag{5.26}$$

In step-by-step methods, differentials are replaced with finite increments of functions. At the step beginning, an "elastic" calculation is assumed that allows obtaining an "elastic" stress increment $\Delta\sigma_e$ and calculating a new active stress $\langle\sigma_e^*\rangle = \sigma^* + \Delta\sigma_e^*$ where $\Delta\sigma_e^* = \Delta\sigma_e$. For a material stable state, an active yield point σ_y^* is constant.

If $|\langle\sigma_e^*\rangle| \leq \sigma_y^*$, then the "elastic" calculation due to conditions (5.7)–(5.8) is confirmed. Otherwise, we take $\langle\sigma^*\rangle = \text{sign}\sigma^*(\sigma_y^*)$ and define the increment of micro-stresses $\Delta r = \langle\sigma_e^*\rangle - \langle\sigma^*\rangle$.

Equation (5.13) in terms of finite increments yields

$$\Delta\varepsilon_p = \frac{\Delta r}{E_p(r, \varepsilon_p)} \tag{5.27a}$$

where a current modulus of plasticity $E_p(r, \varepsilon_p)$ is defined due to Eq. (5.14) using the values of plastic strain ε_p and micro-stress r which have been calculated at the previous step.

A new value of plastic strain is

$$\langle\varepsilon_p\rangle = \varepsilon_p + \Delta\varepsilon_p \tag{5.27b}$$

Using Eqs. (5.14), (5.27a, b), we can improve the values of modulus E_p and plastic strain ε_p by successive approximations.

Quantities $\Delta\sigma$ or $\Delta\varepsilon$ may be specified by a trial program. In the last case, plastic deformation occurs at $|\Delta r| > 0$ where

$$\Delta r = E_t\left(\Delta\varepsilon - \frac{\text{sign}\sigma^*(\sigma_y^*) - \sigma^*}{E}\right) \tag{5.28}$$

Equations (5.14), (5.27a, b) remain valid.

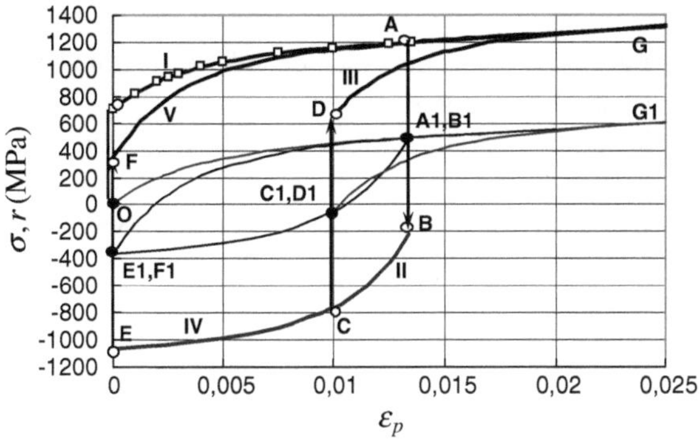

Fig. 5.5 Example of calculation of sign-altering elastoplastic processes using SAH model

5.2.1.3 Numerical Example

In Fig. 5.5, total stresses σ (bold lines) and micro-stresses r (thin lines) versus plastic strain ε_p for several loading processes are calculated by Eq. (5.21). In addition to material characteristics taken above in Fig. 5.4, the value of yield point $\sigma_y^* = 710$ MPa has been taken due to the diagram of Fig. 3.7.

Stage I. Line OA, initial loading up to $\sigma = 1200$ MPa (light cubes) with $r = 490$ MPa (point $A1$) and $\varepsilon_p = 0.0134$. Active stress is $\sigma^* > 0$.
Stage II. Unloading along line AB with constant values of micro-stress r and plastic strain ε_p (points $A1$, $B1$) and next loading in the opposite direction up to $\sigma = -780\,\text{MPa}$ (point C) with micro-stress r increase in opposite direction (to point $C1$). Active stress is $\sigma^* < 0$.
Stage III. Repeated loading along the trajectory C–D–G. Active stress is $\sigma^* > 0$, micro-stresses increase from point $D1$ (coinciding with $C1$) by curve $D1$–$G1$.
Stages IV and V. Repeat the previous two stages II and III, but for larger range of the plastic strain.

The calculated curves reproduce all experimental features of material behavior for sign-changing 1D loading, described above: various rates of plastic strain evolution in dependence on deformation prehistory, gradual transition to the maximum rate of plastic yielding with the minimum value of E_{p0}, and distinction in the diagrams of the first and the subsequent half-cycles even for material stable state.

Instability of a material behavior can also be taken into account with use of an experimental dependence $\sigma_y^*\left(\varepsilon_{p\Sigma}\right)$.

5.2.2 *Non-isothermal Processes*

5.2.2.1 Smooth Change of Temperature

All characteristics of a material presented in Eq. (5.14) depend on a body temperature T. The functions $E_p0(T), r_{\lim}(T)$, and $\beta_p(T)$, together with $E(T)$ and

$\sigma_y^*(T)$, can be found by processing of the primary experimental elastoplastic diagrams $\sigma = f(\varepsilon, T)$, and the secondary diagrams $r = f(\varepsilon_p \cdot T)$ obtained at several constant temperatures.

When temperature smoothly changes and design steps are small enough, material parameters of an element are assumed to remain constant during the current step. Calculation experience shows that it is better to take a maximum temperature for the given step as its key value. Conditions (5.7)–(5.8) must also be checked at this key temperature.

There is no necessity to estimate temperature dependences of material parameters with too great accuracy because they can be determined even at constant and especially varying temperature imprecisely and with significant scatter. Therefore, these dependences must be averaged.

5.2.2.2 Increments of Plasticity Parameters at Varying Temperature

Material characteristics in Eq. (5.13) may be considered as continuous functions of temperature. In this case, small temperature increments of material characteristics are

$$dE_{p0} = \frac{\partial E_{p0}}{\partial T} dT, \quad dr_{\lim} = \frac{\partial r_{\lim}}{\partial T} dT, \quad d\sigma_y^* = \frac{\partial \sigma_y^*}{\partial T} dT.$$

Equation (5.26) with temperature term yields

$$d\sigma = (E^* + E_p) d\varepsilon_p + \operatorname{sign}\sigma^* \left(\frac{\partial \sigma_y^*}{\partial T} + \frac{\partial E_p}{\partial T} \right) dT, \tag{5.29a}$$

where according to Eq. (5.14) with $\partial r''/\partial T = 0$ is

$$\frac{\partial E_p}{\partial T}(\varepsilon_p, T) = \frac{\partial E_{p0}}{\partial T} + \beta_p \frac{\partial r_{\lim}}{\partial T}. \tag{5.29b}$$

Parameter β_p within some temperature and stress ranges can be averaged.
It follows from Eqs. (5.29a, b) that

$$d\varepsilon_p = \frac{1}{E^* + E_p} \left[d\sigma - \operatorname{sign}\sigma^* \left(\frac{\partial \sigma_y^*}{\partial T} + \frac{\partial E_p}{\partial T} \right) dT \right]. \tag{5.30}$$

Equations (5.29a, b)–(5.30) allow analyzing various conditions of plasticity occurrence and each step calculating by use of the theory of non-isothermal plastic yielding of Chap. 3.

A procedure of step-by-step calculation begins from recalculation of all material characteristics onto new temperature $\langle T \rangle = T + \Delta T$ where temperature increment ΔT for a current step is assumed to be known for each element of a body.

Further calculations of a current step are conducted at constant temperature to similarly that as has been specified above in Sect. 5.2.1.2.

5.3 Stabilized Anisotropic Hardening—Multi-axial Loading

5.3.1 Stress and Strain Vectors in the Theory of Plasticity

For multi-axial loading, the concept of total stress, considered as a sum of active and two parts of micro-stresses leads to representation of stress vector in the form

$$\boldsymbol{\sigma}_i = \boldsymbol{\sigma}_i^* + \mathbf{r}_i' + \mathbf{r}_i'' \tag{5.31}$$

Analogously, the concept of total strain, considered as a sum of elastic and plastic parts, leads to the representation of strain vector in the form

$$\boldsymbol{\varepsilon}_i = \boldsymbol{\varepsilon}_i^e + \boldsymbol{\varepsilon}_i^p. \tag{5.32}$$

5.3.1.1 Projections of a Vector and Its Increment onto Any Axis

In application to the theory of plasticity, the following vector characteristics described in common view in Chap. 1 are used.

In solid mechanics, nine-dimensional stress vector $\boldsymbol{\sigma}_i$ with components $(3/2)^{0.5} s_{ij}$ and strain vector $\boldsymbol{\varepsilon}_i$ with components $(2/3)^{0.5} e_{ij}$ are commonly used. For simplicity, $\boldsymbol{\sigma}_i$ is referred to as a stress vector, and $\boldsymbol{\varepsilon}_i$—as a strain vector, though they characterize only deviators of the corresponding tensors. Magnitudes of these vectors are

$$|\boldsymbol{\sigma}_i| = \sqrt{\frac{3}{2} s_{ij} s_{ij}} = \sigma_i \tag{5.33}$$

and

$$|\boldsymbol{\varepsilon}_i| = \sqrt{\frac{2}{3} e_{ij} e_{ij}} = \varepsilon_i, \tag{5.34}$$

which for $\mu \approx 0.5$ coincide with definition of stress σ_i and strain ε_i intensities, respectively.

- Increments of stress $d\boldsymbol{\sigma}_i$ and strain $d\boldsymbol{\varepsilon}_i$ vectors with components $(3/2)^{0.5}ds_{ij}$ and $(2/3)^{0.5}de_{ij}$.
- Magnitudes of these vectors

$$|d\boldsymbol{\sigma}_i| = \sqrt{\frac{3}{2}ds_{ij}ds_{ij}} = (d\sigma)_i \tag{5.35}$$

and

$$|d\boldsymbol{\varepsilon}_i| = \sqrt{\frac{2}{3}de_{ij}de_{ij}} = (d\varepsilon)_i. \tag{5.36}$$

- Small increments of stress $d\sigma_i$ and strain $d\varepsilon_i$ intensities determined as

$$d\sigma_i = d\left(\sqrt{\frac{3}{2}s_{ij}s_{ij}}\right) = \frac{3s_{ij}ds_{ij}}{2\sigma_i} \tag{5.37a}$$

and

$$d\varepsilon_i = d\left(\sqrt{\frac{2}{3}e_{ij}e_{ij}}\right) = \frac{2e_{ij}de_{ij}}{3e_i}. \tag{5.37b}$$

Intensities of active stress $\boldsymbol{\sigma}_i^*$, micro-stress $\mathbf{r}_i$, and plastic strain $\boldsymbol{\varepsilon}_i^p$ vectors and their increments may be also presented in identical forms. A mean value of micro-stress components $r = r_{ii}/3 = 0$, as for plastic strain, therefore the micro-stress parameters $|\mathbf{r}_i| = r_i$, $|d\mathbf{r}_i| = (dr)_i$, and dr_i are defined by formulas analogous to Eqs. (5.34), (5.36), and (5.37b).

In view of the paired relationship of tangential stresses and shear strains, vectors $\boldsymbol{\sigma}_i$ and $\boldsymbol{\varepsilon}_i$ may be defined as 6D, and as they do not depend on mean values σ and ε, they can be reduced to 5D ones (Il'ushin 1963).

Generally, the direction of increment vector $d\boldsymbol{\sigma}_i$ can make an arbitrary angle ϕ_σ with the direction of the vector $\boldsymbol{\sigma}_i$ (Fig. 5.6a). The same is referred to vector $d\boldsymbol{\varepsilon}_i$ which is inclined to direction of the active stress vector $\boldsymbol{\sigma}_i^*$ on an angle ϕ_ε (Fig. 5.6b).

Due to Eq. (1.49) of Chap. 1, a projection of a vector $d\boldsymbol{\sigma}_i$ onto the direction of $\boldsymbol{\sigma}_i$ is

$$(d\boldsymbol{\sigma}_i)_\phi = \frac{3s_{ij}ds_{ij}}{2\sigma_i} \tag{5.38}$$

Comparison Eq. (5.38) with Eq. (5.37a) shows that for a *small* angle ϕ, an increment $(d\boldsymbol{\sigma}_i)_\phi \approx d\sigma_i$. These quantities are scalar unlike a positive value of a vector increment $|d\boldsymbol{\sigma}_i| = (d\sigma)_i$.

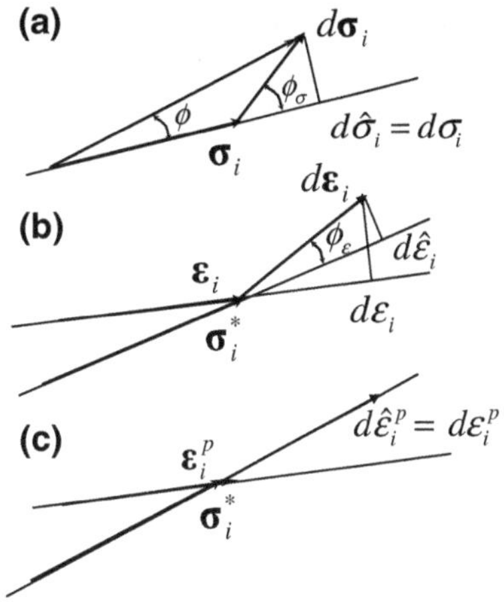

Fig. 5.6 Vectorial presentation of stresses and strains and their increments

In the theory of plasticity, a significant value has the vector projections onto the direction of active stress vector $\boldsymbol{\sigma}_i^*$. Further, a projection of any vector $\mathbf{v}_i$ onto direction of $\boldsymbol{\sigma}_i^*$ will be denoted $\hat{v}_i$. A projection $\hat{v}_i$ of any vector $\mathbf{v}_i$, collinear to the active stress vector $\boldsymbol{\sigma}_i^*$, coincides in quantity with the vector magnitude $|\mathbf{v}_i|$, but unlike, it is a scalar function.

In particular, projections of the vectors $d\boldsymbol{\varepsilon}_i$ and $d\boldsymbol{\varepsilon}_i^p$ onto the direction of the active stress vector $\boldsymbol{\sigma}_i^*$ (Fig. 5.6b) yield

$$d\hat{\varepsilon}_i = \frac{s_{ij}^* de_{ij}}{\sigma_i^*} \tag{5.39a}$$

and

$$d\hat{\varepsilon}_i^p = \frac{s_{ij}^* d\varepsilon_{ij}^p}{\sigma_i^*}. \tag{5.39b}$$

Analogous formulas are true for the projections of active stresses, micro-stresses, and plastic strain and their increments.

5.3.1.2 Conditions of Vectors Collinearity

Collinear vectors are parallel but can be directed opposite, e.g., $d\boldsymbol{\sigma}_i$ and $\boldsymbol{\sigma}_i$ under uniaxial unloading.

The *elastic strain* and total stress vectors are parallel, $\boldsymbol{\varepsilon}_i^e \| \boldsymbol{\sigma}_i$, as far as their increments $d\boldsymbol{\varepsilon}_i^e \| d\boldsymbol{\sigma}_i$, a ratio of their magnitudes is a constant equal to 3G.

In *the deformation theory of plasticity,* it is accepted that $\varepsilon_{ij} = \lambda s_{is}$, and the vector of the total strain is parallel to the stress vector, $\boldsymbol{\varepsilon}_i \| \boldsymbol{\sigma}_i$.

In *the theory of plastic yielding* it is usually accepted that $d\varepsilon_{ij}^{p} = s_{ij} d\lambda$ i.e., the vector of increment of plastic strain is parallel to the stress vector $d\boldsymbol{\varepsilon}_i^p \| \boldsymbol{\sigma}_i$. Experimental data show that for sign-varying loading, it is necessary to take $d\varepsilon_{ij}^{p} = s_{ij}^{*} d\lambda^{*}$, or $d\boldsymbol{\varepsilon}_i^p \| \boldsymbol{\sigma}_i^{*}$ (see Fig. 5.6c). In this case,

$$d\hat{\varepsilon}_i^p = (d\varepsilon^p)_i = \sqrt{\frac{2}{3} d\varepsilon_{ij}^p d\varepsilon_{ij}^p}. \tag{5.40}$$

Condition of vectors $d\boldsymbol{\varepsilon}_i^p$ and $\boldsymbol{\sigma}_i^{*}$ collinearity, due to Eq. (1.50) of Chap. 1, is

$$\frac{d\varepsilon_{ij}^p}{(d\varepsilon^p)_i} = \frac{3 s_{ij}^{*}}{2\sigma_i^{*}}. \tag{5.41}$$

5.3.1.3 Work of Shape Change

An increment of the work of shape change is defined by the expression

$$dW_{sp} = s_{ij} de_{ij}, \tag{5.42a}$$

and an increment of work of plastic shape change is defined by the formula

$$dW_{sp}^{p} = s_{ij} d\varepsilon_{ij}^{p} = \hat{\sigma}_i d\hat{\varepsilon}_i^{p}, \tag{5.42b}$$

where increment $d\hat{\varepsilon}_i^p$ is expressed by Eq. (5.40).

The increment of the work dW_{sp} can be any sign, while the increment of the plastic work dW_{sp}^{p} is always positive, owing to conditions (5.8) and (5.9).

5.3.2 Experimental Data on Multi-axial Loading

Under tests, total stresses (without their partition on active and micro-stresses) and total strains (without partition on elastic and plastic parts) are measured and can be considered as strongly experimental data. Their constituents resulted by the uniaxial tests and some theoretical concepts can be referred to conditionally experimental. Our regard should be given to either of these data.

5.3.2.1 Different Kinds of Tests for Multi-axial Loading

Processes of plastic yielding under combined stress state can be studied in different ways

- When an increment projection $d\hat{\sigma}_i$ of total stress vector onto the CH axis (i.e., onto varying direction of active stress vector) is positive during all test time, $\hat{\sigma}_i d\hat{\sigma}_i > 0$, then the increment of plastic strain intensity $d\varepsilon_i^p = d\hat{\varepsilon}_i^p$ does not change the sign. This leads to permanent growth of plastic strain intensity ε_i^p.

At that, a direction of the CH axis can smoothly vary so that a rate of the axis turn is small, $(d\phi^*/d\varepsilon_p)^2 \ll 1$. A temporary cease of plastic deformation at $d\hat{\sigma}_i = 0$ without axis turn is possible.

This kind of *sign-constant plastic deformation* has been realized in a number of tests.

- If a projection of total stress increment $d\hat{\sigma}_i$ onto the varying direction of the CH axis becomes at some instant negative, $\hat{\sigma}_i d\hat{\sigma}_i < 0$, then plastic deformation temporarily stops.

Active stresses may arbitrarily change their values and directions within the elastic area. This can lead to finite alteration of the direction of the CH axis to the moment when plastic deformation is renewed. Such *sign-changing plastic deformation* tests are more informative, but their data are rather limited.

Some trials are carried out by altering of stress component quantities with control of strain components (sometimes named "soft" tests), while the other ones —by altering of strain and control of stress components ("rigid" tests).

5.3.2.2 Analysis of the Test Results

From analysis of a number of tests with multi-axial non-proportional (out-of-phases) loading, it follows the following:

- Elastic strain vector $\boldsymbol{\varepsilon}_i^e$ always follows the total stress vector $\boldsymbol{\sigma}_i$, while the increment of plastic strain $d\boldsymbol{\varepsilon}_i^p$ follows the active stress vector $\boldsymbol{\sigma}_i^*$. At that, in general cases vectors of strains $\boldsymbol{\varepsilon}_i$, $\boldsymbol{\varepsilon}_i^p$, stresses $\boldsymbol{\sigma}_i$, $\boldsymbol{\sigma}_i^*$, and $\mathbf{r}_i$ are not collinear.
- Under tests with permanent turn of stress vector $\boldsymbol{\sigma}_i$, the direction of total strain vector $\boldsymbol{\varepsilon}_i$ turns also permanently (e.g., along the trajectory in the form of an evolving Archimedes's spiral). In this case, the dependence $\sigma_i = f(\varepsilon_i)$ remains near of the usual generalized deformation diagram.
- After abrupt change of the loading direction and continuing increase of stress vector magnitude σ_i, the growth of strain vector $\boldsymbol{\varepsilon}_i$ proceeds not only in the direction of the newly applied increasing stress, but also in a direction of the previous stress in spite of the fact that this stress component is maintained constant.
- At abrupt change of the loading direction when the stress vector increment $d\boldsymbol{\sigma}_i$ makes an arbitrary angle with the vector $\boldsymbol{\sigma}_i$, the vectors $\boldsymbol{\sigma}_i$ and $\boldsymbol{\varepsilon}_i$ themselves turn gradually but with different rates. At the beginning, the strain vector rotates after the stress vector, and then the strain vector rapidly lags behind, and in the sequel under active loading, both vectors' directions become close again.

- When a loading direction changes with intermediate unloading to $\sigma_i = 0$ (i.e., after transition through the point $\sigma_i^* = 0$), a new initial rate of plastic strain depends on what angle $\Delta\phi^*$ the active stress vector $\boldsymbol{\sigma}_i^*$ turns (Shishmarev and Shcherbo 1983). At $\Delta\phi^* \approx 0$, plastic deformation proceeds with the former rate; at $\Delta\phi^* \approx 0.5\pi$, the rate equals the rate at the beginning of the process (i.e., at $\varepsilon_i^p = 0$); at $\Delta\phi^* \approx \pi$, the rate is twice so as under proportional loading with sign alteration.

In a number of tests, a trajectory of strain vector $\boldsymbol{\varepsilon}_i$ instead of stress $\boldsymbol{\sigma}_i$ is given, which is supported by corresponding regulation of stress components s_{ij}. The observed results can be explained by the same material behavior.

Tests of various materials demonstrate considerable number of individual effects both anisotropic and isotropic features. A desire to simulate the individual features possibly fully leads to the necessity of using a great number of additional theoretical and experimental characteristics and complicated communications between parameters.

Proceeding from the given monograph orientation, only the most common and practically important features of material behavior are further allocated and rather simple mathematical models are offered for their description.

5.3.3 *Theoretical Concept*

5.3.3.1 Common Backgrounds

The presented theory of stabilized anisotropic hardening is based on the following common backgrounds and available experimental data briefly described above.

Isotropic and Anisotropic Material Characteristics

Mechanical characteristics describing *material properties* of an inherently isotropic body are identical *along any direction*, i.e., they have *isotropic* character. In case of stable state of a material, these parameters do not vary and, independently of loading type, remain isotropic up to failure. To such properties, refer elasticity moduli E, G, minimum modulus of plasticity E_{p0}, active yield stress σ_y^* (a half-width of the elastic area), limit magnitude of a nonlinear part of micro-stresses $r_{\lim}$, parameter of hardening stabilization β_p, and also the strength properties.

Other mechanical characteristics describing *material behavior* are linked with *anisotropic* hardening of a material during its deformation in a *certain direction*. They mirror modifications in an inherently isotropic body as a result of its deformation. These properties are as follows displacement of the elastic area, rotation of the micro-stress and plastic strain vectors, and fading rate of the nonlinear micro-stresses part. On the one hand, they have a *qualitative* character; and on the other hand, they can be determined quantitatively as it will be shown below.

Axes of Anisotropic Hardening

Rate of the anisotropic hardening is defined by quantity of the modulus of unsteady plasticity E_p''. It has been shown above that after turn of the stress on 180°, the modulus increases as a result of change of the active stress σ^* direction in relation to the previous direction of the unsteady part of plastic strain r''.

For a multi-axial loading, this effect can be described by a turn of the vector $\boldsymbol{\sigma}_i^*$ concerning the previous direction of the vector $\mathbf{r}_i''$. We denote these directions "the axis CH" and "the axis FH," *respectively*. The axis FH coincides with the direction of the non-stationary micro-stresses vector $\mathbf{r}_i''$ and reflects influence of the plastic strains affected by *former anisotropic hardening*; the axis CH coinciding with the vector of active stress $\boldsymbol{\sigma}_i^*$ shows a direction of *current anisotropic hardening*.

Vector of plastic strain increment $d\boldsymbol{\varepsilon}_i^p$ always follows the CH axis.

When the vector $\boldsymbol{\sigma}_i^*$ turns on 180°, the CH axis also turns after it. In this moment, the vectors $\boldsymbol{\varepsilon}_i^p$, $\mathbf{r}_i''$, and $\mathbf{r}_i''$ do not vary, while their projections onto the CH axis, $\hat{\varepsilon}_i^p$, $\hat{r}_i'$, and $\hat{r}_i''$ alter the signs. After unloading and new loading in the opposite direction, the projection $\hat{r}_i''$ gradually decreases to zero and then changes the sign. In this moment, the FH axis also changes its direction.

Due to available experimental data, it is natural to assume that for proportional sign-changing loading, correlations between projections of all specified vectors onto the CH axis can be presented in a form similar to the uniaxial tension/compression.

A concept of the rotating axes of anisotropic hardening will be used further in the theory of non-proportional loading.

Other Assumptions

Other features of the material behavior under stabilized anisotropic hardening are as follows:

- Tensors' deviators of all parts of stresses and strains, as well as their increments, are represented by vectors in 6D space. Directions of these vectors may be different and alter during a deformation process.
- While active stresses remain within the elastic area, a rotation of the active stress axis (the CH axis) does not affect material behavior.
- The plastic strains with the corresponding micro-stresses develop along the changing direction of the CH axis, and the elastic area moves in the same direction.

- Rotation of the CH axis changes a rate direction of the steady part of the plastic strain and in some cases influences also its magnitude.
- The unsteady part of the plastic strain rate increases depending on the projection $\hat{r}_i''$ of the micro-stress vector $\mathbf{r}_i''$ onto the changed direction of the CH axis,
- Further evolution of plastic strains in a changed direction of the CH axis is accompanied by rotations of micro-stresses and plastic strain vectors.
- If a new direction of loading after rotation remains constant, direction of the plastic strain vector gradually approaches with it and rate of the unsteady part fades to zero.
- The body temperature influences only material isotropic properties, but a modification of temperature pattern of the structure, provoking thermal stresses, can lead to veering of the active stress vector and directions of the anisotropic hardening axes.
- For an unstable condition of a material (in particular, at the beginning of loading), the material properties, remaining isotropic, can change in size depending on normalized accumulated plastic strain (see Temis and Shorr 2011).

5.3.3.2 Proportional Sign-Varying Loading

As it was specified in Chap. 3, proportional (or simple) loading is named a process in which all loads are simultaneously changed proportionally to each other. In the general case of sign-changing proportional loading, the loads can both increase and decrease. Therefore, in any point of a body the sign of the stress component's increment ds_{ij} can coincide or be opposite to sign of the stresses component s_{ij}; however, their ratio during the entire process must be the same for all element's components, $ds_{ij}/s_{ij} = \pm\lambda$, where a constant $\lambda \geq 0$.

Under proportional loading, all constituents of the stress and strain vectors are collinear, so the angle ϕ_ε shown in Fig. 5.6b is always zero, while the angles ϕ_σ and ϕ in Fig. 5.6a may be zero or 180°. Particularly,

$$\hat{\sigma}_i = \frac{3s_{ij}^* s_{ij}}{2\sigma_i^*} \tag{5.43a}$$

and

$$d\hat{\sigma}_i = \frac{3s_{ij}^* ds_{ij}}{2\sigma_i^*}. \tag{5.43b}$$

Analogous expressions are valid for active stresses and micro-stresses.

Projection of the total stress vector can be presented as a sum of projections

$$\hat{\sigma}_i = \hat{\sigma}_i^* + \hat{r}_i \tag{5.44a}$$

and also

$$d\hat{\sigma}_i = d\hat{\sigma}_i^* + d\hat{r}_i. \tag{5.44b}$$

Equations (5.7)–(5.8) for the beginning or renewal of plastic yielding take the form

$$\begin{aligned} \hat{\sigma}_i^* &= \sigma_y^*, \\ d\hat{\sigma}_i &> 0, \end{aligned} \tag{5.45}$$

since the projection $\hat{\sigma}_i^*$ of active stress onto the CH axis (i.e., "on itself") is always positive.

At $\sigma_y^* = \text{const}$, it yields $d\hat{\sigma}_i^* = 0$ and $d\hat{r}_i = d\hat{\sigma}_i$.

Under plastic yielding, as the generalization of Eqs. (5.13)–(5.14), an increment of micro-plastic stresses projection is determined as

$$d\hat{r}_i = E_p(\hat{r}, \hat{\varepsilon}_i^p) d\hat{\varepsilon}_i^p, \tag{5.46}$$

where

$$E_p(\hat{r}_i, \hat{\varepsilon}_i^p) = E_{p0} + \beta_p(r_{\lim} + E_{p0}\hat{\varepsilon}_i^p - \hat{r}_i). \tag{5.47}$$

The increment of micro-plastic stresses projection is assumed to be consisted of two members

$$d\hat{r}_i = d\hat{r}_i' + d\hat{r}'' = [E_{p0} + E_p''(\hat{r}_i, \hat{\varepsilon}_i^p)] d\hat{\varepsilon}_i^p, \tag{5.48}$$

where

$$E_p'' = \beta_p(r_{\lim} - \hat{r}_i''). \tag{5.49}$$

A projection of the nonlinear micro-stresses part varies within the limits $-r_{\lim} \le \hat{r}_i'' \le r_{\lim}$, and the variable modulus E_p''—within the range $0 \le E_p'' \le 2\beta_p r_{\lim}$. It follows from Eqs. (5.46)–(5.48) that

$$d\hat{\varepsilon}_i^p = \frac{d\hat{r}_i}{E_{p0} + E_p''}, \tag{5.50}$$

$$d\hat{r}_i' = \frac{E_{p0} d\hat{r}_i}{E_{p0} + E_p''}, \tag{5.51}$$

and

$$d\hat{r}_i'' = \frac{E_p'' d\hat{r}_i}{E_{p0} + E_p''}. \tag{5.52}$$

Equations (5.27a, b) for multi-axial loading takes the form

$$d\hat{\varepsilon}_i = \frac{d\hat{\sigma}_i}{3G} + d\hat{\varepsilon}_i^p = \left[\frac{1}{3G} + \frac{1}{E^* + E_p}\right] d\hat{\sigma}_i, \tag{5.53}$$

where $E^* = \frac{d\sigma_y^*}{d\varepsilon_i^p}$.

Equation (5.22), represented for multi-axial stress state in projections onto the CH axis, is valid only for a loading section when plastic strains develop in a *constant* direction. In this case,

$$\hat{r}_i' = E_{p0} \hat{\varepsilon}_i^p. \tag{5.54a}$$

and

$$\hat{r}_i'' = r_{\lim} - [r_{\lim} - (\hat{r}_i'')_{in}] \exp\{-\beta_p [\hat{\varepsilon}_i^p - (\hat{\varepsilon}_i^p)_{in}]\} \tag{5.54b}$$

with the initial condition $\hat{r}_i'' = (\hat{r}_i'')_{in}$ at $\hat{\varepsilon}_i^p = (\hat{\varepsilon}_i^p)_{in}$.

5.3.3.3 Non-isothermal Proportional Sign-Varying Loading

Quantities σ_y^*, E, E_{p0}, $r_{\lim}$, and β_p, as all other material characteristics, are temperature-dependent and can be obtained by deformation diagrams. At step-by-step calculations, the specified characteristics can be taken constant for the given step.

Proportional sign-varying loading will be considered further as a particular case of the general SAH theory.

In order that a loading may be considered as proportional, temperature stresses s_{ij}^T arising in non-isothermal problems should satisfy the equations

$$s_{ij}^T = s_{ij}^* \frac{\sigma_i^T}{\sigma_i^*}, \tag{5.55}$$

where $\sigma_i^T = (3 s_{ij}^T s_{ij}^T / 2)^{0.5}$.

In the majority of practical problems this requirement is satisfied only approximately.

5.4 General Model of Stabilized Anisotropic-Ray Hardening

5.4.1 Micro-stresses Modification Due to Turn of Anisotropy Axis

5.4.1.1 Influence of the Turn's Angle

According to the experimental data, a rate of plastic strain depends on a current value of the angle ϕ_r between directions of the vectors $\boldsymbol{\sigma}_i^*$ and $\mathbf{r}_i''$.

Under the angle $\phi_r = 90°$, the previous anisotropic hardening does not render any influence on the rate of the non-stationary part of micro-stresses $\hat{r}_i''$ which starts in new direction with the rate $E_p'' = \beta_p r_{\lim}$. For the angle $\phi_r = 180°$, the starting rate of stresses $\hat{r}_i''$ is duplicated to $E_p'' = 2\beta_p l_{\lim}$. To describe such material behavior for any angle ϕ_r, we assume that $\hat{r}_i''$ in Eq. (5.50) is a projection of micro-stress vector $\mathbf{r}_i''$ onto the change direction of the CH axis, namely

$$\hat{r}_i'' = r_i'' \cos \phi_r. \tag{5.56}$$

If initial directions of the vectors $\boldsymbol{\sigma}_i^*$ and $\mathbf{r}_i''$ coincide (e.g., at the process beginning), the angle ϕ_r equals to the angle ϕ^* of the axis CH rotation for a given step.

A similar approach was considered by Temis (1989).

In Fig. 5.7, a generalized graph of projection $\hat{r}_i''$ onto the CH axis for all angles $0 \geq \phi_r \geq \pi$ is displayed. Point locations of the limit unsteady state $\hat{r}_i'' = r_{\lim}$ for different angles ϕ_r are presented by a semicircle 1–3–5 (generally, in the coordinate system $\sqrt{3/2}s_{ij}$, they lay down on a hyper-sphere of the radius $r_{\lim}$). For the initial direction I, it is assumed $\hat{r}_i'' = r_{\lim}$ (a site 0–1).

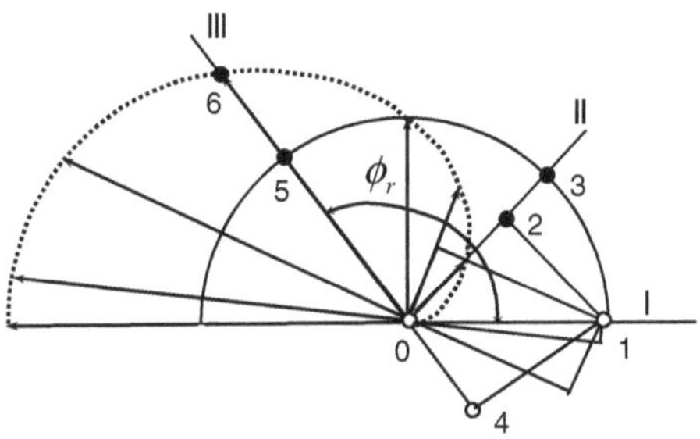

Fig. 5.7 Changes of the micro-stress unsteady part projection $\hat{r}_i''$ and plastic strain rate E_p'' after turn of the stress vector

Site 0–2 in Fig. 5.7 corresponds to the projection $\hat{r}_i''$ onto direction II, site 0–4—onto direction III, etc.

After a turn, a starting rate of micro-stress unsteady part (equaled E_p'' at $\beta = 1$) is defined, due to Eq. (5.50), by a difference $r_{\lim} - \hat{r}_i''$. This difference is presented for direction II by site 2–3, for direction III—by site 4–5, etc., and also is shown by length of the corresponding rays displayed by arrows. The arrow tips organize a dotted curve which each point matches a particular direction of a CH axis after its turn after unloading.

If at the former loading a magnitude $\hat{r}_i''$ takes any position $0 \le \hat{r}_i'' \le r_{\lim}$, then the limit curves, defining the starting rate after the turn, occupy the intermediate locations between curves 3 and 6, shown in Fig. 5.7.

5.4.1.2 Arbitrary Turn of the CH Axis

In Fig. 5.8a, possible positions of a total stress vector $\boldsymbol{\sigma}_i$ and its components $\boldsymbol{\sigma}_i^*$ and $\mathbf{r}_i$ are shown for the condition $\sigma_i^* = \sigma_y^*$ referred to one of the boundaries of the elastic area at some loading instant. The similar picture for total strain vector $\boldsymbol{\varepsilon}_i$ and its components $\boldsymbol{\varepsilon}_i^e$ and $\boldsymbol{\varepsilon}_i^p$ are shown in Fig. 5.8b.

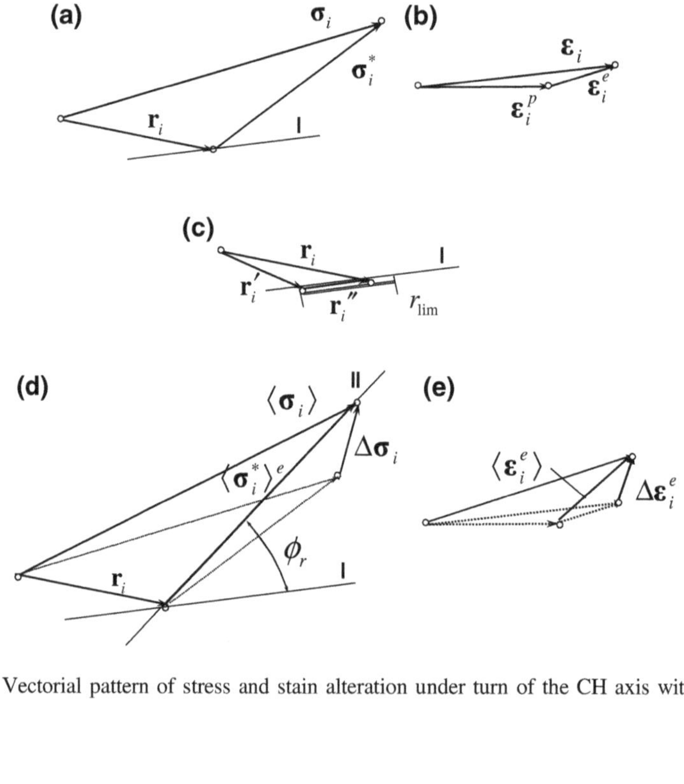

Fig. 5.8 Vectorial pattern of stress and stain alteration under turn of the CH axis within elastic area

Components $\mathbf{r}_i''$ and $\mathbf{r}_i''$ of the micro-stress vector are presented in Fig. 5.8c. We assume the vector of a non-stationary part of micro-stresses $\mathbf{r}_i''$ to be directed along the anisotropy axis FH (line I) and its magnitude to be less than the limiting value $r_i'' < r_{\lim}$.

In Fig. 5.8d, a pattern of stress vectors positions is displayed after the stress vector obtains an increment $\Delta\boldsymbol{\sigma}_i$ and a new total stress vector becomes $\langle\boldsymbol{\sigma}_i\rangle = \boldsymbol{\sigma}_i + \Delta\boldsymbol{\sigma}_i$. The former vectors' positions are shown by dotted lines.

Assuming first the new loading to be elastic with $\mathbf{r}_i = \text{const}$, we find the active stress vector $\langle\boldsymbol{\sigma}_i^*\rangle^e = \langle\boldsymbol{\sigma}_i\rangle - \mathbf{r}_i$. The angle between axes FH (line) I and CH (line II) is indicated by ϕ_r.

If the condition $\langle\sigma_i^*\rangle^e \leq \langle\sigma_y^*\rangle$ is satisfied, the stress/strain state remains within the elastic area, i.e., $\langle\boldsymbol{\sigma}_i^*\rangle = \langle\boldsymbol{\sigma}_i^*\rangle^e$, $\langle\boldsymbol{\varepsilon}_i\rangle = \boldsymbol{\varepsilon}_i + \Delta\boldsymbol{\varepsilon}_i^e$ as shown in Fig. 5.8e, and calculation of the given step comes to an end. In this case, the stress increment vector $\Delta\boldsymbol{\sigma}_i$ can change any way within restriction on size $\langle\sigma_i^*\rangle^e$ specified above.

Otherwise, at $\langle\sigma_i^*\rangle^e > \langle\sigma_y^*\rangle$, it is necessary to accept $\langle\sigma_i^*\rangle = \langle\sigma_y^*\rangle$ and define an increment of the micro-stress vector $\Delta\mathbf{r}_i = \langle\boldsymbol{\sigma}_i^*\rangle^e - \langle\boldsymbol{\sigma}_i^*\rangle$. Since the micro-stresses change along the direction of active stresses, then projections of the specified vectors onto the axis CH will be connected by the relation $\Delta\hat{r}_i = \langle\hat{\sigma}_i^*\rangle^e - \langle\hat{\sigma}_i^*\rangle$, as is shown in Fig. 5.9a.

Projecting the vector $\mathbf{r}_i''$ onto the axis CH, we find a scalar quantity $\hat{r}_i'' = r_i'' \cos\phi_r$ (see Fig. 5.9c) and using Eq. (5.50)—the modulus of unsteady plasticity E_p'' for a new position of the axis CH. An increment of plastic strain intensity $\Delta\hat{\varepsilon}_i^p$ along the axis CH (see Fig. 5.9b) will be defined by Eq. (5.50) and values of

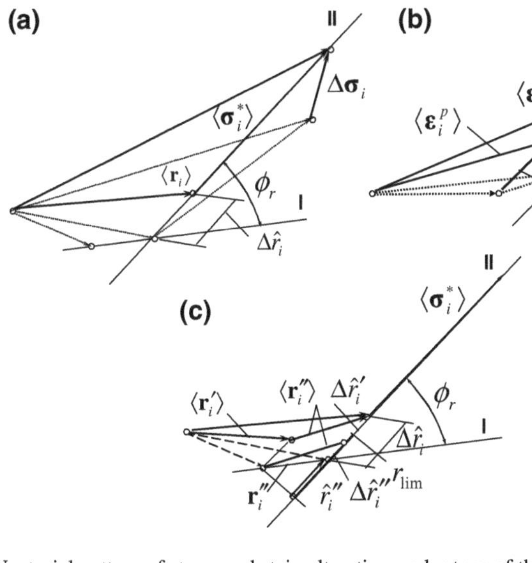

Fig. 5.9 Vectorial pattern of stress and stain alteration under turn of the CH axis with evolution of plastic deformation

increments of stationary and unsteady parts of micro-stresses $\Delta\hat{r}'_i$ and $\Delta\hat{r}''_i$ by Eqs. (5.51) and (5.52). New vectors of these parts will be $\langle \mathbf{r}'_i \rangle = \mathbf{r}'_i + \Delta\mathbf{r}'_i$ and $\langle \mathbf{r}''_i \rangle = \mathbf{r}''_i + \Delta\mathbf{r}''_i$ (Fig. 5.9c).

Turn of the FH Axis

Maximum magnitude of the unsteady micro-stress vector $\langle \mathbf{r}''_i \rangle$ for the position of Fig. 5.9c cannot exceed the limiting value $r_{\lim}$. Since a projection $\hat{r}''_i$ for this position at $\phi_r > 0$ is always less than $r_{\lim}$, growth of unsteady micro-stresses along the axis II will be continued.

Let a calculated position of the vector $\langle \mathbf{r}''_i \rangle$ at any moment referred to the current value of increment $\Delta r''_i$ (site 6–5 in Fig. 5.10) be shown by site 1–5 with its magnitude $\langle r''_i \rangle > r_{\lim}$. Then, the vector $\langle \mathbf{r}''_i \rangle$ must rotate to position 4–5 corresponding to value $\langle r''_i \rangle = r_{\lim}$, while its protection on the previous axis I decrease on a value $(\Delta r''_i)_{ad} = -\Delta_{1-4}$ (site 1–4).

It follows from geometrical relationships that the additional shift Δ_{1-4} can be obtained from the set of equations

$$\begin{aligned} \Delta_{1-4} &= r''_i - r_{\lim}\cos\phi'' + \Delta r''_i \cos\phi_r \\ r_{\lim}\sin\phi'' &= \Delta r''_i \sin\phi_r. \end{aligned} \tag{5.57}$$

For $\phi_r = 0$, there is $\Delta_{1-4} = \langle r''_i \rangle - r_{\lim} = 0$, and for $\phi_r = 0.5\pi$ and $\Delta r''_i = r_{\lim}$, we find a quantity $\Delta_{1-4} = r''_i$.

We assume the decrease of projection $(\Delta r''_i)_{ad} = -\Delta_{1-4}$ to be substituted by occurrence of an additional component of stationary micro-stresses $(\Delta r'_i)_{ad}$ equal to the site Δ_{1-4}. Thus, the sum is $(\Delta r'_i)_{ad} + (\Delta r''_i)_{ad} = 0$. Vector sums of sites (1–6) + (6–5) and (1–4) + (4–5) are the same; therefore, neither total micro-stresses $\mathbf{r}_i$ nor plastic strain $\boldsymbol{\varepsilon}^p_p$ are affected by additional turn of the vector $\mathbf{r}''_i$.

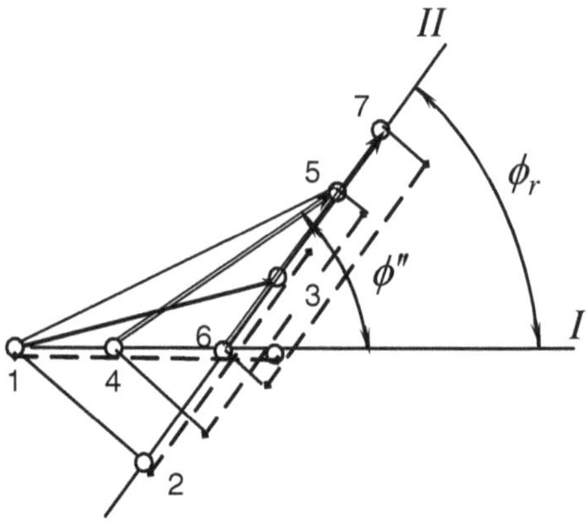

Fig. 5.10 Process of unsteady micro-stress vector evolution under rotation of the active stress axis

In Fig. 5.10 a process of consecutive evolution of vector $\langle \mathbf{r}_i'' \rangle$ from the position 1–3 into 4–5 and then into 6–7 with conservation of equality $\langle r_i'' \rangle = r_{\lim}$ is displayed. In the last case, there is $\langle r_i'' \rangle \to \langle \hat{r}_i'' \rangle \to r_{\lim}$ and increase of the unsteady micro-stresses stops.

5.4.1.3 Particular Cases of the CH Axis Turn

If micro-stress intensity is not equal to zero and the vector of active stress reaches boundaries of the elastic area, any turn of a total stress following by preservation or increase in its magnitude leads to growth of plastic deformation. This demands carrying out of step-by-step calculations with small stress increments. In this case, projections onto new direction of the CH axis do not practically differ from corresponding projections onto the previous direction, in particular $\Delta\hat{r}_i \approx \Delta\hat{\sigma}_i$, and plastic strain evolution induced by stress increase occurs near the same as under proportional loading, namely

$$\Delta\hat{\varepsilon}_i^p = \frac{\Delta\hat{\sigma}_i}{E_p' + E_p''}, \tag{5.58}$$

where the variable modulus E_p'' is defined by Eq. (5.49).

At turn of a total stress vector with reduction of its magnitude, deformation process passes into elastic area.

Neutral Loading

So-called neutral loading is theoretically possible if a total stress vector $\boldsymbol{\sigma}_i$ varies so that the active stress intensity is a constant, $\sigma_i^* = \sigma_y^*$, and the vector $\Delta\boldsymbol{\sigma}_i$ remains all times normal to the vector $\boldsymbol{\sigma}_i^*$. Then, $d\hat{\sigma}_i = 0$ and $d\varepsilon_i^p = 0$. The previous analysis describes this process. However, it is very complicated to ensure an orthogonality of the vectors $\Delta\boldsymbol{\sigma}_i$ and $\boldsymbol{\sigma}_i^*$ with continuous change of the stress vector $\boldsymbol{\sigma}_i$ during a real test. Projections sum of the vectors $\boldsymbol{\sigma}_i^*$ and the finite increments $\Delta\boldsymbol{\sigma}_i$ onto the altering directions of the CH axis will exceed the yield point σ_y^*, so the plastic strain will continue a little growing, as is observed in experiments.

5.4.2 Comparison of Calculated and Experimental Results

5.4.2.1 Tests of Thin-Walled Tubular Specimens

Many researchers, beginning from Hohenemser and Prager (1932), tested thin-walled tubular specimens under various combinations of loads. Pairs of loads: Axial tension with internal pressure and torsion with axial tension or internal pressure that led to plane stress state were usually examined. A vast number of new test results are presented in Zubchaninov et al. (2004).

Processes, in which projections $\hat{\sigma}_i$ and $\hat{\varepsilon}_i$, increasing or decreasing in quantities, did not change the signs, and the accumulated plastic strain $\sigma_{p\Sigma}$ did not surpass size of residual strain Δ under the tension diagram, are the most investigated. Unloadings were mostly examined in the elastic area.

Arutjunjan's (1964) experiments confirmed that at various paths of the tubular specimens loading the center of elastic area were displaced, but the shape of the plastic deformation surface remained almost invariable.

5.4.2.2 Test Example

In Fig. 5.11 a, test results for combined tension and torsion loading of carbon steel specimens are presented. Axial $\sigma_{11} = \sigma_0$ and tangential $\sigma_{12} = \tau_0$ trial stresses versus loading steps t are shown by dotted curves 1 and 2, respectively.

Stress intensity calculated according to Eq. (1.35b) as

$$\sigma_i = \sqrt{\sigma_{11}^2 + 3\sigma_{12}^2},$$

is shown by bold curve 3.

In Fig. 5.11b, longitudinal $\varepsilon_{11} = \varepsilon_0$ and shear $\varepsilon_{12} = 0.5\gamma_0$ strains measured under the test are presented by dotted curves 1 and 2, respectively, and the strain intensity calculated according to Eq. (1.41) as

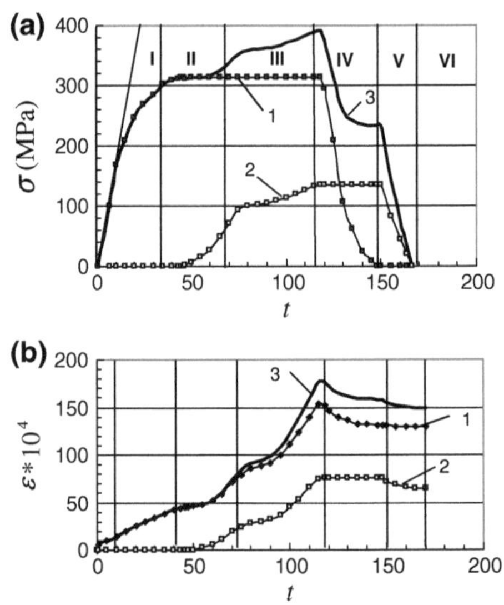

Fig. 5.11 Experimental data of combined tension and torsion loading (Zubchaninov et al. 2004)

$$\varepsilon_i = \sqrt{\varepsilon_{11}^2 + 4\varepsilon_{12}^2/3},$$

is displayed by bold curve 3.

For processing of test results, it was accepted $\mu \approx 0.5$.

The loading sequence consisting of six stages is designated in Fig. 5.11a with Roman numerals. Stage I refers to the elastic axial tension, stage II—to continuation of tension in elastoplastic area, stages III and IV—to increase of the tangential stresses without change of the normal ones, stage V—to the axial unloading up to $\sigma_{11} = 0$ with preservation of tangential stresses, and stage VI—to full unloading up to $\sigma_{11} = 0$ and $\sigma_{12} = 0$.

Comparison of results obtained by numerical calculation to experimental data is presented in charts of Fig. 5.12. Dimensionless parameters $\bar{X}_\sigma = \sigma_{11}/\sigma_y^*$ and $\bar{Y}_\sigma = \sqrt{3}\sigma_{12}/\sigma_y^*$ for stress components and $\bar{X}_\varepsilon = \varepsilon_{11}/\varepsilon_y^*$ and $\bar{Y}_\varepsilon = (4/3)^{0.5}\varepsilon_{12}/\varepsilon_y^*$ for strain ones are used. The Roman numerals concern the same stages of process as in Fig. 5.11.

In Fig. 5.12a, a right-angled path of the end point of the stress intensity vector $\boldsymbol{\sigma}_i$ corresponding to Fig. 5.11a is displayed. The process stages are restricted by light points, double lines mark three sections (at $\sigma_i^* = \sigma_y^*$), where plastic yielding occurs. Calculation results referred to the active and micro-stress vectors evolution are shown by arrows.

Dashed arrows show varying direction and magnitude of the micro-stress vector $\mathbf{r}_i$ and the trajectory of its end defining positions of the center of the elastic area. Solid arrows show directions of the active stress intensity vector $\boldsymbol{\sigma}_i^*$ at various process

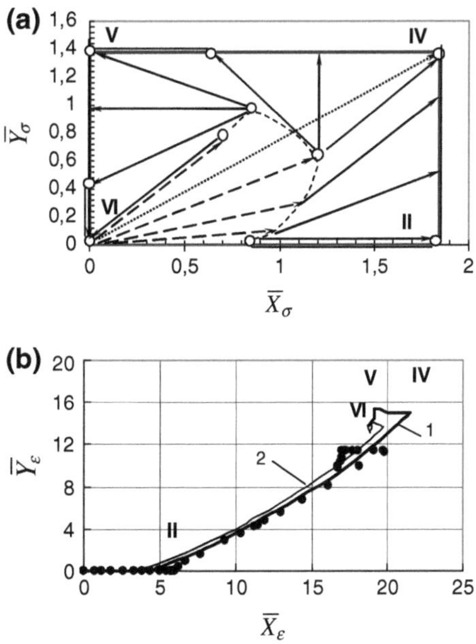

Fig. 5.12 Vector presentation of experimental and numerical results for the test example: **a** stresses, **b** strains

points. The dotted arrow displays a maximum magnitude of the total stress intensity vector $\boldsymbol{\sigma}_i$ at a final action of the coupled extension and torsion (in point IV).

In Fig. 5.12b, measured components of the total strain are displayed by dots, a solid line 1 shows the appropriate calculated curve $\bar{Y}_\varepsilon = f(\bar{X}_\varepsilon)$, and a thin line 2—the calculated curve of the plastic strain components $\bar{Y}^p_\varepsilon = f(\bar{X}^p_\varepsilon)$.

5.4.2.3 Ray Hardening

According to the presented model, the influence of the material anisotropic hardening has a "ray" character and the rate of the plastic yielding depends not only on the stress intensity but also on deviation of a stress from the previous hardening direction. Therefore, the model of the plastic deforming using specified approach is possible be named the "*theory of a stabilized anisotropic-ray hardening*"—the SARH model.

This model adequately enough reflects all the features of the elastoplastic deformation under multi-axis non-proportional loading.

5.4.3 Analysis of Thermoplasticity Problems Using Finite Element Method

5.4.3.1 General

Strength calculations of machine parts on the basis of finite element method (FEM) have obtained last decades a wide circulation in the engineering. The reader is supposed be familiar with the main principles of FEM analysis (see, i.e., Zienkiewicz 1967; Zienkiewicz and Taylor 2000) and constructions of the universal software codes, stated in special managements.

In theory of FEM, a tensor (matrix-column) v and a matrix K is usually designated as $\{v\}$ and $[K]$. However, describing a calculation procedure for an element, we shall use for vector an index notation.

Equations of FEM for step-by-step calculations are represented in terms of stress $\{\Delta\sigma\}$ and total strain $\{\Delta\varepsilon\}$ tensors increments. The latter is equal to the sum

$$\{\Delta\varepsilon\} = \{\Delta\varepsilon_e\} + \{\Delta\varepsilon_p\} + \{\Delta\varepsilon_T\} + \{\Delta\varepsilon_e^T\}, \tag{5.59}$$

where $\{\Delta\varepsilon_e\}$, $\{\Delta\varepsilon_p\}$, and $\{\Delta\varepsilon_T\}$ are tensors increments of the elastic and plastic strains and thermal expansion, respectively.

Temperature-dependent part of the elastic strain (see Chap. 2) is included in Eq. (5.60) as the additional term $\{\Delta\varepsilon_e^T\}$.

In case of isothermal elasticity, an equation inverse to Eq. (2.12) yields

$$\{\Delta\sigma\} = [D]\{\Delta\varepsilon_e\},$$

where $[D]$ is an elasticity matrix

$$[D] = \frac{E}{(1+\mu)(1-2\mu)}\begin{bmatrix} 1-\mu & \mu & \mu & & & \\ \mu & 1 & \mu & & 0 & \\ \mu & \mu & 1 & & & \\ & & & 1-2\mu & & \\ & 0 & & & 1-2\mu & \\ & & & & & 1-2\mu \end{bmatrix} \tag{5.60}$$

Therefore, the basic FEM correlation between stress $\{\Delta\sigma\}$ and total strain $\{\Delta\varepsilon\}$ increments in application to a small element of a body may be presented in a form

$$\{\Delta\sigma\} = [D](\{\Delta\varepsilon\} - \{\Delta\varepsilon_p\} - \{\Delta\varepsilon_T\} - \{\Delta\varepsilon_e^T\}). \tag{5.61}$$

For non-isothermal processes, the matrix $[D]$ must be recalculated in the beginning of each step using the elasticity modulus $E(T)$ referred to a new value of element temperature. Temperature-dependent part of the elastic strain (see Chap. 2) is included in Eq. (5.60) as the additional member $\{\Delta\varepsilon_T^e\}$.

At moderate temperatures, the elasticity modulus can be averaged that allows taking $\{\Delta\varepsilon_T^e\} \approx 0$ and matrix $[D]$—as a constant.

The plastic strain increment $\{\Delta\varepsilon_p\}$ during a given step depends on the stress and total strain increments and also on current values of stress, strain, temperature, and parameters reflecting influence of the previous steps. The increments of the total and plastic strains and also stresses, during a current step, are determined from the FEM solution of the entire structure by successive approximations taking into account the converge conditions (Temis 1980, 1994a). Below, we consider only procedures related to calculation of plastic strain increment during the current step.

5.4.3.2 Solution Procedure for the SARH Model of Thermal Plasticity

For solving a thermal plasticity problem with use of SARH model, the following material characteristics, along with thermoelasticity parameters, must be set: yield point in active stresses σ_y^*, modulus of stationary plasticity E_{p0}, limit magnitude of a nonlinear part of micro-stresses $r_{\lim}$, and parameter of hardening stabilization β_p. For a stable material state considered below, the specified characteristics depend only on temperature.

Before the beginning of a current step of loading and heating, the following parameters are assumed to be known from the proceeding calculations or initial conditions: tensor components of active stresses σ_{ij}^*, steady r'_{ij} and unsteady r''_{ij} parts of micro-stresses, and also of plastic strains ε_{ij}^{p}. Parameters referred to the step termination are parenthesized as $\langle\rangle$.

Step Beginning

At the step beginning, all material characteristics are recalculated for new temperature $\langle T\rangle$ that yields $\langle\sigma_{y}^{*}\rangle$, $\langle E_{p0}\rangle$, $\langle r_{\lim}\rangle$, and $\langle\beta_{p}\rangle$. Then, the following parameters are calculated:

Values of initial step parameters of stress/strain state are as follows:

- components of the micro-stress tensor $r_{ij} = r'_{ij} + r''_{ij}$;
- components of the total stress tensor $\sigma_{ij} = \sigma_{ij}^{*} + r_{ij}$;
- components of the active and total stress deviators

$$s_{ij}^{*} = \sigma_{ij}^{*} - \delta_{ij}\sigma_{ii}^{*}/3,$$
$$s_{ij} = s_{ij}^{*} + r_{ij};$$

- intensities of all stress parts

$$\sigma_{i}^{*} = \left(3s_{ij}^{*}s_{ij}^{*}/2\right)^{0.5}, \quad \sigma_{i} = \left(3s_{ij}s_{ij}/2\right)^{0.5},$$
$$r'_{i} = \left(2r'_{ij}r'_{ij}/3\right)^{0.5}, \quad r''_{i} = \left(2r''_{ij}r''_{ij}/3\right)^{0.5}$$

- intensity of the plastic strain $\varepsilon_{i}^{p} = \left(2\varepsilon_{ij}^{p}\varepsilon_{ij}^{p}/3\right)^{0.5}$.

Iteration Procedures

The subsequent procedures concern each of the FEM iterations for the solution of thermoplasticity problem. During entire step, external loads and temperatures are assumed to be constant.

New Parameters after Input of Stress Increment

Increment components $\Delta\sigma_{ij}$ of the stress tensor are assumed to be set or obtained in the previous calculation step or iteration. New components values are $\langle\sigma_{ij}\rangle = \sigma_{ij} + \Delta\sigma_{ij}$ and increments of the stress deviator components are

$$\Delta s_{ij} = \Delta\sigma_{ij} - \delta_{ij}(\Delta\sigma_{ii})/3.$$

At the beginning, we let strain increment $\Delta\varepsilon_{ij}$ to be elastic, i.e., $\Delta\varepsilon_{ij}^{p}=0$, $\Delta r_{ij}=0$, and $\Delta\sigma_{ij}^{*}=\Delta\sigma_{ij}$. New stress parameters values for "elastic" solution are as follows:

- active stress $\langle\sigma_{ij}^{*}\rangle^{e}=\sigma_{ij}^{*}+\Delta\sigma_{ij}$;
- active stress deviator $\langle s_{ij}^{*}\rangle^{e}=s_{ij}^{*}+\Delta s_{ij}$;
- active stress intensity $\langle\sigma_{i}^{*}\rangle^{e}=[3\langle s_{ij}^{*}\rangle^{e}\langle s_{ij}^{*}\rangle^{e}/2]^{0.5}$.

Verification of the Plastic Yielding Conditions

- If $\langle\sigma_{i}^{*}\rangle^{e}\leq\langle\sigma_{y}^{*}\rangle$, then the assumption on the elastic strain increment is true. In this case,

$$\langle\sigma_{ij}^{*}\rangle=\langle\sigma_{ij}^{*}\rangle^{e},\quad\langle\varepsilon_{ij}^{p}\rangle=\varepsilon_{ij}^{p},\quad\langle r_{ij}'\rangle=r_{ij}',\quad\langle r_{ij}''\rangle=r_{ij}''$$

  and the step calculation is completed.
- Inequality $\langle\sigma_{i}^{*}\rangle^{e}>\langle\sigma_{y}^{*}\rangle$ denotes plastic yielding in a new direction of the active stress $\langle\boldsymbol{\sigma}_{i}^{*}\rangle$ coinciding with the direction of $\langle\boldsymbol{\sigma}_{i}^{*}\rangle^{e}$. In this case, we take

$$\langle\sigma_{i}^{*}\rangle=\langle\sigma_{y}^{*}\rangle\quad\text{and}\quad\Delta\hat{r}_{i}=\langle\sigma_{i}^{*}\rangle^{e}-\langle\sigma_{y}^{*}\rangle.$$

Calculation of a Change of the Micro-stresses Nonlinear Part

A cosine of the angle ϕ_r between the former direction of the vector $\mathbf{r}_i''$ and new direction of the vector $\langle\boldsymbol{\sigma}_{i}^{*}\rangle$ is

$$\cos\phi_r=\frac{r_{ij}''\langle s_{ij}^{*}\rangle^{e}}{r_i''\,\langle\sigma_{i}^{*}\rangle^{e}}$$

and a projection of the vector $\mathbf{r}_i''$ onto new direction is

$$\hat{r}_i''=r_i''\,\cos\phi_r.$$

If $\hat{r}_i''\geq k_r\langle r_{\lim}\rangle$ where $0.95<k_r<1$, we take $\langle\hat{r}_i''\rangle=\langle r_{\lim}\rangle$ and $\langle E_p''\rangle=0$. A value k_r may be specified by the user.

If $\hat{r}_i''<k_r\langle r_{\lim}\rangle$, we define modulus of unsteady plasticity $\langle E_p''\rangle$ and increment of unsteady micro-stresses $\Delta\hat{r}_i''$ from a set of nonlinear equations

$$\langle E_p'' \rangle = \langle \beta_p \rangle [\langle r_{\lim} \rangle - (\hat{r}_i'' + \Delta \hat{r}_i'')],$$
$$\Delta \hat{r}_i'' = \frac{\langle E_p'' \rangle \Delta \hat{r}_i}{\langle E_{p0} \rangle + \langle E_p'' \rangle}.$$

New components of the nonlinear micro-stresses part increment are

$$\Delta \hat{r}_{ij}'' = \Delta \hat{r}_i'' \frac{\langle s_{ij}^* \rangle^e}{\langle \sigma_i^* \rangle^e}.$$

New components of the nonlinear micro-stresses part vector are

$$\langle r_{ij}'' \rangle = r_{ij}'' + \Delta r_{ij}'',$$

and their intensity is $\langle r_i'' \rangle = [2\langle r_{ij}'' \rangle \langle r_{ij}'' \rangle / 3]^{0.5}$.

Calculation of the Plastic Strain Increment

An increment of the plastic strain intensity is

$$\Delta \varepsilon_i^p = \frac{\Delta \hat{r}_i}{\langle E_{p0} \rangle + \langle E_p'' \rangle}.$$

Components of the plastic strain vector's increment are

$$\Delta \varepsilon_{ij}^p = \frac{3 \Delta \varepsilon_i^p \langle s_{ij}^* \rangle^e}{2 \langle \sigma_i^* \rangle^e}.$$

and new values of the components are $\langle \varepsilon_{ij}^p \rangle = \varepsilon_{ij}^p + \Delta \varepsilon_{ij}^p$.

Transition to the Next Iteration

Calculated increments of the plastic strain vectors $\{\Delta \varepsilon_p\}$ for each element are substituted into corresponding equations of FEM code for a structure and the current iteration is carried out. New values of the stress increment vectors $\{\Delta \sigma\}$ allow the next iteration of a current step to be executed.

Step Termination

Upon completion of the iterations, we calculate new values of micro-stress parts.
If a calculated value is $\langle r_i'' \rangle \leq \langle r_{\lim} \rangle$, we take $(\Delta r_{ij}')_{ad} = -(\Delta r_{ij}'')_{ad} = 0$.

If $\langle r_i'' \rangle > \langle r_{\lim} \rangle$, then additional increments of the micro-stresses parts intensities are

$$(\Delta r_i')_{ad} = -(\Delta r_i'')_{ad} = r_i'' - \langle r_{\lim} \rangle \cos \phi'' + \Delta \hat{r}_i'' \cos \phi_r,$$

where

$$\phi'' = \arcsin \phi_r \frac{\Delta \hat{r}_i''}{\langle r_{\lim} \rangle}.$$

Components of the corresponding vectors are

$$(\Delta r_{ij}')_{ad} = (\Delta r_i')_{ad} \frac{r_{ij}''}{r_i''}, \quad (\Delta r_{ij}'')_{ad} = -(\Delta r_{ij}')_{ad}.$$

New values of the micro-stress parts

$$\langle r_{ij}' \rangle = r_{ij}' + \Delta \hat{r}_{ij}' + (\Delta r_{ij}')_{ad}$$
$$\langle r_{ij}'' \rangle = r_{ij}'' + \Delta \hat{r}_{ij}'' + (\Delta \hat{r}_{ij}'')_{ad}.$$

and the values of the last iteration of the active stresses $\langle \sigma_{ij}^* \rangle$ and plastic strains $\langle \varepsilon_{ij}^p \rangle$ tensors are taken as new components for the next step.

5.4.4 Damage Accumulation and Failure

5.4.4.1 Failure Criteria

Failure criteria of an inherently isotropic material usually remain identical in cases when elastoplastic strains develop in different directions. However, quantities of the failure criteria depend not only on material behavior but also on loading conditions.

In Chaps. 3 and 4, the damage accumulation and failure for monotonic increase of the stress intensity and cyclic sign-constant or sign-varying loading under constant orientation of stress tensor have been considered.

For any sign-varying elastoplastic deformation with rotation of the CH axis, two types of damage are possible:

- *Quasi-static* by static damage accumulation when a magnitude of the plastic strain vector $\boldsymbol{\varepsilon}_i^p$ gradually increases in permanently changing directions. This comes to end by static failure after the plastic strain intensity achieved the limit quantity ε_i^p corresponding to the generalized deformation diagram.

- *Low cyclic fatigue* by cyclic damage accumulation due to increase of the number of multiply repeated loading cycles when magnitude and direction of the plastic strain vector $\boldsymbol{\varepsilon}_i^p$ periodically change.

In some cases, both types of damage accumulation can occur simultaneously.

Each portion of 1D sign varying processes between two consecutive changes of stress direction can be considered as *k*th "half-cycle" with a corresponding range of plastic strain. These half-cycles can be conditionally interpreted as "loading/ unloading" (Temis 1989). In Fig. 5.1, half-cycle CD is associated with the plastic strain range $\Delta\varepsilon_p^{(k)}$ shown by line EF.

In case of multi-axial processes, alteration of "half-cycles" occurs in the moment when a projection of total stress increment $\Delta\hat{\sigma}_i$ onto the CH axis

$$\Delta\hat{\sigma}_i = \Delta\sigma_i \cos\phi_\sigma = \frac{3\Delta s_{ij}s_{ij}}{2\sigma_i} \tag{5.61}$$

(see Fig. 5.6a where $d\hat{\sigma} \approx \Delta\hat{\sigma}$ changes the sign). Above, in the chart of Fig. 5.12a, five half-cycles, three from which are elastoplastic, can be found out.

The number of half-cycles for LCF failure can be estimated using the approaches of Chap. 4. However, failure criteria for the multi-axial non-isothermal loading of a material under arbitrary non-proportional paths are studied insufficiently.

5.4.4.2 Damage Accumulation Under Unstable Material State

In common case, isotropic material characteristics σ_y^*, $E_{p0}, r_{\lim}$, and β_p depend not only on the temperature but also on the structural parameter χ. The equality $\chi \approx \varepsilon_{p\Sigma}$ where $\varepsilon_{p\Sigma}$ is the accumulated plastic strain (Odqvist's parameter) as damage parameter is valid only at sign-constant deformation (see Chap. 1) or small number of loading cycles. In common case, other approaches are required. At a great number of half-cycles, a normalized parameter $\chi = \varepsilon_{p\Sigma}/(\varepsilon_{p\Sigma})_d$, in which accumulated plastic strain $\varepsilon_{p\Sigma}$ is referred to the failure quantity $(\varepsilon_{p\Sigma})_d$ under the same average value of the plastic strain range $\Delta\varepsilon_{pm}$, gives better results as it was demonstrated by Temis and Shorr (2011) (see Chap. 4).

5.4.5 Complementary Notes

5.4.5.1 Partition of Elastic and Plastic Strains

A great number of different approaches to plastic deformation modeling under combined stress state and complex loading are considered in the literature.

A concept of trajectories of loading and deforming proposed by Il'ushin (1963) is extensively used by his followers. The trajectories are described by movement of the ends of the vectors $\boldsymbol{\sigma}_i$ and $\boldsymbol{\varepsilon}_i$ in five-dimensional space. The length of a trajectory is taken as "time" parameter. A number of experimental works showed that experimental stress–strain dependences for monotonic increase of plastic strain can be rather adequately described by a choice of suitable design parameters using material characteristics obtained by particular tests processing.

However, the use of full strain vector $\boldsymbol{\varepsilon}_i$, without its partition to elastic and plastic parts, does not allow taking into account the different rate of plastic evolution after change of loading sign and direction, which conflicts with real material behavior. Besides, a study of deformation for given trajectories is useful for substantiation of theoretical approaches to the analysis of plasticity problems. But as applied to practical aims, the loading and deforming trajectories are not known in advance and can be established only during the design.

In the so-called endochronical theory of plasticity proposed by Valanis (1971), only full strains are also inherently considered. But the further development of this theory demands, as it was shown by Kadashevich and Mosolov (1990), to come back in this or that form to notion of elastic and plastic strain areas partition.

5.4.5.2 Development of the SAH Theory

Uniaxial Stress State

On the basis of Ishlinsky's (1954) and Kadashevich and Novojilov's (1958) concept that plastic strains are accompanied by development of residual micro-stresses, some variants of the theory of anisotropic plastic hardening have obtained a further elaboration as stabilized anisotropic hardening. By Shorr (1975) and also in Birger and Shorr (1975), Chaboche et al. (1979), Bondar' (1982, 1987), Temis (1989), and partly others were established that in the governing Eq. (5.13)

$$dr = E_p(\varepsilon_p, r)d\varepsilon_p \tag{a}$$

a varying modulus of plasticity E_p should have a trinomial form

$$E_p(\varepsilon_p, r) = C_0 + (C_1\varepsilon_p - C_2 r)\,\mathrm{sign}\sigma^* \tag{b}$$

Equations (5.13) and (5.14) coincide with Eqs. (a) and (b), respectively, taking $C_0 = E_{p0} + \beta_p r_{\lim}$, $C_1 = \beta_p E_{p0}$, and $C_2 = \beta_p$.

After integrating within the loading section with $\mathrm{sign}\sigma^* = \mathrm{const}$, Eq. (a) leads to Eq. (5.21) and for initial conditions $r_{in} = 0$ at $\varepsilon_{p,in} = 0$ and $\sigma^* > 1$ takes the form of Eq. (5.23).

An opportunity of micro-stresses r connection with plastic deformation ε_p in the form of Eq. (c) was specified by Novojilov (1964).

Chaboche (1977) with the reference to Amstrong and Frederick (1966) proposed the equation (in our notations)

$$dr = Cd\varepsilon_p - \beta_p r d\varepsilon_{p\Sigma}, \tag{c}$$

which can be considered as a particular case of Eqs. (a and b) with $C_1 = 0$, $r_{\lim} = 0$, and $d\varepsilon_{p\Sigma} = (\text{sign}\sigma^*) d\varepsilon_p$.

A solution based on Eq. (c) leads to fading of anisotropic deformation with a strong overestimate of the plastic strain that is in variance with experimental data. Therefore, in Chaboche et al. (1979), Chaboche (1989), it was offered to represent r in the form of a series of functions. Introduction in Eq. (c) of an additional term has led to the equation

$$dr = Cd\varepsilon_p + \beta_p (C' \varepsilon_p - r) d\varepsilon_{p\Sigma}, \tag{d}$$

which coincides with Eq. (b) at $C = C_0 = E_{p0} + \beta_p r_{\lim}$, $C' = E_{p0}$, and $d\varepsilon_{p\Sigma} = (\text{sign}\sigma^*) d\varepsilon_p$.

The equation of type (d) was also considered by Bondar' (1982, 1987) who demonstrated that this equation can be interpreted as concretization of the tensor linear theory offered by Kadashevich (1967). Equations of this type were successfully used by Bondar' and Danshin (2008) for computational modeling of many experimental stress and strain trajectories at various loading conditions and also were dealt in some other publications. Conception on a limited quantity $r_{\lim}$ of micro-stresses fading part was also used by Temis (1989).

Some difference of Eq. (d) from Eqs. (a, b) consists in the following. In Eq. (d), the term $(\text{sign}\sigma^*) r d\varepsilon_p$ is represented in the form of $r d\varepsilon_{p\Sigma}$ but a quantity $\varepsilon_{p\Sigma}$ should not be identified with accumulated plastic strain since reversible anisotropic hardening, to which specific equations concern, cannot be connected with irreversible strain. Occurrence of the signσ^* in Eq. (b), as it follows from its deduction, concerns the micro-stress r and strain ε_p, and not the increment $d\varepsilon_p$. At change of a signσ^*, the initial conditions r_{in} and $\varepsilon_{p,in}$ for a new direction should be accepted by the final condition for the previous stage.

Combined Stress State

At a combined stress state under *proportional* loading, the stress and strain components in the equations given above are replaced with their intensities. Using parameter projections onto direction of the active stress vector (see Sect. 5.3.3.2), an affect of loading sign change is considered in the equations automatically.

Accounting that the CH axis rotates with the active stress vector, we obtain from Eqs. (5.46)–(5.47) the equation

$$d\hat{r}_i = [E_{p0} + \beta_p (r_{\lim} + E_{p0} \hat{\varepsilon}_i^p - \hat{r}_i)] d\hat{\varepsilon}_i^p, \tag{e}$$

which is valid for multi-axial sign-varying loading and does not include the term with irreversible strain $\varepsilon_{p\Sigma}$.

To describe the effect of anisotropic hardening at a combined stress–strain state under *non-proportional* (out-of-phase) loading, various approaches in a number of works were proposed. Detailed insights into the problem are given in the special literature, see Mroz's (1969), Chaboche (1989, 2008), Temis (1989), Zubchaninov (2002), Bondar' and Danshin (2008), Volkov and Korotkikh (2008), etc. To this end, a system of governing equations usually becomes considerably complicated, and new material characteristics which demand the undertaking of additional experiments are introduced.

An approach to analysis of elastoplastic multi-axial processes under arbitrary non-proportional loading on the base of the concept proposed in Shorr (2013) and has been conditionally named above "the theory of stabilized anisotropic-ray hardening" (the SARH theory) is described here in detail seemingly for the first time. In spite of simplicity, this approach is well experimentally substantiated and reveals the most important features of material behavior in manifold complex conditions.

References

Amstrong PJ, Frederick CO (1966) A mathematical representation of the multiaxial Bauschinger effect. C.E.G.B. Report No. RD/B/N 731

Arutjunjan RA (1964) About cyclic loading of elastoplastic medium. Izv. AN SSSR, Mekh. i Matem. 4:89–92 (in Russian)

Birger IA, Shorr BF (eds) (1975) Thermal strength of machine parts. Mashinostroenie, Moscow (in Russian)

Bondar' VS (1982) Solutions of nonlinear problems for combine loading of rotation shells. In: Proceedings of strength and stiffness designs, vol 4. Mosstankin, Moscow, pp 85–95 (in Russian)

Bondar' VS (1987) Mathematical modal of a material inelastic behaviour and damage accumulation. In: Applied problems of strength and plasticity, Gor'kov Universitet, pp 24–28 (in Russian)

Bondar' VS, Danshin VV (2008) Plasticity. Proportional and non-proportional loadings. Fizmatlit, Moscow (in Russian)

Chaboche JL (1977) Viscoplastic constitutive equations for the descriptions of cyclic and anisotropic behaviour of metals. In: Proceedings of 17th Polish conference on mechanics of solids, vol 25. Szczyrk. Bul. de l'Acad. Polonaise des Scinces, Serie Sc et Techn., p 33

Chaboche JL (1989) Constitutive equations for cyclic plasticity and cyclic viscoplasticity. Int J Plast 5:247–302

Chaboche JL (2008) A Review of some plasticity and viscoplasticity. Constitutive Equ Int J Plast 24:1642–1693

Chaboche JL, Dang-Van K, Cordier G (1979) Modelization of the strain memory effect on the cyclic hardening of 316 stainless steel. In: Proceedings of the 5th international conference on SMiRT, Div.L, Berlin

Grüning M (1926) Die Tragfähigkeit statisch unbestimmter Tragwerke aus Stahl bei beliebig häufig wiederholter Belastung, Berlin

Hohenemser K, Prager W (1932) About mechanics of steel plastic behaviour. ZAMM, Bd.12, H.1:1–14
Il'ushin AA (1948) Plasticity. OGIZ, Moscow-Leningrad (in Russian)
Il'ushin AA (1963) Plasticity. Foundation of general mathematical theory. AN SSSR, Moscow (in Russian)
Ishlinsky AY (1954) General theory of plasticity with linear hardening. Ukrainsky Matem J 6 (3).314–324 (in Russian)
Kadashevich YI (1967) About various tensor-linear relations in the theory of plasticity. Investigations on elasticity and plasticity. Izd.LGU, Leningrad: 6:39–45 (in Russian)
Kadashevich YI, Mosolov AB (1990) The relationship between the endochronic theory of plasticity and the "new" measure of internal time. Prikl.Matem.Mechan. 54(4):571–576 (trans 1991, Pergamon Press, in Russian)
Kadashevich YI, Novojilov VV (1958) Theory of plasticity accounting of residual microstresses. J PMM XXI(I)(1):78–89 (in Russian)
Koiter WT (1960) General theorems for elastic-plastic solids. In: Sheddon I, Hill R (eds) Progress in solid mechanics, vol 1. North-Holland publishing company, Amsterdam
Lemaitre J, Chaboche JL (1990) Mechanics of solid materials. Cambridge University Press, Cambridge
Lemaitre J, Chaboche JL (2004) Mecanique des materiaux solides, 2nd edn. Dunod, Paris
Mroz Z (1969) An attempt to describe the behavior of metals under cyclic loads using a more general workhardening Model. Acta Mech 7:199–212
Novogilov VV (1964) On complex loading and perceptivities of phenomenological approach to study of micro-stresses. J Prikladnaja matematica I mekanika 28(3):393–400 (in Russian)
Odqvist FKG (1933) Verfestigung von flussisenahnlichen Korpern. ZAMM 13(5):360–363
Prager W (1957) Shakedown in elastic, plastic media subjected to cycles of load and temperature. In: Symposium su la plasticitá nella scienza delle costrutioni in onore de Arturo Danusso, Varenna, Sept 1956, Bologna, pp 239–244
Reuss A (1930) On account of the elastic strains in the theory of plasticity Zeitschr für Angev Math und Mech 10(3):226–274
Shishmarev OA, Shcherbo AG (1983) Image of a loading process for strain double-section broken trajectories with break angle more 90°. Izv. AN SSSR, MTT, no. 5 (in Russian)
Shorr BF (1975) To theory of non-isothermal plastic deforming at arbitrary loading lows. In: Proceedings of KuAI on mechanics, 8. Kuybyshev, pp 107–114 (in Russian)
Shorr BF (2013) About some problems and restrictions in the theory of thermal plasticity. In: Vasin RA, Kiyko IA (eds) Bondar' VS. Modern concepts of thermoviscoplasticity in the applied problems of design analysis and high parameter technologies. Works of the VI School-seminar. Mosk. Gos. Universitet Mashinostroenija, Moscow, pp 125–143
Temis YM (1980) Method of successive approximations with error correction in geometrically nonlinear elastic problems. In: Proceedings of applied problems of strength and plasticity. Algorithmic solution procedures in elasticity and plasticity problems. Izd.Gor'kovskogo Gos. Universiteta. Go'kiy 16:3–10 (in Russian)
Temis YM (1989) Plasticity and creep of turbo engines components at cyclic loading. In: Strength and dynamics problems in avia-engines building, CIAM Proceedings 1237, No. 4, Moscow, pp 32–50 (in Russian)
Temis YM (1994a) Step-by-step and iterative algorithms and principles of software formation for mathematical modelling of heat-stressed structures. In: Frolov KV et al (eds) Machinery, encyclopedia, vol 1–3, no 1. Mashinostroenie, Moscow: 256–262 (Kolesnikov KS (ed), in Russian)
Temis YM (1994b) Theory of non-isothermal plastic yielding with isotropic and anisotropic hardening. In: Frolov KV et al (eds) Machinery, encyclopedia, vol 1–3, no 1. Mashinostroenie, Moscow, pp 227–231 (Kolesnikov KS (ed), in Russian)
Temis YM, Shorr BF (2011) About a role of accumulated plastic strain in the theory of plasticity. NNGU, No 4, part 4, Nijnij Novgorod, pp 1879–1881 (in Russian)

Valanis KC (1971) A theory of viscoplasticity without a yield surface. Arch Mech Stosow 23 (4):517–551

Volkov IA, Korotkikh YuG (2008) Constitutive equations of viscoelastoplastic media with damages. Physmatgiz, Moscow (in Russian)

Zienkiewicz OC (1967) The finite element method in structural and continuum mechanics. McGraw-Hill, London

Zienkiewicz OC, Taylor RL (2000) The finite element method, 5th edn. Butterworth-Heinemann, Oxford

Zubchaninov VG (2002) Mathematical theory of plasticity. TGTU, Tver' (in Russian)

Zubchaninov VG, Okhlopkov NL, Garanikov VV (2004) Experimental plasticity, no 2. Processes of combine loading, Tver's State Technical University, Tver' (in Russian)

Chapter 6
Creep and Long-Term Strength at Stationary Stresses and Temperatures

Abstract The developed computational models for creep and long-term strength of structural materials under multi-axial proportional loading associated with steady or monotonically varying mechanical and thermal disturbances are justified using numerous experimental data. Specialized and generalized models for non-isothermal creep and ranges of applicability are discussed. A complete three-stage creep diagram is also presented. The latter may be described by six material parameters depending upon stress and temperature. Creep of statically determinate and indeterminate mechanical systems is considered along with computational methods oriented to the prescribed kinematic laws of steady creep velocity distribution over a body. Several examples for various creep conditions show that the stress field resulting from external loading, and especially induced by non-uniform thermal expansion, may deviate considerably from the field corresponding to purely elastic calculation and also from that related to the average body temperature. In case of creep, temperature variation makes a major impact on stress redistribution caused by stress relaxation as well as on the magnitude of residual stresses originated from unloading and cooling. The possibility for predicting structure durability in case of long-term maintenance is addressed, including equivalent trials under heavier stress and temperature conditions.

The first investigations on metal creep belong to Bailey (1929), Norton (1929), Tapsell (1931), Odqvist (1933). Investigations have obtained further development in the second half of the last century. Backgrounds of creep and durability theory as a partition of solid mechanics are stated by Rabotnov (1948, 1960, 1966), Odqvist (1974), and in various works. Many of them contain extensive bibliography material (Lemaitre and Chaboche 1990, 2004; Hosford 2005; Yao et al. 2007, and others).

Basic experimental and theoretical results presented below are most interesting and useful in science and engineering.

B.F. Shorr, *Thermal Integrity in Mechanics and Engineering*,
Foundations of Engineering Mechanics, DOI 10.1007/978-3-662-46968-2_6

6.1 Time-Dependent Mechanical Properties

6.1.1 Experimental Backgrounds of Creep and Durability

6.1.1.1 Creep and Durability

Structure elements fabricated out of stable materials and operated in a range of moderate temperatures and stresses can save their dimensions and working capacity for a very long time. In these conditions, the time describes only the sequence of loading and heating processes. However, at temperatures exceeding the "threshold" value for a given material, the duration of exposition to a load, i.e., the *operating time,* starts to directly influence the mechanical behavior of a construction. For metal alloys, the threshold value is usually 30–40 % Kelvin temperature of melting.

There are a great number of investigations in which temporal physical processes of deformation and damage connected with material composition, structure, phase changes, and other analogous factors are studied. At the phenomenological approach within the frame of material representation as a solid, the development of deformation and damage appears in two directions:

- In gradual resizing of a body which is commonly named *creep*;
- In formation of micro-cavities and cracks which are local in the beginning, but then transmit into a global (through thickness) crack that leads to structure failure due to *exhaustion of durability.*

These effects appear at variable and constant stresses and temperatures.

As a result of creep, a construction can obtain an inadmissible change in size. This can, for example, lead to contact between a rotating wheel and stator elements in air engines. Durability exhaustion can happen as a creep completion process under considerable strains, and also at small size changes, which in itself do not represent danger.

In literature, one can sometimes meet, instead of the term "durability," other terms with the same meaning such as creep rupture strength, and long-term or long-time strength. We will mainly use the most widespread term *long-term strength.*

Basic creep characteristics are obtained by tests at constant stresses and temperatures. These data are then approximately generalized for analysis processes at smoothly varying stresses and temperatures. Creep and durability under alternating and sign-varying stresses and temperatures have specific features which are discussed in Chap. 7.

Creep and plastic strains, and also the characteristics of short and long-term strength of a material, are linked between themselves. These effects were studded by some researchers, including Sizova, whose data presented in Birger and Shorr (1975), will be in part used further in this chapter.

6.1.1.2 Creep Diagrams in Tension

The technique and instrumentation for specimen tests on creep and long-term strength are standardized, see, e.g., Evans and Wilshire (1985). The temperature expansion $\varepsilon_T = \alpha(T - T_0)$ is usually excluded from analysis. The initial loading from stress $\sigma = 0$ to a chosen value of σ is assumed to be considered as "instant." Thus, in the creep test beginning at $t = 0$ an elastoplastic part of a strain $\varepsilon_\sigma = \sigma/E + \varepsilon_p$, as a function of applied stress and temperature is found. In the course of the further testing, the total strain $\varepsilon(t)$ versus time is measured. A creep part of the total strain increasing in time is

$$\varepsilon_c(t) = \varepsilon(t) - \varepsilon_\sigma. \tag{6.1}$$

The Sizova's tests of a wrought heat-resistant nickel alloy are shown in *creep diagrams* in the form of families of curves $\varepsilon_c(t, \sigma)$ at temperatures 900 °C as is displayed in Fig. 6.1a where $\varepsilon_c(0, \sigma) = \varepsilon(0, \sigma) - \varepsilon_\sigma = 0$.

Time prior to specimen failure t_d (*rupture time*) depends on the magnitude of operating stress σ and temperature T.

A magnitude of stress $\sigma_d(T, t_d)$, at which rupture occurs at the instant t_d and temperature T, is named *a limit of long-term strength* (or in abbreviated form *a long-term strength*). A corresponding magnitude of creep strain $\varepsilon_{c,d}$ is sometimes named *a long-term plasticity*.

From the assaying of creep diagrams, some additional characteristics can be detected.

The stress $\sigma_{\varepsilon_c/t_c}$ at which the creep strain at temperature T reaches a specified value ε_c for the set time t_c is named a *limit of creep*. In manuals, it is usually accepted to take $\varepsilon_c = 0.2\ \%$ or $\varepsilon_c = 1\ \%$ and $t_c = 10^n$ hours $(n = 2–4)$. The notation $\sigma_{0.2/100}$ corresponds to $\varepsilon_c = 0.2\ \%$ and $t_c = 100$ h.

A strain $\varepsilon_c(t)$ reflects a whole process of creep evolution, while an instant growth of creep strain is determined by *creep rate* $v_c = \dot{\varepsilon}_c$. The layout view of dependences $v_c(t, \sigma)$ for the basic types of creep diagrams is displayed in Fig. 6.1b.

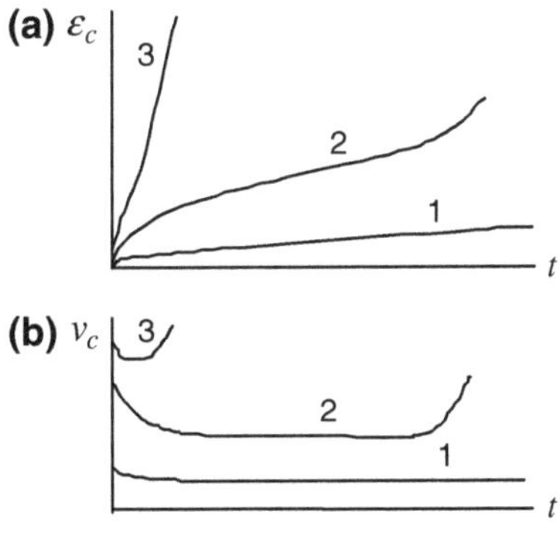

Fig. 6.1 Sketches of creep diagrams for different stresses; $\sigma_1 < \sigma_2 < \sigma_3$

The creep diagrams obtained at different temperatures are presented in the form of tables or graphs $\varepsilon_c(\sigma, T, t)$ and $v_c(\sigma, T, t)$.

Creep diagrams differ by great varieties, including test duration—from several minutes at short-term creep to tens of thousands of hours. In unique tests carried out by Robinson (1943) within 11 years and also by Wilson et al. (1971), the duration reached hundred thousand hours.

6.1.1.3 Full Creep Diagram and Long-Term Strength

In typical full creep diagrams (curves 2 in Fig. 6.1), there are three distinguishable stages. The first primary phase occurs immediately after the application of constant stress. Here, creep has *unsteady* (*transient*) character when rate v_c decreases from initial $v_{c0}(\sigma, T, 0)$ to minimum $v_{c,\min}(\sigma, T)$ value for the given stress and temperature. This effect reflects hardening of a material owing to the development of inelastic deformation.

Within the *secondary*—*steady* (*stabilized*)—stage, creep rate remains almost constant during some time. At the third (sometimes called *tertiary*) stage, the creep rate again increases up to the moment of the specimen rupture. The failure can be caused by the accumulation of material damage, and also a reduction of the specimen's cross-sectional area in tension leading to increase of real stresses.

In practice, the specified stages of creep can have no accurate differentiations. At small stresses (curve 1 in Fig. 6.1), the damage can occur under the secondary creep stage without appreciable macro-strains. At great stresses and temperatures (curve 3), the primary stage directly transmits into the tertiary one. In some cases, the tertiary stage occurs alone.

If a specimen, stressed lower than the yield point, is completely unloaded, its residual strain remains equal to the accumulated creep. In metals and their alloys, the creep strain remains usually without change. In polymeric materials (and more rare in metals), partial or complete recovery of the initial sizes with time appears (effect of restitution). This problem is studied in detail in the theory of viscoelasticity (see Rabotnov 1966; Shevchenko and Terekhov 1982; Lemaitre and Chaboche 2004, and others).

Test results show considerable scatter in magnitude of creep strains and time at long-term rupture. The maximum and minimum durations of the identical specimens can differ by several times (see, e.g., Kablov and Golubovsky 1998). Figure 6.2 displays the ranges of creep curves of a heat-resistant cast alloy, obtained by tests of 10 samples at each of two stress levels 1 ($\sigma = 300$ MPa) and 2 ($\sigma = 400$ MPa) at identical high temperature (due to Sizova' data specified above). Such scatter must be taken into consideration for the assignment of long-term strength margins (see Chap. 8), and also for the approximation of experimental data by analytical functions.

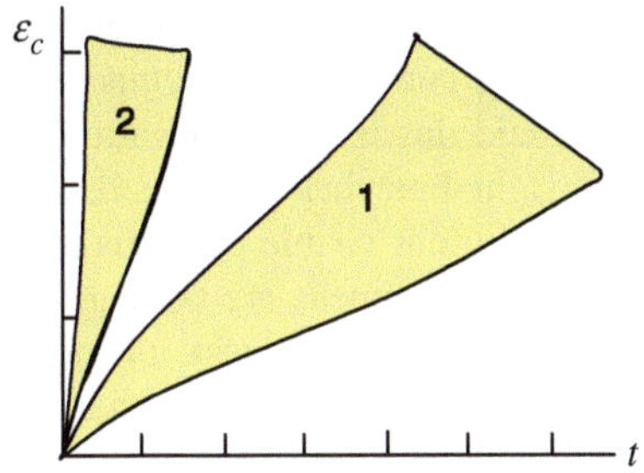

Fig. 6.2 Example of scatter of creep strains and long-time rupture at two stress levels: $\sigma_1 < \sigma_2$

As time-dependent damage of many materials in operating conditions occurs at a very small creep strain, special tests for the study of long-term strength are carried out.

The dependence of a long-term strength material limit σ_d on rupture time t_d at different temperatures is usually represented in double-logarithmic coordinates $\log \sigma_d = f(\log t_d)$, as is shown in Fig. 6.3 for a typical heat-resistant chrome–nickel alloy (Sizova's data).

Parts of horizontal lines at different stress levels display significant ranges of time scatter before rupture at corresponding stresses. Sometimes half-logarithmic coordinates $\sigma_d = f(\log t_d)$ are also used.

Tests using large specimen numbers showed that the time prior to long-term rupture submitted to the logarithmic normal law of distribution (within two–three mean square deviations).

In Fig. 6.3, the function $\log \sigma_d = f(\log t_d)$ is presented in the form of two sites of almost straight lines. An appreciable breaking between these lines is typical enough for many structural materials.

The breaking reflects the transition from one type of damage to another. At the left branches of curves at high stress levels, there is a considerable creep strain, and

Fig. 6.3 Long-term strength diagram for a heat-resistant alloy at different temperature: $T_1 = 600\,°C$, $T_2 = 700\,°C$, $T_3 = 800\,°C$, $T_4 = 900\,°C$

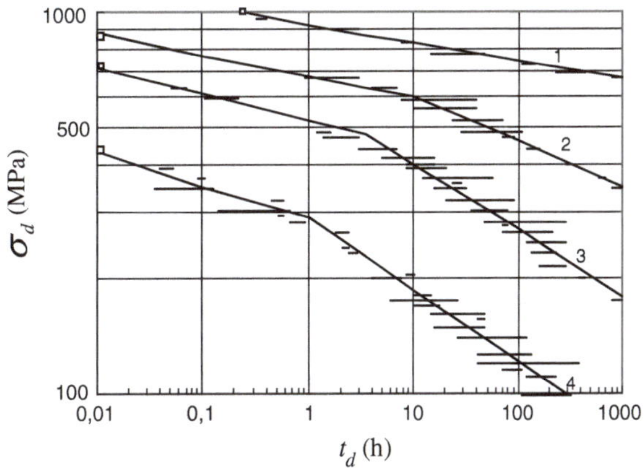

a rupture observed at small times has viscous character as well as at short-term tests. Short-term strength points at the corresponding temperatures, displayed in Fig. 6.3 by white small squares, are laid down on prolongation of stress rupture curves at small test periods. At the right branches where stresses are moderate or small, rupture is fragile and occurs without visible previous macro-strains.

At some long-time tests, the specimens are not carried out to failure. Exposing non-ruptured specimens to short-time tensile test, it is possible to obtain an ultimate strength and a relative elongation or cross-sectional area reduction at rupture. Changes of the specified parameters in comparison with their values for new specimens give information on damage accumulation in time and also can be used for predicting the remaining design resource.

6.1.1.4 Creep and Plastic Strains Under Simultaneous Action

Short-Term Creep

At a very high temperature appreciable creep strain can occur even during short-term tests where durability is measured in tens of seconds. Experimental researches of such processes, named *short-term creep,* demand the use of special installations and measurement methods (Mileyko and Telenkov 1962; Rabotnov and Mileyko 1970). At the appearance of "instantaneous" plastic deformations, they interact with consequent creep strains leading to rapid growth of common inelastic strains over time.

Long-Term Creep

At some tests with moderate rate of loading, creep can grow simultaneously with an increase in plastic deformations. As a result, real experimental deformation diagrams can differ from the ideal diagrams corresponding to instantaneous loading. In Fig. 6.4, deformation diagrams for a heat-resistant alloy obtained at a rapid rate of loading ($\dot{\varepsilon}_p = 8.5 \times 10^{-2}\ \%/s$, curves 1) and at a more slow rate ($\dot{\varepsilon}_p = 4 \times 10^{-4}\ \%/s$, curves 2) are presented for two test temperatures (Sizova's data). In the given case, creep at the elevated temperature with slow loading rate reduces the visible yield point. In addition, a value of measured static Young's modulus occurs less than its dynamic quantity (see Chap. 2).

Analogous changes of stress/strain diagrams due to creep have been marked in a number of tests, e.g., by Manson et al. (1971).

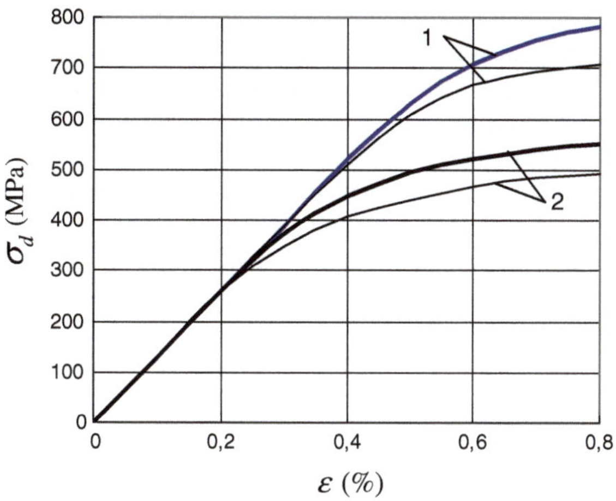

Fig. 6.4 Deformation diagrams of a heat-resistant alloy at different rates of loading; *1* $T = 20$ °C, *2* $T = 500$ °C

6.1.1.5 Creep and Plastic Strains Under Sequential Loading

Investigations show that in most cases the long-duration exposure of the stressed specimens in creep conditions leads to lowering of material plasticity, measured at the subsequent test for short-term strength (embrittlement phenomenon). A preliminary plastic deformation generates "the strain hardening" of a material that reduces its long-term strength.

However, poor experimental data do not often permit using these phenomena in calculations. In cases when plastic strains do not exceed 0.2–0.5 %, it is admissible to assume that creep curves and limits of long-term strength are not dependent on preliminary plastic strains. The diagrams of plastic deformation and limits of short-term strength are also assumed to be unaffected by accumulated creep strain.

6.1.2 Experimental Data on Creep and Long-Term Strength at Various Stressed States

6.1.2.1 Creep and Long-Term Strength in Compression

Investigations on creep at compression were carried out by a number of researchers at various time (Sully et al. 1948; Torshenov 1961; Straub 1995, and others).

Tests on creep of cylindrical specimens at axial compression demonstrate that the time dependence of creep under compression and tension at identical temperatures and stresses within small strains (0.3–0.5 %) usually coincide close enough to allow the use of common creep characteristics in their calculations.

At significant strains, the creep at compression usually grows slower than at tension.

A small axis misalignment of the application of the compressing force may lead to progressive curving of the long specimen, when stress distribution over its cross sections becomes non-uniform. Therefore, the ratio of specimen length to diameter must be no more than 3:1, and it is necessary to take specimen with tapered heads or with special shoulders.

The time when strains begin to rapidly increase may be considered as the time of long-term failure at compression. A conditional limit of long-term strength at compression referring to the identical time of failure can be 40–50 % above than at tension. Under the identical stresses, the ratio of fracture times at compression and tension is usually $t_{comp}/t_{tens} \approx 2.5-3$.

However, in some tests creep under compression developed faster than under tension.

6.1.2.2 Creep and Long-Term Strength in Torsion

Long-term tests at torsion, like short-time, are usually carried out on thin-walled tubular specimens when a uniform distribution of tangential stresses τ over cross section is ensured until the failure and dependence of creep shear strains on time $\gamma_c = f(t)$ is easily defined.

The common features of creep development and the link of failure time with the operating stress and temperature at torsion and tension are similar (Oding and Tuljakov 1958; Johnson 1960). However, the accelerated growth of creep at the third stage is affected only by the damage accumulation in the material as the specimen cross section remains practically invariable. This fact makes tests in torsion "purer" (Crossland et al. 1971).

The criteria of long-term strength and creep at torsion are implied from the common criteria at a combined stress state.

6.1.2.3 Creep at Combined Stressed States

For long-time tests at a plane stress state, thin-walled tubular samples are exposed not only to a tension/compression and torsion, but also to internal pressure. Thus, a tension/compression in two directions with various correlations of stresses in magnitude can be obtained. It is extremely difficult to conduct direct tests at a 3D stressed state.

Creep at combined (often named multi-axial) stress states was studied experimentally by Bailey (1935), Norton (1939), Johnson and Tapsell (1940), Oding and Tuljakov 1958), Johnson et al. (1962), and others. Many works were analyzed in detail by Rabotnov (1966). Despite inevitable differences in various materials' behavior and a large scatter of test results, experiments on creep allow the following experimental results to be justified for use as the base of design methods.

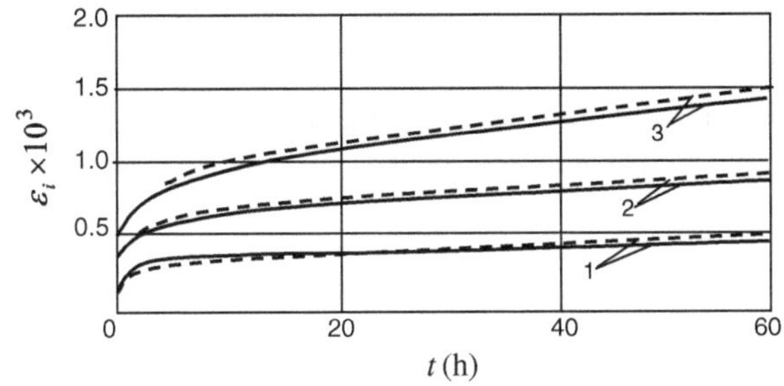

Fig. 6.5 Creep diagrams of a titanium alloy at combined stress state: *solid lines* tension, *dotted lines* torsion; *1* $\sigma_i = 588$ MPa, *2* $\sigma_i = 637$ MPa, *3* $\sigma_i = 686$ MPa (Samarin and Klebanov 1994)

- Volumes of body elements do not change;
- At any stress state, the same three stages of creep, as at uniaxial tension, are observed to some extent, and the tertiary stage is terminated by material damage.
- Creep strains, as well as short-term plastic ones, are defined by shear processes in a material; therefore, at any stress state the development of inelastic strains depends on stress intensity σ_i or the maximum tangential stress $\tau_{\max}$. In Fig. 6.5, experimental creep diagrams at tension and torsion of a titanium alloy, as dependence of total strain intensity ε_i versus time t, for several values of the stress intensity σ_i are reduced. The diagrams illustrate validity of specified criteria.

Specified experimental results have been used by a number of authors for elaboration of the general theory and calculation methods on creep.

6.1.2.4 Long-Term Strength at Combined Stressed States

Most tests on long-term strength were carried out using thin-walled tubular samples and also disks. In the last cases, failure began at a surface where there was also a plane stress state.

The tests of some authors (Johnson and Frost 1955; Sdobyrev 1959; Rabinovich 1966, and others) have shown that at small creep strains, a material fracture has character nearing fragile. Therefore, the maximum tensile stress (i.e., the first principle stress) σ_1 could serve as a criterion of long-term strength σ_d of materials.

However, for the materials whose damage is preceded by considerable creep strains, the stress intensity σ_i can also be treated as a criterion of long-term strength. Rather close results gives the use of the maximum tangential stress

$$\tau_{\max} = 0.5(\sigma_1 - \sigma_3), \tag{6.2a}$$

to which corresponds the equivalent normal stress

$$\sigma_{eqv} = 2\tau_{\max} = \sigma_1 - \sigma_3. \tag{6.2b}$$

Some combinations of the specified parameters with the introduction of additional experimental factors are also used as equivalent stresses, for example (Kablov and Golubovsky 1998)

$$\sigma'_{eqv} = \beta\sigma_m + (1-\beta)\sigma_i. \tag{6.3a}$$

where σ_m is a mean value of stress components.

Sdobyrev (1959) proposed $\beta \approx 0.5$, i.e.,

$$\sigma'_{eqv} = 0.5(\sigma_1 + \sigma_i). \tag{6.3b}$$

For some alloys at elevated temperatures in the conditions of joint operation of tension and torsion, Eqs. (6.3a, b) yielded more satisfactory results.

More complex criteria with several experimental factors found application (see Yao et al. 2007).

At the multi-axial stress state, fracture can begin in the depth of a body and essentially depends on character of loading and temperature conditions.

If a material undergoes an all-round non-uniform tension at $\sigma_2 > 0$ and $\sigma_3 > 0$, creep deformations develop poorly and the condition $\sigma_d = \sigma_1$ may be approximately taken as the failure criterion. However, some tests show that long-term durability at double-axis tension $\sigma_d = \sigma_1$ is lesser than at uniaxial.

At a main stress $\sigma_1 > 0$ and negative stresses $\sigma_2 < 0$ and $\sigma_3 < 0$, creep deformations amplify and there are foundations to accept $\sigma_d = \sigma_i$ as the failure criterion.

6.1.3 Approximation of Experimental Data on Creep and Long-Term Durability

Direct experimental data on creep and long-term durability are usually determined using a limited quantity of specimens for some constant temperatures and stresses, and are represented in a tabular or graphic form. The intermediate values of the characteristics necessary in designing can be obtained by numerical interpolation or analytical approximation formulas.

6.1.3.1 Description of Primary and Secondary Parts of Creep Diagrams

Stabilized (Steady) Creep

A constant rate $v_{c,st} = \dot{\varepsilon}_{c,st}$ of the stabilized creep is calculated as

$$v_{c,st} = \frac{\varepsilon_c(t_2) - \varepsilon_c(t_1)}{t_2 - t_1} \tag{6.4}$$

where $\varepsilon_c(t_1)$ and $\varepsilon_c(t_2)$ are values of creep strain at the time points t_1 and t_2.

To describe the function $v_{c,st} = f(\sigma)$, any expressions transferring fast increase of rate with the growth of stress and satisfying condition $v_{c,st} = 0$ at $\sigma = 0$ are valid.

To this goal, a power law proposed by Norton (1929)

$$v_{c,st}(\sigma, T) = k(T)\sigma^{n(T)}, \tag{6.5a}$$

where the exponent accepts values from $n = 1.5–2$ to $n = 10$ and above, is widely used.

Equation (6.5a) can be presented in the form

$$v_{c,st}(\sigma, T) = v_{c0}(T)\left(\frac{\sigma}{\sigma_0}\right)^{n(T)}, \tag{6.5b}$$

where v_{c0} is a creep rate at a reference stress σ_0.

Having written down Eq. (6.5a) for two values of stresses σ_1 and σ_2 at the given temperature T and having divided one into another, we find

$$n\,(T) = \frac{\log v_{c2} - \log v_{c1}}{\log \sigma_2 - \log \sigma_1}, \tag{6.5c}$$

after which it is possible to define the factor $k(T)$.

Analogously, other suitable functions are used, e.g., an exponential one

$$v_{c,st} = k(T)\{\exp[n(T)\sigma] - 1\} \tag{6.6a}$$

or a hyperbolic sine

$$v_{c,st} = k(T)\sinh[n(T)\sigma]. \tag{6.6b}$$

For physical reasons (see, e.g., Rabotnov 1966), the dependence of creep rate on temperature ought to contain an exponential member of a kind

$$v_{c,st}(T) \sim \exp(-b/T) \tag{6.7}$$

where $b > 0$ and T is Kelvin temperature. A function $v_c(T)$ can be non-monotonic and dependent on stress.

Linear creep dependence in time

$$\varepsilon_c(t) = \varepsilon_c(t_1) + v_{c,st}(t - t_1) \tag{6.8}$$

is valid beginning from a time moment $t_1 \neq 0$ after the initial transient site terminates.

Stepped Representation of a Creep Diagram

At long work duration and moderate stresses, analysis of the machine parts behavior within the initial transient site does not represent interest. Then, a process may be conditionally considered as stabilized since the moment $t_1 = 0$, believing that at this instant the strain had already some inherent value $\varepsilon_c = \varepsilon_{c0}$. This quantity reflects a final effect of unsteady creep (Odqvist 1974). Then, we obtain from Eq. (6.8) the approximated description of the creep curve in a stepped linearized form

$$\varepsilon_c(\sigma, T, t) = \varepsilon_{c0}(\sigma, T) + v_{c,st}(\sigma, T)t. \tag{6.9}$$

An initial plastic strain can also be included in the inherent strain value ε_{c0}.

Isochronous Curves of Creep

Another way of creep diagram approximation taking into account a transient site is a use of so-called isochronous curves. To this goa,l the experimental curves of creep obtained at constant stress and temperature $\varepsilon(\sigma, T, t)$ are formally reconstructed in the curves $\sigma(\varepsilon, T, t)$, which represent dependence between stress σ and total strain ε, varying during the operating time t. At the initial moment $t = 0$, an isochronous curve $\sigma(\varepsilon, T, 0)$ coincides with the curve of the usual diagram of short-term elastoplastic deformation $\sigma_0(\varepsilon, T)$.

If we conditionally accept the obtained curves $\sigma(\varepsilon, t)$ for the real curves of elastoplastic deformation at a fixed time t, the calculation under creep conditions can be reduced to the elastoplastic calculation stated in Chap. 3.

Further simplification is possible if the dependence $\sigma(\varepsilon, t)$ is presented in the form of a product:

$$\sigma(\varepsilon, t) = \sigma_0(\varepsilon)f(t). \tag{6.10}$$

In this case, for the full description of a material behavior at creep, it is enough to find an experimental function $f(t)$ for a single stress–strain curve $\sigma_0(\varepsilon)$ as all creep curves at various temperatures are believed similar.

The model of a material whose behavior can be written using Eq. (6.10) is sometimes named "time hardening material." When the stress steadily increases, such an approach does not contradict some experiences.

Model of a Strain-Hardening Material

Reduction of creep rate under strain development at stable state of a material must be connected with changes of material structure due to deformation but not with operating time. The magnitude of accumulated creep strain can serve itself as an indicator of such changes. Then, a general dependence of creep rate $\dot{\varepsilon}_c$ on the strain ε_c at constant stress σ and temperature T should look like

$$\dot{\varepsilon}_c = f(\varepsilon_c, \sigma, T). \qquad (6.11a)$$

It is convenient to present Eq. (6.11a) as a product

$$\dot{\varepsilon}_c = f_\varepsilon(\varepsilon_c) f(\sigma, T), \qquad (6.11b)$$

where a component of creep rate $v_c = f(\sigma, T)$ can be described by Eqs. (6.5a, b, c)–(6.6a, b).

Equations (6.11a, b) describe a "strain-hardening material."

Rabotov's Model

Rabotnov (1948, 1966) suggested to accept Eq. (6.11b) in a form

$$\dot{\varepsilon}_c = \varepsilon_c^{-\alpha} f(\sigma, T), \quad 0 < \alpha < 1, \qquad (6.12a)$$

Integrating Eq. (6.12a) at $\sigma = \text{const}$ and $T = \text{const}$ yields

$$\varepsilon_c(t) = \left[\frac{f(\sigma, T)}{m}\right]^m t^m \qquad (6.12b)$$

where $0 < m = 1/(1+\alpha) < 1$.

Having chosen at a creep diagram two points concerning the time moments t_1 and t_2, we find

$$m = \frac{\log \varepsilon_c(t_2) - \log \varepsilon_c(t_1)}{\log t_2 - \log t_1}$$

and corresponding value $\alpha = (1-m)/m$.

According to Eq. (6.12a), the rate of creep at $\alpha > 0$ and $m < 1$ decreases from $\dot{\varepsilon}_c = \infty$ at $t = 0$ to zero at $t \to \infty$, i.e., process is described as completely unsteady with fading creep rate that is true only at very small stresses. The point at $t = 0$ has no special value, but Eq. (6.12a) must be changed to describe transition to the stabilized creep. To this goal, it was suggested (Shorr 1958) to take

$$\dot{\varepsilon}_c = \begin{cases} \varepsilon_c^{-\alpha} f(\sigma, T) & \text{at } \varepsilon_c < \varepsilon_{c,*} \\ \varepsilon_{c,*}^{-\alpha} f(\sigma, T) & \text{at } \varepsilon_c \geq \varepsilon_{c,*} \end{cases}, \tag{6.13}$$

where $\varepsilon_{c,*}$ is a creep strain at which the established process starts.

Using Eq. (6.5b) and an elastoplastic diagram in a kind of

$$\varepsilon_\sigma = \varepsilon_{\sigma 0}\left(\frac{\sigma}{\sigma_0}\right)^{n_p}, \tag{6.14}$$

Odqvist (1953) represented the rate of total strains $\varepsilon = \varepsilon_\sigma + \varepsilon_c$ at $\sigma = \text{const}$ as

$$\frac{d\varepsilon}{dt} = \varepsilon_{\sigma 0}\frac{d}{dt}\left[\left(\frac{\sigma}{\sigma_0}\right)^{n_p}\right] + v_{c0}\left(\frac{\sigma}{\sigma_0}\right)^{n_{sc}}. \tag{6.15}$$

Another way of transient creep inclusion in calculations will be considered in Chap. 7.

6.1.3.2 Link of Long-Term Strength and Durability

Due to the curves shown in Fig. 6.3, experimental dependence between constant operating stress σ and time prior to failure $t_d(\sigma, T)$ can be represented in logarithmic coordinates as a linear function

$$\log \sigma = C(T) - [m(T)]^{-1} \log t_d(\sigma, T), \tag{6.16}$$

Equation (6.16) yields

$$t_d(\sigma, T) = B(T)\sigma^{-m(T)}, \tag{6.17}$$

where $C(T)$ and $B(T) = 10^{C(T)}$ are empirical functions.

Equation (6.17) shows how a material durability t_d reduces with the growth of operating stress σ. The greater the exponent $m(T)$ is, the faster the time t_d falls versus increasing stress σ.

The invert correlation to Eq. (6.17) gives a dependence of a long-term strength $\sigma_{d/t} = \sigma$ on time t

$$\sigma_{d/t}(t, T) = \left(\frac{B(T)}{t}\right)^{1/m(T)}. \tag{6.18}$$

The smaller the exponent $m(T)$ is, the faster the stress $\sigma_{d/t}$ falls versus increasing time t.

Approximate formulas are valid within those limits in which they have been obtained by processing of the original experimental data. For diagrams $\sigma_d = f(t_d)$

with a slope change, as shown in Fig. 6.3, approximate dependences may be applied only separately to each of the branches.

Other methods of approximating the empirical dependence $f(\sigma, T, t) = 0$ are described below.

Larson–Miller Parameter

Creep strain $\varepsilon_{c,d}$ which is reached at tests to the moment of destruction t_d is determined as

$$\varepsilon_{c,d}(\sigma, T, t_d) = \int_0^{t_d} v_c(\sigma, T, t)dt = v_{c,m}(\sigma, T)t_d, \quad (6.19a)$$

where $v_{c,m}(\sigma, T)$ is an averaged creep rate over the entire process.

Due to Eqs. (6.7) and (6.11b), the expression for rate $v_{c,m}(\sigma, T)$ must have a form of equation

$$v_{c,m} \approx Kf_1(\sigma)\exp\{-[A - f_2(\sigma)]/T\} \quad (6.19b)$$

where $f_1(\sigma)$ and $f_2(\sigma)$ are positive functions monotonously increasing with the increase of stress, and K and A are constants.

Some experiments show that in certain ranges of stresses and temperatures, the creep strain quantity $\varepsilon_{c,d}$ at long-term fracture does not greatly vary (Monkman and Grant 1956). Having multiplied both parts of Eq. (6.19b) in time t_d and taking its logarithm, one can come at $\varepsilon_{c,d} \approx$ const to the equation with structure

$$T[\log t_d + F_1(\sigma_d)] = F_2(\sigma_d). \quad (6.20)$$

Equation (6.20) approximately links long-term strength σ_d and time at long-term fracture (durability) t_d with operating temperature T.

Larson and Miller (1952) have offered to use in the left side of Eq. (6.20) the function $F_1(\sigma_d) = C_L \approx$ const. The Larson–Miller parameter

$$P_L(T, t_d) = T[\log t_d + C_L] \quad (6.21)$$

is usually calculated by Kelvin temperature T and time t_d expressed in hours.

A dependence between long-term strength σ_d and parameter P_L, due to Eq. (6.20), yields

$$\sigma_d = F_L[P_L(T, t_d]. \quad (6.22)$$

The function F_L should decrease with the growth of parameter P_L.

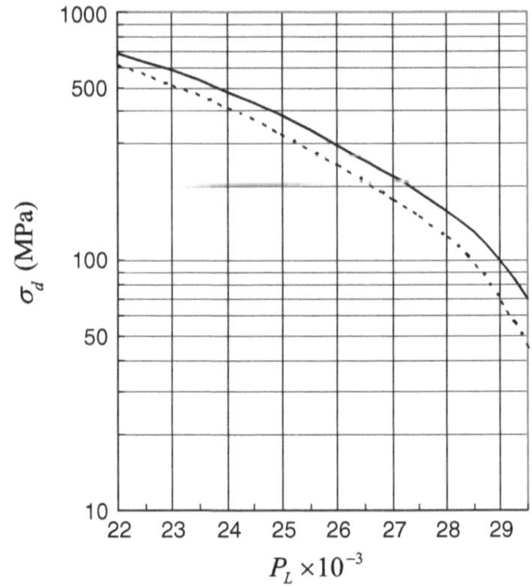

Fig. 6.6 Typical Larson–Miller diagram for a heat-resistant Ni–Cr alloy

Tests of some materials have shown that function F_1 in Eq. (6.20) is not a constant but depends on stress and temperature. The factor C_L changes within a range from 15 to 30. However, for assessed calculations taking into account considerable scatter of experimental data, it is possible to accept $F_1 \approx C_L \approx 20$ as averaged factor for many materials. A concrete type of Eq. (6.22) is defined by the processing of long-time test results. In Fig. (6.6) a typical averaged dependence $\log \sigma_d = f(P_L)$ for a heat-resistant cast alloy due to Sizova's data (1975) obtained at various temperatures is displayed by a continuous line.

In the presence of sufficient experimental data, it is possible to construct at Larson–Miller diagram the curves approximately concerning certain probability of failure at given stress and temperature. An example of such curves for failure probability $P = 0.01$ is shown in Fig. 6.6 by a dashed line. The continuous line corresponds to the probability $P = 0.5$.

Various sites of the generalized Larson–Miller diagrams, similar to that presented in Fig. 6.6, can be constructed by results of tests at various temperatures. It is necessary to be careful using such diagrams for the forecasting of design durability out of temperature limits, within which they were experimentally obtained (especially, toward higher temperatures).

Analogous diagrams connecting various limits of the creep σ_c (see Sect. 6.1.1.2) with corresponding time t_c

$$\sigma_c = F_{Lc}[P_{Lc}(T, t_c)] \tag{6.23}$$

can be built and used for creep calculation.

Other Parameters

Manson and Haferd (1953), based on empirical trials, found that the dependence of long-term strength on time and temperature can be presented as a function

$$\sigma_d = F_M[P_M(T, t_d)], \tag{6.24a}$$

where

$$P_M(T, t_d) = \frac{T - T_a}{\log t_d - \log t_a}. \tag{6.24b}$$

Further tests have shown that values T_a and t_a depend on temperature, but in a moderate temperature range, these values can be considered for a concrete material as constants.

Some other generalized dependences containing those or other experimental factors were also offered in a number of works. Thus, in Trunin II (1969), Trunin et al. (1978) and others, the time prior to destruction is defined as

$$t_d = A_1 T^m \sigma^{-n} \exp\left(\frac{A_2 - A_3\sigma}{T}\right). \tag{6.25}$$

To find material parameters of Eq. (6.25) from tests, Kablov and Golubovsky (1998) have used probabilistically statistic methods.

The more experimental factors a formula contains, the more precisely it can describe results of a separate test, but its use in other cases becomes more problematic.

6.1.3.3 Description of Full Creep Diagram

Experimental dependence of creep deformation in time at constant values of stress σ and temperature T, covering all three stages of creep, can be presented by the equation

$$\bar{\varepsilon}_c(\bar{t}) = \bar{\varepsilon}_{c,0} f_0(\beta_0 \bar{t}) + \bar{\varepsilon}_{c,st}\bar{t} + \left[1 - (\bar{\varepsilon}_{c,0} + \bar{\varepsilon}_{c,st})\right] f_d(\beta_d \bar{t}), \tag{6.26}$$

where the current creep strain and time are referred to strain and time at long-term failure, namely

$$\bar{\varepsilon}_c = \varepsilon_c/\varepsilon_{c,d}, \quad \bar{t} = t/t_d, \quad \bar{\varepsilon}_{c,0} = \bar{\varepsilon}_c(0) \text{ and } \bar{\varepsilon}_{c,st} = v_{c,st} t_d/\varepsilon_{c,d}.$$

The functions $f_0(\beta_0 \bar{t})$ and $f_d(\beta_d \bar{t})$ should satisfy conditions

$$f_0 = \begin{cases} 0 & \text{at } \bar{t} = 0 \\ 1 & \text{at } \beta_0 \bar{t} \gg 1 \end{cases}, \quad f_d = \begin{cases} 0 & \text{at } \bar{t} = 0 \\ 1 & \text{at } \bar{t} = 1 \end{cases} \tag{6.27}$$

In particular, these functions may be taken as

$$\begin{aligned} f_0 &= 1 - \exp(-\beta_0 \bar{t}), \\ f_d &= \{\exp[-\beta_d(1 - \bar{t})] - \exp(-\beta_d)\} \end{aligned} \tag{6.28}$$

with $\beta_0 \gg 1$ and $\beta_d \gg 1$ satisfying conditions (6.27).

The equation of relative creep rate $\bar{v}_c = v_c / v_{d,m}$ following from Eqs. (6.26) and (6.28) is:

$$\bar{v}_c(\bar{t}) = \dot{\bar{\varepsilon}}_c(\bar{t}) = \bar{v}_{c0} \exp(-\beta_0 \bar{t}) + \bar{v}_{c,st} + \bar{v}_{c,d} \exp[-\beta_d(1 - \bar{t})]. \tag{6.29}$$

Here, $v_{d,m} = \varepsilon_{c,d}/t_d$ is a mean rate of creep during the time t_d, $\bar{v}_{c,st} = v_{c,st}/v_{d,m} = \bar{\varepsilon}_{c,st}$ is a relative steady creep rate and the parameters $\bar{v}_{c0} = \beta_0 \bar{\varepsilon}_{c,0}$ and $\bar{v}_{c,d} = \beta_d \left[1 - \left(\bar{\varepsilon}_{c,0} + \bar{\varepsilon}_{c,st}\right)\right]$ are excesses of creep rate over stationary values in the beginning of the first stage (at $\bar{t} = 0$) and in the end of the third stage (at $\bar{t} = 1$), respectively.

Six material characteristics $\varepsilon_{c,0}, \varepsilon_{c,d}, \beta_0, \beta_d, v_{c,st}, t_d$ are stress and temperature dependent. The definition of quantities $\varepsilon_{c,0}\beta_0, v_{c,st}$ for the first and second creep stages has been considered above; values $\varepsilon_{c,d}, \beta_d$ are defined similarly using the third-stage creep curve. The conditions $\beta_0 \gg 1$ and $\beta_d \gg 1$ are practically always fulfilled.

Evans–Wilshire's Model

In literature, various variants of the equations of types Eqs. (6.26)–(6.29) were considered. In Evans and Wilshire (1985), the equation

$$\varepsilon(\theta, t) = \theta_1[1 - \exp(-\theta_2 t)] + \theta_3[\exp(\theta_4 t) - 1] \tag{a}$$

was used to describe creep processes in which the steady stage does not play a significant role.

A rate of the creep process due to Eq. (a) is

$$\dot{\varepsilon}(\theta, t) = \theta_1 \theta_2 \exp(-\theta_2 t) + \theta_3 \theta_4 \exp(\theta_4 t) \tag{b}$$

with a minimum at

$$t_* = \frac{1}{\theta_2 + \theta_4} \ln \frac{\theta_1 \theta_2^2}{\theta_3 \theta_4^2}. \tag{c}$$

It has been shown that dependences $\theta_k(\sigma, T)$ for $k = 1{-}4$ can be approximated by the equations

$$\log \theta_k = a_k + b_k T + c_k \sigma + d_k \sigma T, \tag{d}$$

whose coefficients are defined from creep curves.

The calculation approach based on Eqs. (a)–(b) has been named by Evans "The theory of 4θ" and used to project the long-term creep damage duration (Evans 2000).

6.2 Design Analysis of Stresses and Strains at Creep

6.2.1 Statement of Design Problems in Creep

6.2.1.1 Different Time Scales

Dynamic Creep

As characteristic times of dynamic processes in mechanical systems are incommensurably less than the time of creep development, the influence of inertial forces on creep design is usually negligibly small. However, stress components of a high frequency which are imposed on constant or slowly changing stresses can in certain cases rather appreciably increase the creep rate. The creep in such conditions is sometimes named *dynamic creep* (Rabotnov 1966) and needs special experimental investigation.

Creep and Plastic Yielding

At durability calculations of mechanical structures, the representative rates of creep and plastic yielding under varying loads at transient regimes are usually noncomparable, which makes their description as a unique process unreasonable.

However, theoretical and experimental study of inelastic strains at creep and plastic yielding as a uniform process is of certain scientific and, in more rare cases, of engineering interest.

To this goal, the creep test data can be presented in total strains, using isochcronous diagrams, or uniting strain terms of creep and plastic flow in a common state equation as in Odqvist's "total creep deformation theory" (1953), see Eq. (6.15) above.

In some works, the creep and plastic flow are assumed to run simultaneously in real time (Palley 1964; Shevchenko and Terekhov 1982, and others). Problems arising under this approach will be discussed in Chap. 8.

Prehistory of Loading

Experimental data show that the material behavior at a current time instant is defined by two groups of factors: (a) by local material state depending on the entire "life" prehistory of the given body and (b) by stress and temperature operating at the present instant. The material state can depend on operating time at elevated temperature and time-out duration, current and accumulated strains and their aspects (elastoplastic, creep), micro-cracks, and so forth.

When external loads and temperatures are constant or become constant starting at some instant of time, creep passes to the fully stabilized state and creep rate takes a minimum value independent of time. The creep of the body as a whole cannot become entirely stabilized until the creep process has been stabilized at all points of the body. In some cases, a stabilized creep stage can be considered in an expanded sense as the material state, at which a previous loading and heating history weakly influences running creep rate. The rate can smoothly vary assuming dependence only on running stress and temperature. This approach makes calculation appreciably simpler.

Steady and Unsteady Loading and Heating

Creep can be affected by steady (stationary) loads at a constant temperature field, and also by non-stationary loads and temperatures, varying with time. Certainly, any real steady process begins with an initial transitive stage, but such a process, in which operating time is essentially longer than the transitive stage, can be understood as stationary. In the case of unsteady processes, it is expedient to separate them under smoothly varying, stepwise and cyclic loads and temperatures. In this chapter, the steady and smoothly varying processes are considered, and the rest—in Chaps. 7 and 8.

6.2.1.2 The Simplest Theories of Creep

In stationary problems, in which stresses and temperatures remain constant despite creep development, creep experimental curves can be directly used for calculation, interpolating them on intermediate values of stresses and temperatures. Thus, it is not required to mark out separate stages of creep, and time appears as the uniquely defining factor of this process.

Theory of Aging

If to assume that under unsteady smoothly varying processes a time remains the same unique governing factor, we come to the *theory of aging*. The previously described isochronous curves are based on this concept.

Flow Theory

Generalizing experimental curves of creep rate $v_c = f(\sigma, T, t)$ obtained at fixed stresses and temperatures (Fig. 6.1b), on unsteady problems we come to the *flow theory,* which is formally close to the theory of plastic yielding considered in Chap. 3. For calculation, the approximated analytical representations of experimental curves are usually used.

The differential form of the flow theory when strain increments $\Delta\varepsilon_c = v_c(\sigma, T, t)\Delta t$ are calculated at each time step makes it more suitable for analysis of unsteady problems. A well-founded application of such an approach is also limited by conditions of smoothly varying stresses and temperatures.

Theory of Hardening in Creep

Generalizing Eq. (6.10) on unsteady problems, we come to *the theory of hardening.* Depending on an accepted aspect of function $v_c = \dot{\varepsilon}_c = f(\sigma, T, \varepsilon_c)$, various variants of the theory of hardening are offered.

As shown further in Chap. 8, this theory can be limitedly used in some cases for calculations that include unloading.

6.2.2 Statically Determinable Problems

6.2.2.1 Small Strains

If in a statically determinable system outer loads are known, the stresses can be calculated from conditions of static equilibrium independently of strain character, including creep. After this, one can calculate total and creep strains using material experimental characteristics or their approximated representation.

If only displacements or thermal strains are set in such a load-free system, the stresses can be calculated using the material characteristics. The calculations in a linear approach are carried out on inherent system geometry.

6.2.2.2 Finite Strains

As creep strains increase in time, the equilibrium equations should refer to the deformed state of the system. If this process occurs gradually and steadily without a stress sign alteration, the equation of the creep flow theory in increments for step-by-step calculation may be used in the form

$$\Delta\varepsilon_c(t) = v_c[\sigma(t), T, t]\Delta t. \quad (6.30a)$$

At significant deformation, creep usually passes in the steady stage when rates of strains cease to depend on time and the last equation becomes

$$\Delta\varepsilon_c(t) = v_c[\sigma(t), T]\Delta t. \quad (6.30b)$$

In statically determinable problems at constant external loads, the redistribution of stresses between elements of a body occurs only under large strains when it demands calculation in a nonlinear statement.

6.2.3 Statically Indeterminable Problems at Steady Creep

6.2.3.1 Application of Steady Creep Analysis to Unsteady Processes

In statically indeterminable problems, creep initiates change of stresses in an entire body or redistribution of stresses between some parts of a body already at small strains. The magnitude of a strain rate in these conditions usually changes, i.e., creep has unsteady but smoothly altering character.

In the creep conditions for statically indeterminable problems at constant outer loads and temperature field, a redistribution of stresses in comparison with their distribution at the initial moment occurs rather gradually. When creep passes into the stabilized stage, redistribution of stresses stops.

Structural parts, designed for a long service life under constant loads and temperature, may operate under conditions of stabilized creep for the greater part of the lifetime. The long-term strength of such parts cannot be found in terms of the initial stresses, since it is affected by the stresses existing for a long period of time. The latter can be calculated using the theory of stabilized creep.

Kinematic Laws of Creep Rate Distribution

In some group of problems, kinematic laws of distribution of total strains and also of total strain rates are known (e.g., the plane cross-sectional hypothesis for rod tension and bending or the hypothesis of the straight-line normal for plate bending). The same kinematic law establishes a distribution of stabilized creep rate over a non-uniformly heated body (Shorr 1958). Having resolved the governing equations concerning stresses, we can find the stress distribution in the creep conditions.

6.2.3.2 Extension of a Non-uniformly Heated Rod

Let us consider a rod of cross-sectional area A, which is extended by a constant longitudinal force F and affected by a steady-state temperature distribution $T(x, y)$ that does not change along the length z of the rod. The cross section and temperature distribution are assumed to be double-symmetrical.

According to the kinematic hypothesis of plane sections, the total longitudinal strain ε and also its rate $v = \dot{\varepsilon}$ are constant at all points of any cross section. Under stabilized stage of the creep at the above conditions, we obtain $v = v_c = \text{const}$.

Using Eqs. (6.6a) and (6.7) in the form

$$v_c(\sigma, T) = k\exp(-\beta/T)[\exp(n\sigma) - 1] \tag{6.31}$$

and solving Eq. (6.31) for stress distribution $\sigma(x, y)$, we obtain

$$\sigma(x, y) = \frac{1}{n}\ln\left[\bar{v}_c \exp\left(\frac{\beta}{T(x, y)}\right) + 1\right], \tag{6.32}$$

where $\bar{v}_c = v_c/k$ is a relative value of stabilized creep rate.

From the conditions of the rod equilibrium it follows that

$$\int\limits_A \sigma dA = F. \tag{6.33}$$

The rate $\bar{v}_c$ can be found from the equation

$$\varphi(\bar{v}_c) = F - \frac{1}{n}\int\limits_A \ln\left[\bar{v}_c \exp\left(\frac{\beta}{T(x, y)}\right) + 1\right] dA = 0, \tag{6.34}$$

which has to be solved numerically. Substituting $\bar{v}_c$ into Eq. (6.32), we obtain the distribution of stresses $\sigma(x, y)$.

In the specific case, when the function $\exp(n\sigma) \gg 1$ is valid at all points of the cross section, this yields

$$\sigma(x, y) = \sigma_m + \beta\left[\frac{1}{T(x, y)} - \left(\frac{1}{T}\right)_m\right] \tag{6.35a}$$

where $\sigma_m = F/A$ and

$$\left(\frac{1}{T}\right)_m = \frac{1}{A}\int\limits_A \frac{dA}{T(x, y)}. \tag{6.35b}$$

The following features of the stress distribution in a rod in tension under stabilized creep conditions can be shown up:

- In a uniform temperature field, the stresses are constant.
- In a non-uniform temperature field, the stresses drop off at the hotter points of a cross section and increase at the colder points (in comparison to their mean value σ_m).
- If only thermal stresses have been induced under absent of external forces, then Eq. (6.34) is satisfied by $\sigma(x, y) = 0$. This result does not depend on any creep law because the kinematics condition $v_c = \text{const}$ demands the constant sign of steady stresses that is possible at $F = 0$ only by $\sigma(x, y) = 0$.

Below it will be shown that the thermal stresses indeed completely vanish (relax) in conditions of stabilized creep.

- Stress distribution at stabilized creep does not depend on initial stress distribution at the instant $t = 0$.

Stresses for stabilized creep in a rod with rectangular cross section $B \times H$ induced by tensile force, under the parabolic temperature distribution over the rod width, where calculated in Shorr (1962). It was taken $\sigma_m = 250$ MPa and the relative temperature difference $\Delta\bar{T} = (T_{\max} - T_{\min})/T_m = 0.17$, see Fig. 6.7a, where $\bar{T} = T/T_m$.

For a heat-resistant alloy, it was also taken $n = 0.0285\ (\text{MPa})^{-1}$ and $\beta = 5.34 \times 10^4$ K. In Fig. 6.7b, the calculated stress distribution $\bar{\sigma} = \sigma/\sigma_m$ along the width of the cross section $-0.5 \leq \bar{y} \leq 0.5$, where $\bar{y} = y/H$, is denoted by the bold curve $\sigma(\infty)$. Other curves related to a transient process of creep stabilizing will be described further.

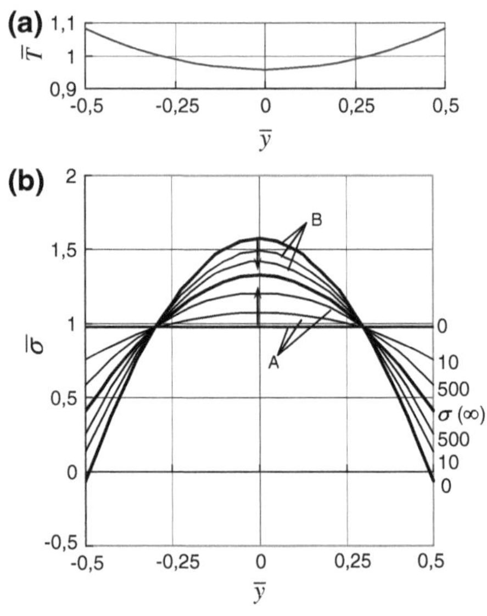

Fig. 6.7 Temperature (**a**) and stress (**b**) distribution along width of a non-uniformly heated rod in tension at different creep processes

6.2.3.3 Plane Bending of a Non-uniformly Heated Beam

In the literature, solutions of many problems on stabilized creep (mainly, in a uniform temperature field) obtained in various ways are shown. An example of such a problem, reflecting the influence of a non-uniform temperature field on stress distribution under creep, is shown below (Shorr 1964).

A beam of a rectangular cross section with a linear temperature distribution along the width (Fig. 6.8a) is loaded by a bending moment M. Material characteristics, mean temperature, and its non-uniformity are the same as that in the previous problem.

On the basis of the hypothesis of flat cross sections, creep rate should vary linearly with the width of the cross section of the beam $v_c(\bar{y}) = a + b\bar{y}$, and its sign coincides with the stress sign at the same point. In this case, Eq. (6.32) must be recorded in the form

$$\sigma(x, y) = \frac{\text{sign}(\bar{v}_c)}{n} \ln\left[|\bar{v}_c| \exp\left(\frac{\beta}{T(x, y)}\right) + 1\right], \tag{6.36a}$$

where $\bar{v}_c = \bar{a} + \bar{b}\bar{y}$.

Constants $\bar{a}$ and $\bar{b}$ are defined from equilibrium conditions

$$\int_A \sigma dA = F = 0, \quad \int_A \sigma y dA = M. \tag{6.36b}$$

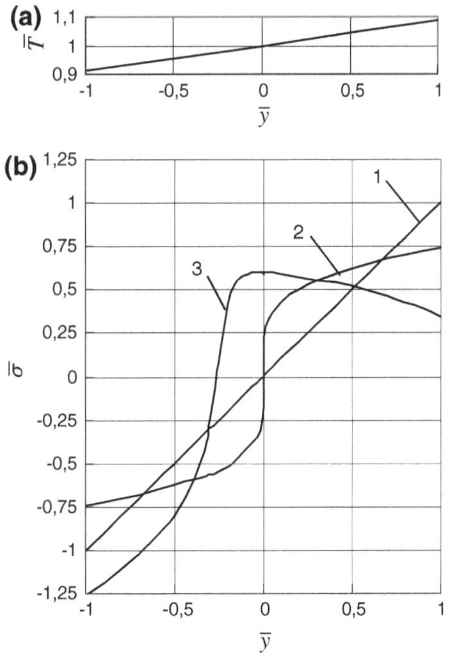

Fig. 6.8 Temperature (**a**) and stress (**b**) distribution along width of a non-uniformly heated beam in pure bending at stabilized creep

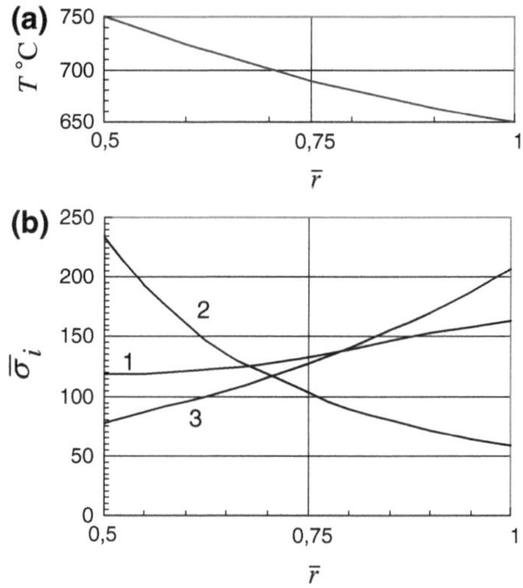

Fig. 6.9 Distribution of stress intensity along the radius of a non-uniformly heated cylinder affected by inner pressure

Calculation results $\bar{\sigma} = f(\bar{y})$ are presented in Fig. 6.9b where line 1 corresponds to elastic calculation with the maximum stress $\sigma_{\max} = 250$ MPa, curve 2—to stress distribution at constant mean temperature $T_m = 700$ °C, and curve 3—to stress distribution at a linearly changed temperature.

The stress distribution for stabilized creep at a non-uniform temperature field significantly differs not only from the results of elastic calculation, but also from the stress distribution at mean temperature.

6.3 Creep Calculation at a Combined Stress–Strain State

6.3.1 Theoretical Backgrounds

6.3.1.1 Creep at Proportional Loading

In Odqvist (1934), Boley and Weiner (1960), Rabotnov (1966) and in a number of other monographs, backgrounds of theory of creep are expounded on the base of the general theory of solids. With reference to practical problems of mechanics and engineering, these backgrounds can be represented as follows:

- Full strain is presented as a sum of elastic, plastic, creep strains, and thermal expansion

$$\varepsilon_{ij} = \varepsilon_{ij}^{e} + \varepsilon_{ij}^{p} + \varepsilon_{ij}^{c} + \varepsilon_{ij}^{T}, \tag{6.37a}$$

  and analogously in their increments

$$d\varepsilon_{ij} = d\varepsilon_{ij}^{e} + d\varepsilon_{ij}^{p} + d\varepsilon_{ij}^{c} + d\varepsilon_{ij}^{T} \tag{6.37b}$$

Below, it is assumed initial plastic strain to be absent, $\varepsilon_{ij}^{p} = 0$.

- Creep deformation occurs with practically no change in volume, so a mean creep strain is zero

$$\varepsilon^{c} = \varepsilon_{ii}^{c}/3 = 0 \tag{6.38}$$

and the strain tensor components coincide with that of deviator ones

$$\varepsilon_{ij}^{c} = e_{ij}^{c}. \tag{6.39}$$

- The analogous expressions are valid for creep rate:

$$v_{ij} = \dot{\varepsilon}_{ij} = \dot{\varepsilon}_{ij}^{e} + \dot{\varepsilon}_{ij}^{c} + \dot{\varepsilon}_{ij}^{T}, \quad v^{c} = v_{ii}^{c}/3 = 0, \quad v_{ij}^{c} = \dot{\varepsilon}_{ij}^{c} \tag{6.40}$$

- At stationary loading and heating considered in this chapter, a stressed state of a body does not vary in time. In these conditions, the loading is proportional (simple), at which tensors of stress and strain keep constant directions. As at analogous conditions of plasticity, components of creep rate tensor have the same direction as components of stress deviator $s_{ij} = \sigma_{ij} - \delta_{j}\sigma$, e.g.,

$$v_{ij}^{c} = \lambda s_{ij}, \tag{6.41a}$$

where λ is a scalar that may depend on stresses, temperature, and some structure parameters.

In terms of increments, it is

$$d\varepsilon_{ij}^{c} = v_{ij}^{c} dt = \lambda s_{ij} dt. \tag{6.41b}$$

Using the equality $v_{ij}^{c} v_{ij}^{c} = \lambda^{2} s_{ij} s_{ij}$ and usual expressions for stress and strain rate intensities

$$\sigma_{i} = \sqrt{\frac{2}{3} s_{ij} s_{is}} \text{ and } v_{i}^{c} = \dot{\varepsilon}_{i}^{c} = \sqrt{\frac{3}{2} v_{ij}^{c} v_{ij}^{c}} = \sqrt{\frac{3}{2} \dot{\varepsilon}_{ij}^{c} \dot{\varepsilon}_{ij}^{c}}, \tag{6.42}$$

we determine

$$\lambda = \frac{3}{2} \frac{v_{i}^{c}}{\sigma_{i}} \tag{6.43}$$

and

$$v_{ij}^{c} = \frac{3}{2}\frac{v_i^c}{\sigma_i} s_{ij}. \tag{6.44a}$$

or in creep strain increments

$$d\varepsilon_{ij}^{c} = \frac{3}{2}\frac{v_i^c}{\sigma_i} s_{ij} dt. \tag{6.44b}$$

In numerical calculations Eq. (6.44b) is taken as

$$\Delta\varepsilon_{ij}^{c}(t) = \frac{3}{2}\frac{v_i^c(t)}{\sigma_i(t)} s_{ij}(t)\Delta t. \tag{6.45}$$

Nonlinearity of a creep process is described by a nonlinear dependence of scalar λ on stress intensity and also on temperature and other structural parameters.

In case of tension by stress σ_0, when $\sigma_i = \sigma_0$ and $v_i^c = v_0^c$, creep rate v_0^c is linked with stress, temperature, and structural parameters by a function

$$v_0^c = f(\sigma_0, T, q_1, q_2, \ldots, q_n). \tag{6.46a}$$

Assuming that this function is valid for any combined stress–strain state (SSS) if σ_i and v_i^c are taken instead of σ_0 and v_0^c, we obtain

$$v_i^c = f(\sigma_i, T, q_1, q_2, \ldots, q_n). \tag{6.46b}$$

6.3.1.2 Strain-Hardening Creep

To describe the unstable creep and its transition to a stabilized stage at combined SSS and non-isothermal problems, one may operate with any of the approaches discussed above in Sect. 6.1.2. Here, we take temperature smoothly varying with time.

One of the forms of function (6.46b) is

$$v_i^c(t) = f[\sigma_i(t), T(t), q(t)], \tag{6.47}$$

where a single structural parameter q due to the theory of hardening may be determined in different ways:

- as an *accumulated creep strain*

$$q(t) = \int_0^t v_i^c(t_1)\, dt_1; \tag{6.48}$$

- as a corresponding *work of stresses at creep deformations*

$$q(t) = \int_0^t \sigma_i(t_1) v_i^c(t_1) dt_1. \tag{6.49}$$

In the first case, we have

$$\dot{q}(t) = \dot{\varepsilon}_i^c(t) = f\left[\sigma_i(t), T(t), \varepsilon_i^c(t)\right]. \tag{6.50}$$

- Equation (6.47), due to Eqs. (6.12a, b)–(6.13), can be taken in the form

$$\dot{q}(t) = f[\sigma_i(t), T(t)] \left(\frac{q(t)}{q_*(t)}\right)^{-\alpha(q)} \tag{6.51}$$

where $q_*(t) = f_*[\sigma_i(t), T(t)]$, $\alpha(t) = f_\alpha[\sigma_i(t), T(t)]$, and

$$\alpha(q) = \begin{cases} \alpha(t) & \text{at } q/q_* < 1; \\ 0 & \text{at } q/q_* \geq 1. \end{cases} \tag{6.52}$$

After the accumulated creep strain reaches some limiting value q_*, a process under constant stress and temperature will be stabilized.

Within definite limits of stresses and temperatures, the parameter α at $q/q_* < 1$ may be replaced by its mean constant value $\alpha \approx \alpha_m$. Then, Eq. (6.50) yields

$$\dot{\chi}(t) = \Phi(\sigma_i, T), \tag{6.53}$$

where

$$\chi(t) = \frac{\varepsilon_i^c(t)^{\alpha_m+1}}{\alpha_m + 1} \tag{6.54}$$

and

$$\Phi(\sigma_i, T) = f(\sigma_i, T)[q_*(\sigma_i, T)]^{\alpha_m}. \tag{6.55}$$

As for the function $f(\sigma_i, T)$, any forms from Sect. 6.2 are acceptable, e.g.,

$$f(\sigma_i, T) = k \exp(-\beta/T)[\exp(n\sigma_i) - 1]. \tag{6.56}$$

Equations (6.53)—(6.56) were used in some calculations for unstable creep at smoothly changing stresses by Rabotnov (1957, 1960, 1966), Danilova (1959), Shorr (1958, 1959, 1964) and others.

6.3.1.3 Use of Divers Creep Models

Comparison of the stabilized creep equations to the equations of the deformation theory of thermal plasticity shows their similarity. Therefore, solution methods of thermal plasticity can be applied to the stabilized creep problems of non-uniformly heated bodies.

In particular, by analogy to the elastoplastic deformation curve, it is possible to induce a tangential and secant moduli of creep rate curve and to present a dependence $f(\sigma_i, v_{ci}) = 0$ in the analogous form of elasticity for an incompressible body with a variable modulus. This allows the stabilized creep problems to be solved by consecutive approximations using the method of variable elasticity parameters or the other methods of Chap. 3.

In commercial software packages of finite-element analysis for approximating primary creep curves, formulas in the form of the polynomials containing a series of factors C_k are usually used, for example,

$$\varepsilon_{ci} = C_1 \sigma_i^{C_2} T^{C_3} t^{C_4}. \tag{6.57}$$

In engineering literature, there are some other approaches to describing creep using different material models, in particular, the nonlinear theory of heritable medium (Rabotnov 1966) and structural model of a medium (Masing 1923; Besseling 1958; Shevchenko and Marina 1976), and others.

Possibilities of describing creep behavior under non-proportional loading are considered in Chap. 8.

6.3.2 Redistributions of Stresses at Non-uniform Temperature Field—Residual Stresses

6.3.2.1 Redistribution of Stresses in Time

Under stationary loading and heating in a non-uniform temperature field, significant stress redistribution in time can occur (Shorr 1958). Thus, the stresses sometimes change signs. However, in view of smooth stress alteration, the calculations can approximately be carried out using equations of stabilized creep.

As a typical example, let us consider results of creep calculation of a long hollow cylinder with a ratio of internal to external radii $\bar{r}_1 = r_1/r_2 = 0.5$, loaded by constant inner pressure $p_0 = 180$ MPa (Fridman et al. 1961; Shorr 1964). The stationary temperature field along the radius is displayed in Fig. 6.9a. The end faces of the cylinder are assumed to be closed, its material is a high-resistant alloy whose characteristics due to Eq. (6.29) are $k = 43\ \mathrm{h}^{-1}, \beta = 2.2 \times 10^4$ K, and $n(T) = 0.15 + 0.001(T - 973)$.

Distributions of relative stress intensity $\bar{\sigma}_i = \sigma_i/p_0$ along radius $\bar{r} = r/r_2$ for three design cases are displayed in Fig. 6.9b. Curves 1 and 2 refer to the initial time instant, accordingly, with and without an account of thermal stresses. Curve 3 shows the stress distribution in the conditions of the stabilized creep.

It is visible that creep leads to considerable stress redistribution in comparison with the initial stress field.

A smooth stress change from the initial value $\sigma(0) = \sigma_0$ corresponding to elastoplastic calculation, to a value $\sigma(\infty)$ at the stabilized creep, can be approximated by the function

$$\sigma(t) = \sigma_0 \exp(-\kappa t) + \sigma(\infty)[1 - \exp(-\kappa t)]. \tag{6.58}$$

The exponent κ can be estimated on a value of stress $\tilde{\sigma}$ at one of the intermediate time moments using any of the methods specified above.

6.3.2.2 Residual Stresses

In continuing the problem of Sect. 6.3.2.1, *residual* (sometimes named *back*) stresses that occur in a cylinder after cooling and unloading were found (Fig. 6.10).

Residual stresses can be calculated as a difference between the stabilized creep stresses and the corresponding initial elastic ones. It must be kept in mind that at unloading the residual stress intensity σ_i cannot be determined as a difference of stress intensities; here, it can only be calculated by the components σ_r, σ_θ, and σ_x of the residual stresses.

Curves 1 and 2 in Fig. 6.10 present a distribution of the residual stress intensity $\bar{\sigma}_i$ for partial unloading, namely curve 1 refers to the case when pressure is taken off but the non-uniform temperature field remains and curve 2 is refers to the opposite case. A complete unloading with cooling is shown by curve 3.

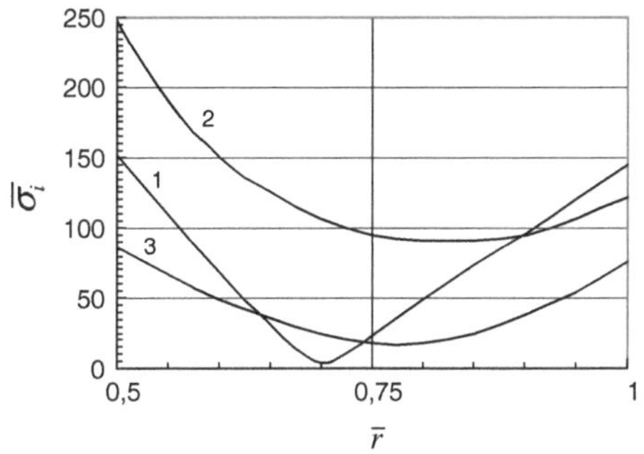

Fig. 6.10 Distribution of residual stress intensity along the radius of non-uniformly heated cylinder for different cases of unloading and cooling

The presented results demonstrate that creep produces considerable stress redistribution during operation and leaves large residual stresses after unloading and cooling the structure.

6.4 Creep and Durability at Steadily Changing Stresses

6.4.1 Stress Relaxation

6.4.1.1 Two-End Fixed Rod

Steadily (monotonically) changing stresses in an elastic rod, originally extended to initial strain ε_0 and then fixed at both ends in this position, can be approximately analyzed using equations of stabilized creep despite the real creep process being unsteady.

Total strain of a two-end fixed rod at constant temperature is $\varepsilon(t) = \varepsilon_0 = \sigma_0/E = \text{const}$, and its rate under creep will be zero

$$\dot{\varepsilon}(t) = \dot{\sigma}(t)/E + \dot{\varepsilon}_c(t) = 0. \tag{6.59}$$

Let us take a law for the stabilized creep in the form of Eq. (6.5a), then

$$\dot{\sigma}(t) + Ek\sigma^n(t) = 0. \tag{6.60}$$

Having separated variables σ and t and integrated Eq. (6.60) within the limits from $t = 0$ to t, we come at $n > 1$ to the equation

$$\frac{\sigma}{\sigma_0} = \frac{1}{\sqrt{[n-1]1 + (n-1)Ek\sigma_0^{n-1}t}}. \tag{6.61}$$

Typical dependence $\sigma/\sigma_0 = f(t)$, where $n = 3$, a product $Ek = 8.5 \times 10^{-6}$ $(\text{MPa}^2 \times \text{h})^{-1}$, and time t is taken in hours, is displayed in Fig. 6.11 for several values of initial stress σ_0.

Due to creep, the elastic strains and stresses diminish.

The fall of stress under creep at a fixed total strain has received a name *stress relaxation*. In practice, such phenomenon results in a decrease of tightness of preliminary stressed joints, and to change of frequency characteristics of mechanical systems and in other undesirable consequences.

A relaxation effect was investigated theoretically and by tests in many works, e.g., by Trumpler (1941), Davis (1943), Malinin (1948), Johnson (1949).

The stress relaxation does not depend on whether the initial tensile stress σ_0 has been induced by an external force with the subsequent fixing of the rod ends at constant temperature, or by gradual cooling on ΔT of a rod with initially fixed ends, when tensile thermal stresses $\sigma_0 = E\alpha_T\Delta T$ arise in it.

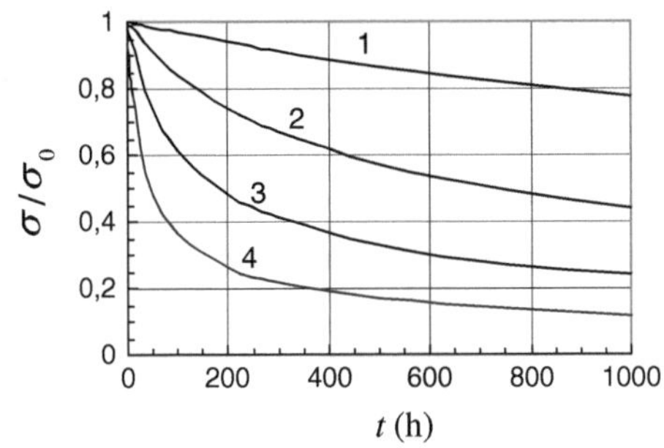

Fig. 6.11 Typical relaxation curves in a two-end fixed rod: *1* $\sigma_0 = 20$ MPa, *2* $\sigma_0 = 50$ MPa, *3* $\sigma_0 = 100$ MPa, *4* $\sigma_0 = 200$ MPa

In a rod fixed at the ends and then rapidly heated, there arise compression thermal stresses. These stresses eventually decrease (relax) just as tension but the constants k and n in Eq. (6.61) should be changed according to the compression diagrams (see Sect. 6.1.2.1).

6.4.1.2 Elastically Fixed Rod

When an elastic rod is fixed at one end through a spring with a stiffness c_{spr}, Eq. (6.59) yields

$$\dot{\sigma}(t)/E_r + \dot{\varepsilon}_c(t) = 0, \tag{6.62}$$

where $E_r = E(1 + E/c_{str})^{-1}$.

Equation (6.61) remains valid by substituting into it a reduced modulus of stiffness E_r. A stiff spring ($c_{spr} \gg E$) does not change the stress relaxation process, but with a very soft spring ($c_{spr} \ll E$), the change of stresses becomes insignificant.

6.4.2 Creep Calculation Under Steadily Varying Stresses

6.4.2.1 Approximate Approaches

In general cases, creep calculations under varying stresses, even at smooth changes of loads and temperatures, should be carried out by numerical methods including step-by-step procedures corresponding to Eq. (6.30a). Solutions of many specific problems may be found in the literature (e.g., Malinin 1952; Shorr 1964; Boyle and Spence 1983).

At moderate steadily changes of stresses and temperatures, the “isochronous” curves of deformation and other approximate modes may be used in approximate calculations.

6.4.2.2 Non-uniformly Heated Extended Rod

Coming back to the problem of redistribution of stresses at creep in a non-uniformly heated extended rod (see Sect. 6.2.3.2), we will consider the process of stress change in time. Initial rapid transition from a cold rod to non-uniformly heated one, at which initial temperature stresses have arisen, is here neglected.

The equation of the theory of hardening in the form of Eq. (6.13) at $f(\sigma, T) = k\exp(-\beta/T)[\exp(n\sigma) - 1]$ has been accepted as a creep law, within strain area $\varepsilon_c \geq \varepsilon_{c*}$.

In Fig. 6.7b, given above, the stress distribution on the rod width at various time instants for two variants of calculation was displayed:

Curves A. Stretching by a constant force without initial temperature stresses, i.e., at thermal expansion factor $\alpha_T = 0$;
Curves B. Stretching by the same force taking into account initial temperature stresses induced by the preset non-uniform temperature field (at $\alpha_T = 2 \times 10^{-5}\,1/°\mathrm{C}$).

The directions of stress evolution during time for both variants of calculation are displayed by arrows.

It is seen that unsteady creep essentially changes the stress field, especially during the first hours of operations. In the course of time, the curves of both calculation variants approach (but from the opposite sides) at the curve $\sigma(\infty)$ of the stabilized creep. In the absence of the temperature stresses (curves A), the tensile stresses in hot points $\bar{y} = \pm 0.5$ decrease as a result of the higher rate of creep at these points. Accounting temperature stresses (curves B), which also gradually decrease, the total stresses in the most hot points increase in time (in the given example).

6.4.3 *Prediction of Damage Development of Creep and Durability*

6.4.3.1 Damage Development at Creep

Progressive development of creep strain occurring at the third stage of loading by a constant force can be provoked by two reasons:

- Real strength of a material in stressed state at elevated temperatures reduces due to micro-cavities or micro-cracks arising and gradually growing over operating time. This effect can be estimated in the simplest way by the parameter of material damaging $0 \le D \le 1$ (Kachanov 1958; Rabotnov 1959). With respect to tension, it is determined as a ratio of a part of the cross-sectional area A_{cr} damaged by cavities to the inherent total area A_0 and usually indicated as $\omega = D = A_{cr}/A_0$. Then, the "real" varying stress $\sigma(t)$ at an undamaged part of the cross-sectional area induced by constant longitudinal force F will be

$$\sigma(t) = \frac{F}{A_0 - A_{cr}} = \frac{\sigma_0}{[1 - \omega(t)]}. \tag{6.63}$$

In reality, an empirical function $\omega(t)$ depends not on the operating time but on the creep strain accumulated under stress and temperature action.
Assuming a smoothly varying creep rate $v_c(t)$ to be determined by Eq. (6.5a) through the real stress

$$v_c(t) = k\left\{\frac{\sigma_0}{[1 - \omega(t)]}\right\}^n, \tag{6.64}$$

we see that the creep rate $v_c(t)$ unrestrictedly increases for the damaging factor $\omega \to 1$.
- Stresses in a stretched rod increase due to reduction of the cross-sectional area induced by lateral contraction even without material damage. In order to demonstrate this effect let us take the rate of full strains in lateral directions of an elastic rod in tension at $\varepsilon \ll 1$ as

$$\dot{\varepsilon}_l(t) = -\mu E^{-1}\dot{\sigma}(t) - 0.5\dot{\varepsilon}_c(t), \tag{6.65}$$

where the mean inelastic strain at creep is $\varepsilon_{c,m} = \left(\varepsilon_c + 2\varepsilon_{c,l}\right)/3$, while plastic deformation is assumed zero.

The rate of cross-sectional area change at the creep rate $\dot{\varepsilon}_c = k\sigma^n$ will be defined as

$$\dot{A} = 2A\dot{\varepsilon}_l = -2\mu A E^{-1}\dot{\sigma} - Ak\sigma^n. \tag{6.66}$$

On the other hand, $A(t) = F/\sigma(t)$, which yields at $\sigma \neq 0$

$$\dot{A} = -F\sigma^{-2}\dot{\sigma} = -A\sigma^{-1}\dot{\sigma}. \tag{6.67}$$

This leads to nonlinear differential equation

$$\left(1 - 2\mu E^{-1}\sigma\right)\dot{\sigma} - k\sigma^{n+1} = 0. \tag{6.68}$$

with the initial condition $\sigma = \sigma_0$ at $t = 0$.

Taking into account that under elastic strains the ratio $\sigma/E \ll 1$ and having integrated Eq. (6.68), we find:

$$t = \frac{1}{n v_{c,0}}\left[1 - \left(\frac{\sigma_0}{\sigma}\right)^n\right], \tag{6.69}$$

where $v_{c,0} = k\sigma_0^n$ is a creep rate at the initial stress σ_0.

From the reverse dependence, it follows

$$\frac{v_c(t)}{v_{c,0}} = \left(\frac{\sigma}{\sigma_0}\right)^n = \left(1 - \frac{t}{t_d}\right)^{-1}, \tag{6.70}$$

where

$$t_d = (n v_0)^{-1} \tag{6.71}$$

is a theoretical time of a rod rupture due to cross-sectional area contraction (Hoff 1953).

Accounting for the development of plastic deformations, Eq. (6.65) should be presented in the form

$$\dot{\varepsilon}_l(t) = -\mu_s[E_s(\sigma)]^{-1}\dot{\sigma}(t) - 0.5\dot{\varepsilon}_c(t). \tag{6.72}$$

where $E_s(\sigma)$ is a secant modulus of the plastic deformation diagram.

Assuming that this modulus does not depend on creep, we arrive, instead of Eq. (6.68), to the nonlinear equation

$$\left\{1 - 2\mu_s[E_s(\sigma)]^{-1}\sigma\right\}\dot{\sigma} - k\sigma^{n+1} = 0. \tag{6.73}$$

Until the full strain remains small ($\varepsilon \ll 1$), Eqs. (6.63)–(6.67) are valid. Otherwise, it is necessary to define a true strain as $\varepsilon^* = \ln(1 + \varepsilon)$. A correct solution of this problem is rather complicated (Rabotnov 1966).

Equations of the previous subsection show initial stages of theoretically possible stress increase at constant load. A real rupture of a rod will occur when the maximum stress in the weakened cross section reaches the magnitudes of the long-term strength for the given temperature.

Under tension of sufficiently plastic materials, a sample rupture is accompanied by a neck formation, as at short-time tests, however, the value of cross-sectional area reduction factor $\psi = A_{neck}/A_0$ after long-time rupture is usually less than at short-time test.

Development of damaging of more brittle materials can occur without any outer visible indications.

6.4.3.2 Prediction and Validation of Durability

A number of the structures intended for operation at elevated temperatures is projected on a resource 100,000 h and more (for example, steam pipelines and other members of power plants). It is impossible to confirm such long durability by direct design tests in operating conditions. Also it is almost impossible to establish characteristics of material for such durability by direct specimen tests. As an alternative, durability prediction is based on relatively short-term tests in 1000–2000 h but at stresses and temperatures, higher, than under operation. In this case, the test results must be extrapolated in time for two and more orders.

Reliability of extrapolation requires a detailed studying of a concrete material's physical and mechanical properties. In the absence of such data, some generalized approaches and formulas are used (Larson and Miller 1952; Orr et al. 1954; Pisarenko and Krivenjuk 1990, and others).

Use of the Primary and Generalized Characteristics of a Material Durability

Primary curves $\varepsilon_c(t)$ and $\sigma_d(t)$, constructed using the results of specimen tests, sometimes are extrapolated over limits of time, within which they have been experimentally obtained. To this goal, approximated formulas such as Eq. (6.14) are usually used. It is visible from Fig. 6.3 that such extrapolation from very short-term tests is not always possible.

With the same purpose, the generalized characteristics of material durability are used. In particular, Eqs. (6.20)–(6.23) concerning Larson–Miller parameter $P_{L,d} = T(C_L(T) + \log t_d) = f(\sigma)$, yield an approached dependence

$$t_d = 10^{\frac{P_L(\sigma)}{T} - C_L(T)}. \tag{6.74}$$

The curves $\sigma(P_l)$ such as those shown above in Fig. 6.6 can be obtained in the course of not too prolonged tests under a high temperature, which must not exceed a maximum permissible for the given material temperature T_{lim}. At lower temperatures $T < T_{\text{lim}}$ and the same stress, i.e., at $P_L = \text{const}$, the estimated time before destruction will quickly increase according to the equation:

$$\frac{t_d(T)}{t_d(T_{\text{lim}})} = 10^{\frac{P_L(\sigma)}{T}} \Delta\bar{T} \tag{6.75}$$

where $\Delta\bar{T} = (T_{\text{lim}} - T)/T_{\text{lim}}$ is a relative decrease in temperature; for moderate values of $\Delta\bar{T}$ is taken $C_l \approx \text{const}$.

For typical values of ratio $P_L(\sigma)/T = 20{-}30$, a temperature decrease of 10 % can raise an estimated time before destruction in 2–3 orders.

The stress reduction leading to growth of parameter P_L also rises time t_d, but this effect becomes usually significant only if the stress decreases several times.

Use of Creep Characteristic

Monkman and Grant (1956) showed that a product of the stabilized creep rate $v_{c,st}$ for a time t_d before destruction at a given temperature has an approximately constant value

$$v_{c,st} t_d = \varepsilon_{c,d} \approx \text{const.} \tag{6.76}$$

This quantity is specific for a material and is not dependent on stress.

Similar tests of Sizova (1971) and others have confirmed this correlation for different materials.

In the cases when a creep rate is stabilized at different stresses in rather short time, Eq. (6.76) allows one to find the value of limit strain $\varepsilon_{c,d}$ using the quantities of $v_{c,st}$ and t_d obtained at a high stress and then estimate a considerably greater durability at smaller stresses.

Equation (6.76) can be considered as a consequence of Eqs. (6.5a) and (6.17) for power exponents $m(T) \approx n(T)$.

With reference to materials with unexpressed stabilized stage of creep, Evans (2000) has shown that Eq. (6.76) is approximately satisfied should a strain $\varepsilon_{c,d}$ be defined by the minimum creep rate $v_{\min}$ (see Sect. 6.1.3.3).

Resource Prediction of a Machine Parts

Characteristics of material durability are necessary for calculations of in-service life (resource) of machine parts.

To provide a real long operation of structures, taking into account the approached possibilities of prediction in a combination with scatter of material properties, preservation of rather large margins in time is demanded (see Chap. 9).

It is also necessary to carefully supervise loads and temperature of operating regimes and watch the output of resource of machine parts using the methods of technical diagnostics.

For indirect verification of predicted resources, stressed tests of a structure or its parts are conducted (Kuznetsov and Tseytlin 1976). These tests provide accumulating during shorter time the damages “equivalent” to damages at long operation.

Instead of special tests, an operating experience of machines in more severe conditions can be used. Thus, power plants for gas-transfer stations maintained with the large resource at lower temperatures can be created on the basis of the aviation engines having the limited resource at very high temperatures.

References

Bailey RW (1929) Creep of steel under simple and compound stresses and the use of high initial temperature in steam power plant. In: Transactions Tokyo sectional meeting of the world power conference, Oct–Nov, Tokyo

Bailey RW (1935) The utilization of creep test data in engineering design. Proc Inst Mech Eng 131

Besseling JF (1958) A theory of elastic, plastic and creep deformation of an initially isotropic material showing anisotropy strain hardening, creep recovery and secondary creep. J Appl Mech ASME 25(4):529–536

Birger IA, Shorr BF (eds) (1975) Thermal strength of machine parts. Mashinostroenie, Moscow (in Russian)

Boley BA, Weiner JH (1960) Theory of thermal stresses. Wiley, New York

Boyle JT, Spence J (1983) Stress analysis for creep. Bafterworths, London

Crossland B, Patton RC, Skelton WJ (1971) The comparison of torsion and tension creep data for 0.18 percent carbon steel. In: Smith AI, Nikolson AM (eds) Advances in creep design. Applied Science Publishers Ltd., London

Danilova IN (1959) To calculation of stress state in a rotor at instabilized creep. Izv. Akad. Nauk SSSR, Mekhanika i Mashinostroenie, 5 (in Russian)

Davis EA (1943) Creep and relaxation of oxygen free copper. J Appl Mech 10(2)

Evans RW (2000) A constitutive model for the high-temperature creep of particle-hardened alloys based on the θ projection method. In: Proceedings of mathematical physical and engineering sciences, vol 456, no 1996. Royal Society, London, pp 835–868

Evans RW, Wilshire B (1985) Creep of metals and alloys. Cariton House Terrace, London

Fridman JB et al (1961) Some problems of thermal strength in reactor engineering. Atomnaya Energia 10(6):606–619 (in Russian)

Hoff NJ (1953) The necking and rupture of rods subjected to constant tensile loads. J Appl Mech 20(1):105–108

Hosford WF (2005) Mechanical behavior of materials. Cambridge University Press, Cambridge

Johnson AE (1949) Creep and relaxation of metals at high temperatures. Engineering 168(4362)

Johnson AE (1960) Complex stress creep of metals. Metall Rev 5(20):447–506

Johnson AE, Frost NE (1955) Fracture under combined stress creep conditions of a 0.5 percent Mo steel. Engineering 191

Johnson AE, Henderson J, Khan B (1962) Complex stress creep, relaxation, and fracture of metallic alloys. H. M. Stationery Office (National Engineering Laboratory), Edinburg

Johnson AE, Tapsell HJ (1940) Creep under combined tension and torsion. Engineering 150(3887, 3889, 3892, 3894)

Kablov EN, Golubovsky ER (1998) High-temperature strength of nickel alloys. Mashinostroenie, Moscow (in Russian)

Kachanov LM (1958) About failure time in creep conditions. Izv.AN SSSR, ONT (8) (in Russian)

Kachanov LM (1960) Theory of creep. (1967, trans). National Lending, Library for Science and Technology, London (in Russian)

Kuznetsov ND, Tseytlin VI (1976) Equivalent researches of gas-turbines engines. Mashinostroenie, Moscow (in Russian)

Larson RF, Miller J (1952) A time-temperature relationship for rupture and creep stresses. Trans Am Soc Metals 74(5):765–775

Lemaitre J, Chaboche JL (1990) Mechanics of solid materials. Cambrodge University Press, Cambridge

Lemaitre J, Chaboche JL (2004) Mecanique des materiaux solides, 2nd ed. Dunod, Paris

Malinin NN (1948) Design foundation on creep. Mashgiz, Moscow (in Russian)

Malinin NN (1952) Design on creep. In: Ponomarev S (ed) Foundations of contemporary methods of strength design in machinery. Mashgiz, Moscow (in Russian)

Malinin NN (1975) Applied theory of plasticity and creep. Mashinostroenie, Moscow (in Russian)

Manson SS, Haferd AM (1953) A linear time-temperature relation for extrapolation of creep and stress rupture data. NACA Tech. Note 2890, Washington
Manson SS, Halford GR, Spera DA (1971) The role of creep in high temperature low cycle creep fatigue. In: Smith AI, Nicolson AM (eds) Advances in creep design (The A.E. Johnson memorial volume). Applied Science Publishers Ltd, London, pp 229–249
Masing G (1923) Wissenschaftliche Veroffentlichungen aus dem Siemens Kon-zern. H. 1. Bd. 111
Mileyko ST, Telenkov VI (1962) Short-time creep of aluminum alloys. J Izv Math Theor Phys (5) (in Russian)
Monkman FC, Grant NJ (1956) An onempirical relationship between rupture life and minimum creep rate in creep-rupture tests. Proc ASTM V56:593–620
Norton FH (1929) Creep of steel at high temperature. McGraw-Hill, New York
Norton FH (1939) Creep in tubular pressure vessels. Trans ASME 61
Oding IA, Tuljakov GA (1958) Creep of an austenitic steel under combine stress state. Izv. AN SSSR, ONT 1:3–10 (in Russian)
Odqvist FKG (1933) Verfestigung von flussisenahnlichen Körpern. ZAMM 13(5):360–363
Odqvist FKG (1934) Theory of creep under combined stresses. In: Proceedings of 4th international congress for applied mechanics, p 228
Odqvist FKG (1953) Influence of primary creep on stresses in structural parts. Transactions of the Royal Institute of Technology, Stockholm (66)
Odqvist FKG (1974) Mathematical theory of creep and creep rupture, 2nd edn. Clarendon Press, Oxford (1st edn 1966)
Orr RL, Sherby OD, Dorn JE (1954) Correlation of rupture data for metals at elevated temperatures. Trans ASM 46:113–128
Palley IZ (1964) On relationships of theory of plasticity connecting stresses and strains at varying temperature conditions. Proc RIIGVF Riga 34:3–16 (in Russian)
Pisarenko GS, Krivenjuk VV (1990) A new approach to prediction of metal long-term strength. Dokl. Akad. Nauk SSSR, Mekhanika 312(3), 558–562 (in Russian)
Rabinovich VP (1966) Strength of turbine disks. Mashinostroenie, Moscow (in Russian)
Rabotnov YN (1948) Designing machine parts for creep. Izv. Akad. Nauk SSSR, OTN. (6) (in Russian)
Rabotnov YN (1957) About possibilities of unsteady creep description with application to rotors creep investigation (in Russian). Izv. Akad. Nauk SSSR, OTN., No.5
Rabotnov YN (1959) About mechanism of long-term destruction. Problems of Material and Structure Strength. Izv, AN SSSR Moscow (in Russian)
Rabotnov YN (1960) The theory of creep and its applications. In: Plasticity. Pergamon Press, Oxford
Rabotnov YN (1966) Creep of construction elements. Physmatgiz, Moscow. (1969) Creep problems in structure members (trans), Amsterdam (in Russian)
Rabotnov YN, Mileyko ST (1970) Short-term creep. Nauka, Moscow (in Russian)
Robinson EL (1943) 100,000 hour creep test. Mech Eng
Samarin JuP, JaM Klebanov (1994) Generalized models in theory of creep. Samarsky Gos. Techn. Universitet, Samara (in Russian)
Sdobyrev VP (1959) Long-life criterion for some high-temperature alloys at complex stress-strain state. Izv. AN SSSR Mekhanika i Mashinostroenie 6:93–99 (in Russian)
Shevchenko YN, Marina VY (1976) Structural modal of medium at nonisothermal loading process. J Appl Mech 12(12):19–27 (in Russian)
Shevchenko YN, Terekhov RG (1982) Physical equations of thermo-viscous-plasticity. Naukova Dumka, Kiev (in Russian)
Shorr BF (1958) Effect of nonuniform heating under creep conditions on the change in a stress state. Dokl Akad Nauk SSSR 123(5):809–812 (in Russian)
Shorr BF (1959) Unstabilized creep calculation of nonuniformly heated rods of an arbitrary cross section. Izv. Akad. Nauk SSSR, Mekhanika i Mashinostroenie 1:89–96 (in Russian)

Shorr BF (1962) Fundamentals of creep calculations on nonuniformly heated parts. In: Fridman JB (ed) (1964) Strength and deformation in nonuniform temperature fields gosatomizdat, Moscow; (trans). Consultants Bureau, New York (in Russian)
Shorr BF (1964) Fundamentals of creep calculations on nonuniformly heated parts. In: Fridman JB (ed) (1962) Strength and deformation in nonuniform temperature fields (trans). Consultants Bureau, New York
Sizova RN (1971) Approximate ways for producing of creep design characteristics. In: Heat stresses in structure elements, vol 10. Naykova Dumka, Kiev, pp 136–145 (in Russian)
Straub S (1995) Verformungsverhalten und Mikrostruktur warmfester marten-sitischer 12 %-Chromstähle. Dissertation. Universität Erlangen-Nürnberg, Fortschr.-Ber. VDI Reihe 5(405) Düsseldorf
Sully AH, Cale GN, Willoughby G (1948, Sept) Creep of metals subjected to compression stress. Nature (162)
Tapsell HJ (1931) Creep of metals. Oxford University Press, Oxford
Temis YM (1994) Theory of non-isothermal plastic yielding with isotropic and anisotropic hardening "Machinary, Encyclopedia". In: Frolov KV et al (eds) Machines dynamics and strength vol 1–3, B 1. Kolesnikov KS (ed). Mashinostroenie, Moscow, pp 227–231 (in Russian)
Temis YM (2005) Plasticity and creep modeling of turbo-engines constructive materials. In: Proceedings of 49th international conference AAI, Part 2, MAMI, Moscow, pp 25–76 (in Russian)
Torshenov NG (1961) Creep of aluminum alloy D-16T under compression. J PMTF 6:158–160 (in Russian)
Trumpler WE (1941) Relaxation in metals at high temperatures. J Appl Phys 12(3)
Trunin II (1969) Determination of long-time strength characteristics of heat-re-sistant materials with great durability. J Problemy Prochnosti 6:3–8 (in Russian)
Trunin II, Levitsky JuV, Shaban GA (1978) Prediction of strength and plasticity characteristics and build-up of the generalised high-temperature strength diagram. J Zavodskaya Laboratoriya 10:1260–1264 (in Russian)
Wilson DJ, Freeman JW, Voorhees HR (1971) Creep rupture testing of aluminium alloy to 100000 hours. J Materials 6(4):981–1000
Yao HT, Xuan FZ, Wang Z, Tu ST (2007) A review of creep analysis and design under multi-axial stress states. Nucl Eng Des 237:1969–1986

Chapter 7
Cyclic Creep: Thermocyclic Durability

Abstract Numerical methods are considered for cyclic creep, in particular, for thermocyclic creep. In case of a stable material state and periodically varying loading and temperature field, the developed methodology enables straightforward evaluation of limiting cycle parameters without the calculation of intermediate transient cycles. Interaction of plastic strain and creep and also a role of long-term exposure during cycle parts under high temperatures are described. Numerical examples indicate the possibility of considerable variation of stress and strain ranges induced by periodic external loads and thermal expansion in case of cyclic creep. This makes a drastic effect on the evaluation of the cyclic durability of a structure.

7.1 Cyclic Creep

7.1.1 Cyclically Changing Stresses

7.1.1.1 General Approach

In some cases, creep is a result of a great number of repeated loadings and heatings. The ranges of stresses and creep rates under these conditions can be not quite identical. However, if stress alternates periodically, then a *cyclic creep* arises.

In some works, the term "cyclic creep" refers to processes when stresses alternate two–three times, but we shall further assume they change up to 10^3 and more times, which, for example, can be related to the number of "flight cycles" of aviation engines.

Results of creep tests at cyclically changing loadings (even at constant temperature) have considerable variety, but nevertheless, the most typical experimental features can be established (Coffin 1954; Shorr and Dul'nev 1968, and others). These allow the formulation of some theoretical approach to cycle creep calculation.

B.F. Shorr, *Thermal Integrity in Mechanics and Engineering*,
Foundations of Engineering Mechanics, DOI 10.1007/978-3-662-46968-2_7

To obtain, in general cases, an overall pattern of strain evolution, calculations at cyclic stresses and temperatures must be carried out using a step-by-step method. However, for a large number of cycles, such an approach is for an extreme duration. Besides, this demands us to take into consideration the poor accuracy of material properties referring to initial cycles and assumes that at the beginning of operational loading, the structure was unstressed. The latter does not always take place. Therefore, a direct stresses and strains calculation for a stable cycle of which its parameters depend only on typical operating steady conditions is of practical interest and can thereby give somewhat more objective estimation of working capacity of a machine.

The general approaches to investigate the cyclic creep problems, taking into account adaptability at creep, were considered by Shorr (1958, 1964, 1966), Rabotnov (1966), and in a number of others publications.

7.1.1.2 Soft Cycles

At "soft" cyclic loading, the tests are mostly carried out using a constant cycle $\sigma_{\max} \to 0 \to \sigma_{\max}$. In Fig. 7.1, the dependence of total strain $\varepsilon = \varepsilon_e + \varepsilon_c$ versus time for specimens of an aluminum alloy at $\sigma_{\max} = 280$ MPa and $T = 150$ °C is shown (Kennedy 1962). At such a maximal stress level, there is some evolution of creep curves in initial cycles, and then, creep rapidly passes into the cyclically established state (this area is shown in the figure by envelopes).

In these tests, almost full creep restitution at repeated unloadings was observed. Without, or with incomplete, restitution, creep at stable cycles $\sigma_{\max} \to 0 \to \sigma_{\max}$ can gradually increase.

Results of Sizova's stepwise tests of a heat-resistant alloy are displayed in Fig. 7.2. Specimens were heated to temperature 900 °C, loaded by compression stress $\sigma < 0$, and exposed during "hot" half-cycle t_h, while unloaded and cooled. Then, this procedure was repeated by tension stress $\sigma > 0$, and so forth. Figure 7.2a

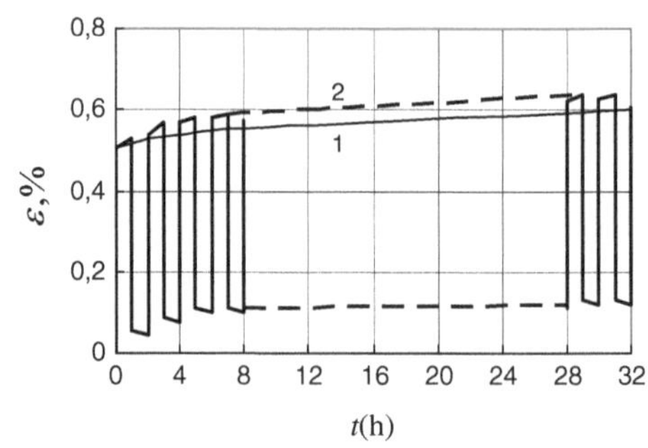

Fig. 7.1 Change of total strain at constant stress $\sigma_{\max}$ (curve 1) and at cycle loading $\sigma_{\max} \to 0 \to \sigma_{\max}$ (curve 2)

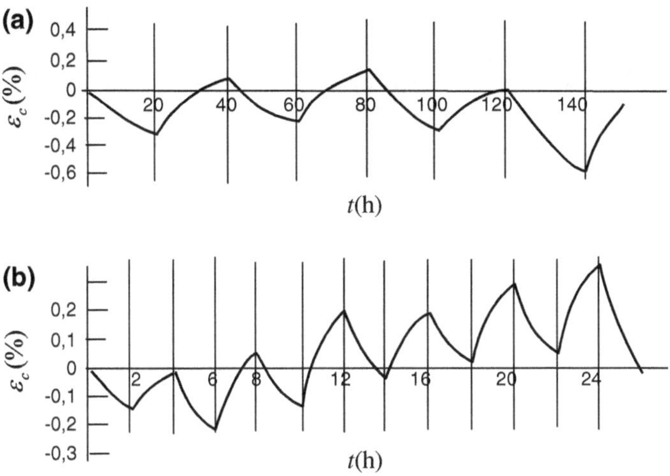

Fig. 7.2 Feature of stepwise creep at sign-alteration stresses $\sigma_{\max} = |\sigma_{\min}|$ Sizova's data in Birger and Shorr (1975)

corresponds to stress amplitude $\sigma = \pm 300$ MPa and exposure t_h = 20 h, and Fig. 7.2b—to $\sigma = \pm 400$ MPa and t_h = 2 h.

Beginning from the second half-cycle, the creep rate considerably exceeds its initial value at the first loading and strain ranges are leveled (though not exactly). A shift of the curves in Fig. 7.2b is connected, apparently, with various speed of creep at tension and compression.

7.1.1.3 Rigid Cycles

The creep tests at varying heat are the most easy to conduct by rigid fastening of the specimen borders when the total strain in it is zero, which corresponds to a rigid loading cycle in elastoplastic tests regarded in Chap. 4. In this case, the mechanical part of the strain is equal to the thermal expansion (with opposite sign)

$$\varepsilon_e + \varepsilon_p + \varepsilon_c = -\varepsilon_T \tag{7.1}$$

and the arising stresses are purely thermal.

Since Coffin (1954), many tests for thermal fatigue have been conducted using this principle. Methods of laboratory testing of creep at varying temperatures are described in some works, e.g., Shorr and Dulnev (1964), Evans and Wilshire (1985), and others. At "rigid" cyclic loading and moderate stresses, when a constant range of total strains is set, after several initial transitive cycles, the stable cycle of stress alteration is also usually established.

Gradual stabilization of stress or strain cycles is an important feature of creep at cyclic loadings. This effect reflects the ability of a material to "adapt" to conditions of loading (see also Chaps. 4 and 5). As a rule, the adaptability phenomenon takes place for typical operating stresses and temperatures which allows the designs to work in such conditions at a great number of repeated loading and heating. At stresses and temperatures when adaptability is not observed, the problem of maintenance of long-time reliable operation of details of the given alloy becomes problematic.

In Fig. 7.3a, a typical pattern of temperature, and in Fig. 7.3b—thermal stress changes during several consecutive cycles, are shown for a test of a heat-resistant alloy under rigid loading, exposure for each cycle at the maximum temperature, and cooling [Dul'nev's data in Birger and Shorr (1975)].

In the beginning of the first cycle (curve 1) at a temperature rise from 20 to 800 °C , there exists large compression stresses exceeding a hot yield point and leading to the occurrence of plastic compression strain. At the maximum temperature and constant total strain, an intensive stress relaxation, induced by creep, decreases compression stresses by almost double. After cooling at the termination of the cycle, large residual tension stresses exceeding a cold yield point appear. The second cycle (curve 2) begins with unloading from the specified residual stresses; therefore, the hot compression stresses increase to some smaller size. The subsequent cycles, beginning from the fourth (curve 3), become almost stable up to destruction of the specimen from thermal fatigue.

In the process of increasing the number of cycles, their asymmetry can change, and the cycle moves into the area of tensile stresses as is shown in Fig. 7.4 for a heat-resistant alloy tested within the temperature range $T_{\min} \Leftrightarrow T_{\max} = 100 \Leftrightarrow 700$ °C (Dul'nev's data) where curves 1 and 2 concern accordingly $T_{\min}$ and $T_{\max}$ and a dashed line 3—average values of stresses in cycles.

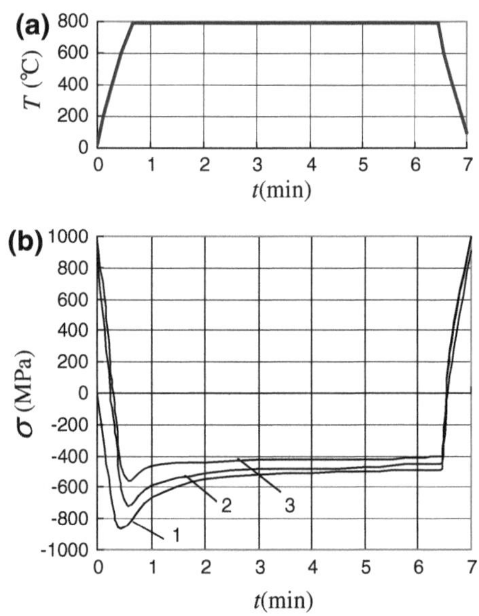

Fig. 7.3 Thermal stresses at cyclic loading of a specimen under constant total strain with exposure at maximal temperature and cooling

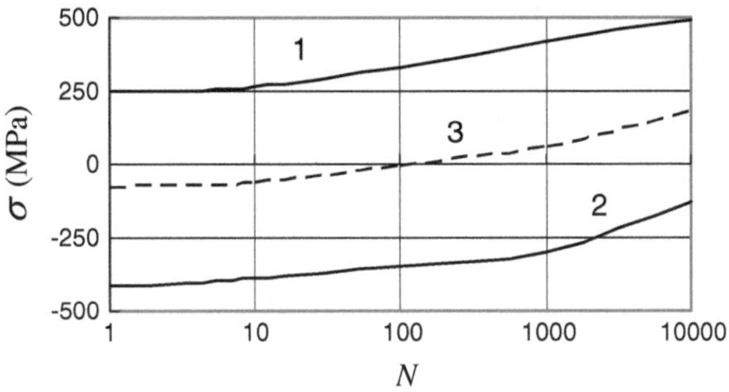

Fig. 7.4 Possible shift of a stress cycle at non-isothermal tests

7.1.1.4 Dynamic Creep

In some conditions, stress oscillation of high frequency with rather small amplitude σ_v is imposed onto constant or slowly changing average stress σ_m, sufficient for creep development. Such conditions are sometimes referred to as *dynamic creep* (Greenwood 1949; Rabotnov 1966). In these and other works, it is shown that for structurally stable materials at a ratio $\sigma_v/\sigma_m < 0.5$, creep rate usually increases, sometimes considerably, although there can be exceptions.

When oscillation amplitudes grow and a stress cycle approaches symmetry, the role of damage from low cycle fatigue increases.

In Chap. 4, the problems of low cyclic elastoplastic deformation at a constant moderate temperature whilst neglecting creep have been considered. However, tests show that at high temperatures and rather slow cycles, the influence of cyclic creep can also become significant. The time before failure will then depend on both the number of cycles and the accumulation of a long-term damage.

When high frequency oscillations with moderate stress amplitudes occur in an elastic area and failures are induced by high cycle fatigue, creep plays a secondary role.

7.1.2 Periodic Processes

7.1.2.1 Periodicity Conditions

The design of a stable cycle with the period t_{cl} is based on a condition that stresses and temperatures at each point of a body change periodically

$$\sigma(t + t_{cl}) = \sigma(t), \tag{7.2a}$$

$$T(t + t_{cl}) = T(t). \tag{7.2b}$$

A process could be periodic in the conditions of creep if outer loads, temperatures, and varying boundary conditions change periodically, and material properties are stable or become stable after some initial cycles. The material characteristics must not depend on common operation time and parameters that unlimitedly increase in time, particularly, on one-side growing inelastic deformations. However, unsteady creep can occur within each cycle, and strains can slowly monotonously increase from cycle to cycle.

Since a temperature field changes slower than a stress state, the period of a stationary non-isothermal cycle should include the time necessary for the temperature field to return to the condition at the cycle beginning. Account of cycle duration is one of the differences of cyclic creep from plastic deformations at low cyclic loading considered in Chap. 4.

To obtain a stably repeating cycle (i.e., a limiting cycle of stress and strain rate), creep characteristics should satisfy some conditions.

This requirement is always satisfied under slowly changing loads and temperatures when the unsteady part of deformation can be neglected considering that $\dot{\varepsilon}^c \approx \dot{\varepsilon}^c_{st} = f(\sigma, T)$.

In Shorr (1966), it was shown that the strain rate can alter periodically

$$\dot{\varepsilon}^c(t + t_{cl}) = \dot{\varepsilon}^c(t) \tag{7.3}$$

when a function $\dot{\varepsilon}^c(t)$ is presented due to the theory of heritable influence (Rabotnov 1966) in a form

$$\dot{\varepsilon}^c(t) = f[\sigma(t), T(t)](1 + A) - A\beta \int_0^t \exp[-\beta(t - \theta)] f[\sigma(\theta), T(\theta)] d\theta, \tag{7.4}$$

where $0 \le \theta \le t$ and A, β are material factors.

Equation (7.4) passes at $A = 0$ into the usual equation of steady creep.

A creep model of stabilized anisotropic hardening (see further Chap. 8) also allows periodic solutions.

Periodic solutions are typically found through the use of numerical methods, and only the simplest problems can be solved analytically. Cyclic change of a temperature field is assumed to be known.

7.1.2.2 Design Situations

Three design situations are possible:

- A cyclic change in stresses can be either obtained or set in advance. This allows the cyclic creep rates and strains to be calculated directly using the relations, deduced from experimental characteristics or theoretical generalizations presented in Chap. 6 (condition of "soft" loading).
- Cyclic change of strains (a condition of "rigid" loading, e.g., provoked by thermal cycling) is set.
- Cyclic change of outer loads and boundary conditions are set.

In the last two cases, the solutions must be found by solving the total system of equations of solid mechanics, which usually demands iteration.

The purpose of calculations of cyclic creep is to estimate inelastic strain $\Delta\varepsilon_{cl}$ or work ΔW_{cl} accumulation within each cycle of loading/heating and also the average rate of cyclic creep $v_{cl} = \Delta\varepsilon_{cl}/t_{cl}$. These results are necessary for estimating the number of cycles before destruction.

7.1.3 Cyclically Changing Temperature

7.1.3.1 Temperature-Dependent Creep Parameters

The equations describing creep in the conditions of alternating stress and fixed temperature can be spread to a case of variable temperature if we take temperature as a varying parameter of functions $v_{st}(\sigma_i, T)$, $\varepsilon_{\lim}(\sigma_i, T)$, and $\beta(\sigma_i, T)$ considering that these dependences are discovered from appropriate experiences at a fast temperature increase.

The rate of steady creep $v_{st}(\sigma_i, T)$ for structurally stable materials does not depend on loading history and is zero at $\sigma_i = 0$. It may be also taken zero at any stress if temperature is smaller than some threshold value $T_{\min}$. Dependence $\varepsilon_{\lim}(\sigma_i, T)$ mirroring the limited unsteady creep is usually defined during uniaxial tests by rapid stress increase from $\sigma_1 = 0$ to a magnitude σ_2 at a fixed temperature T. In order to experimentally find an analogous function $\tilde{\varepsilon}_{\lim}(\sigma_i, T)$ by a rapid increase of a specimen temperature from $T_1 \approx T_{\min}$ to T_2, supporting a fixed stress value σ over entire specimen is very difficult. Therefore, it is necessary to assume that $\tilde{\varepsilon}_{\lim}(\sigma_i, T) \approx \varepsilon_{\lim}(\sigma_i, T)$.

Let us mark, that at trials of some materials at repeated heating/cooling (so-called cyclic high-heat treatment), a size change even in the absence of exterior loads was observed. Assuming that at such trials, there were no temperature stresses and this resizing was a consequence only of creep, and in principle, it is possible to define a corresponding function $\tilde{\varepsilon}_{\lim}(\sigma_i, T)$. Examples of calculating such a function are presented in Shorr and Dul'nev (1968).

In non-isothermal problems, it is necessary to take into account that strains and stresses alter simultaneously.

7.1.3.2 Example of an Analytical Solution

Let us consider a 1D problem of stress σ change versus time, induced by cyclic creep in a viscoelastic rod rigidly fixed at a temperature T'. The temperature alternates step wisely between T' and $T'' > T'$ (Fig. 7.5a) with period $t_{cl} = \tau$ (Shorr and Nafikov 1963). A cyclic thermal expansion of a rod initiates cyclic strains with constant range and stresses varying due to creep.

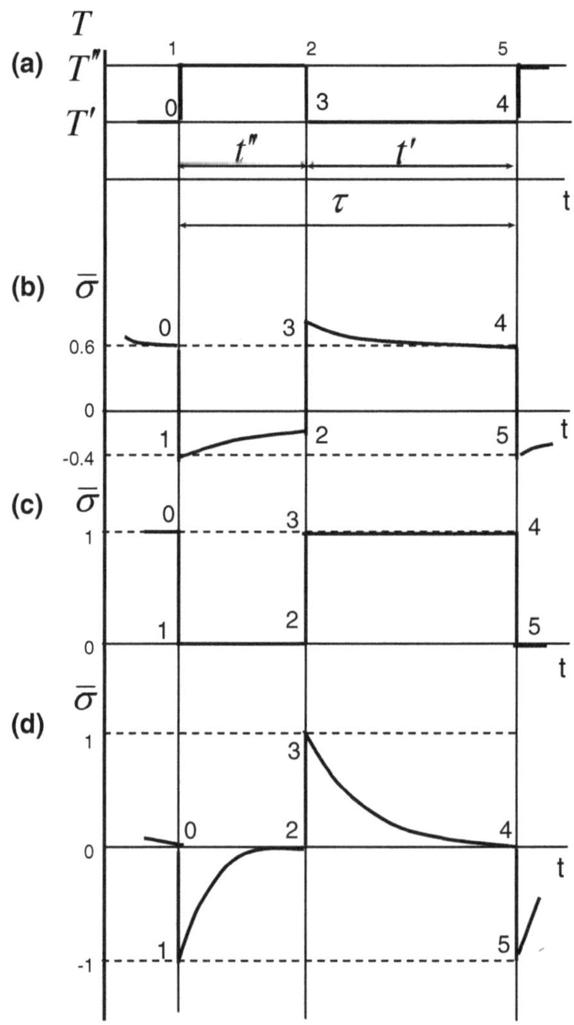

Fig. 7.5 Example of analytical solution of a cyclic creep problem

Neglecting unsteady creep rate increasing during rapid stress change at temperature alteration, we assume that the creep rate at both step portions has steady character and is represented by the dependence

$$\dot{\varepsilon}_c = A\exp(-bT)[\exp(k|\sigma| - 1)]\mathrm{sign}\sigma \tag{7.5}$$

This model supposes the existence of periodic solutions at creep. The relative duration of a portion with temperature T'' is $\xi = t''/\tau$. At rigid fixing of the rod ends the rate of the complete strain is zero, i.e.,

$$\dot{\sigma} + E\{A\exp(-bT)[\exp(k|\sigma| - 1)]\text{sign}\sigma + \alpha_T\dot{T}\} = 0. \tag{7.6}$$

Having entered a new variable $z = \exp(k|\sigma|)$, we reduce Eq. (7.6) to

$$\dot{z} + \varphi(t)z + \psi(t)z^2 = 0, \tag{7.7}$$

where

$$\begin{aligned} \varphi(t) &= kE[\alpha_T\dot{T}\text{sign}\sigma - A\exp(-bT)], \\ \psi(t) &= kEA\exp(-bT) \end{aligned} \tag{7.8}$$

For the step portion with constant temperature ($\dot{T} = 0$), functions $\varphi(t)$ and $\psi(t)$ take fixed values marked as φ', ψ' and φ'', ψ'' where $\psi/\varphi = -1$. The solution of Eq. (7.7) for each portion is

$$z(t_1) = \frac{z(0)\exp\psi t_1}{1 - z(0)(1 - \exp\psi t_1)}, \tag{7.9}$$

where time t_1 is separately counted for each portion.

Having returned to stress $|\sigma(t_1)| = [\ln z(t_1)]/k$ and marked stresses in the characteristic points of a cycle in Fig. 7.5a as $\sigma^{(1)} - \sigma^{(5)}$, we express sequential stresses $\sigma^{(2)} - \sigma^{(5)}$ through $\sigma^{(1)}$, to obtain four equations, including a relative site duration ξ as parameter. Stress jumps between sites are

$$\sigma^{(3)} - \sigma^{(2)} = \sigma^{(4)} - \sigma^{(5)} = E\alpha_T(T'' - T'). \tag{7.10}$$

The requirement of stress periodicity $\sigma^{(5)} = \sigma^{(1)}$ provides the fifth equation. The solution of the system obtained leads to the quadratic equation $ax^2 + bx + c = 0$ in $x = \exp(k|\sigma^{(1)}|)$; $x \geq 1$ where coefficients a, b, c depend on material characteristics and cycle parameters. The sign of stress $\sigma^{(1)}$ is defined by the condition for the first portion $\sigma^{(1)} = -E\alpha_T(T'' - T') < 0$.

One of the typical calculation results is displayed in Fig. 7.5b where $\bar{\sigma} = \sigma/E\alpha_T(T'' - T')$.

The analytical solution reveals some features of cyclic creep at stepwise heating.

- In the given problem, there occurs neither creep nor stresses without thermal expansion (at $\alpha_T = 0$).
- If at lower temperature T' creep is lacking ($\psi' = 0$), stresses at higher temperature T'' drop to zero, while stresses at T' become tensile $\bar{\sigma}^{(3)} = \bar{\sigma}^{(4)} = 1$, as shown in Fig. 7.5c.
- At very long cycle duration, a complete stress relaxation proceeds at each site, and at instant of temperature change, there arises temperature stresses having the same magnitude as that at the beginning of the process, as shown in Fig. 7.5d.

7.1.4 General Approach to Cyclic Creep Analysis

7.1.4.1 Direct Calculation of a Stabilized Cycle

Direct calculation of a stabilized cycle assumes that its characteristics have to be found from a periodicity condition of Eq. (7.2) without stress and strain calculation for preceding loading and heating cycles (Shorr 1966, 1970a, b).

To this end, we present a stress tensor for any point of a body at a current time t_1 in the form

$$\sigma_{ij}(t_1) = \sigma_{ij}(0) + \Delta\sigma_{ij}^{\circ}(t_1) + \int_0^{t_1} \dot{\sigma}_{ij}^{c}(\theta)d\theta; \quad 0 \le \theta \le t_1 \le \tau \tag{7.11}$$

where $t_1 = t - t_{in}$ is a time from the beginning of the current cycle τ.

Here, $\sigma_{ij}(0)$ is a stress at the beginning of the cycle, $\Delta\sigma_{ij}^{\circ}(t_1)$ is a change of stress from external loads and temperature field to the time t_1 calculated in absent of the creep (to the whole period $\Delta\sigma_{ij}^{\circ}(\tau) = 0$), and $\dot{\sigma}_{ij}^{c}(\theta)$ is the rate of additional self-counterbalanced stresses induced by the creep. The values of $\Delta\sigma_{ij}^{\circ}(t_1)$ and $\dot{\sigma}_{ij}^{c}(\theta)$ in a point of the body are obtained from calculation of the whole system. Therefore, the rate of the stress $\dot{\sigma}_{ij}^{c}(\theta)$ can be presented in the form of an operator

$$\dot{\sigma}_{ij}^{c}(\theta) = L[\dot{\varepsilon}_{ij}^{c}(\theta)] \tag{7.12}$$

which should be calculated by solving a corresponding linear problem of the theory of elasticity.

Fields of stress and strain rates are time dependent, but their correlation due to Eq. (7.11) remains for the given problem invariable.

From a periodicity condition (7.1), it follows that

$$\int_0^{\tau} \dot{\sigma}_{ij}^{c} d\theta = 0. \tag{7.13}$$

For some typical problems, it is possible to present dependence (7.12) in a closed form. The operator $L[\dot{\varepsilon}_{ij}^{c}(\theta)]$ can be obtained using mathematically similar solutions of the problems on temperature stresses and also the elastoplastic solutions based on a method of additional deformations (see Chap. 5).

In the general case, an application of FEM is possible.

As a whole, the cyclic creep problem demands a joint solution of Eqs. (7.11)–(7.13) and the constitutive equation of the cyclic creep theory $\dot{\varepsilon}_{ij}(t_1)$. The solution results in an iteration procedure starting from the initial approach with $\dot{\sigma}_{ij}^{c}(\theta) = 0$ where any time moment can be accepted as the period beginning.

A principle feature of iterative calculation on cyclic creep is that the size of stress in the end of the previous iteration is accepted as the value of stress at the beginning of the next iteration. Thus, the periodicity condition is only approximately satisfied in consecutive iterations; however, precision of accomplishment of the periodicity condition increases in process of magnification of iterations number. Convergence of the computing process is ensured with known approaches (see, e.g., Birger 1956 and also Shorr 1970b).

The universal FE computation codes using step-by-step methods of calculation do not consider material models from the point of view of adaptability requirements. Therefore, it is necessary to use so-called user's creep models whose algorithms can be developed on the basis of stabilized anisotropic hardening creep theory (see Chap. 8).

7.1.4.2 Integral Correlations

There are at least two substeps with a different sign of stress rate $\dot{\sigma}$ at cyclic processes. For this case, an equation of cyclic creep can be presented in integral form (Shorr 1966)

$$\varepsilon_*(t) = \varepsilon_{*in} \exp[-\beta(t - t_{in})] + \beta \int_{t_{in}}^{t} \varepsilon_{*\infty}(\theta) \exp[-\beta(t - \theta)] d\theta, \tag{7.14}$$

where $\varepsilon_{*in} = \varepsilon_*(t_{in})$ is an unsteady creep at the beginning of the set step, $\varepsilon_{*\infty}(\theta) = f[\sigma(\theta), T(\theta)]$, and a parameter β is a constant.

In the case of a symmetrical cycle for a material without restitution, creep occurs only under increasing stress value.

Different variants of creep integral equations are considered in Rabotnov (1966), Shorr and Dul'nev (1968), Shorr (1970a, b) and have been used preferably for analytical solutions. So, a rate of creep $\dot{\varepsilon}^c(t)$ can be written using Eq. (7.14) in the form

$$\dot{\varepsilon}^c(t) = f(t) + F_0(t) - \beta \int_{t_{in}}^{t} F_0(\theta) \exp[-\beta(t - \theta)] d\theta, \tag{7.15}$$

where $f(t) = v_{st}[\sigma(t), T(t)]$ is a steady creep rate, $F_0(t) = \beta\varepsilon_{*0}[\sigma(t), T(t)]$ is unsteady creep rate at the moment of rapid stress application, and a function $F_0(\theta)$ reflects the influence of prior states.

For a cyclic creep analysis, Eq. (7.15) must be integrated within a single cycle. An example of the cyclic creep direct calculation is resulted below.

7.1.4.3 Cyclic Creep of a Beam at a Bend

Computation (Shorr 1970b) has been executed with reference to a beam with one free and other fixed end. The beam was subjected to a cyclic force F applied at the free end. The force repeatedly alternated in steps from $F_{\min} = 600$ N to $F_{\max} = 900$ N due to the program shown in Fig. 7.6a. The length of the beam was 1000 μm, and the size of the rectangular cross section was 20 × 32 mm. The beam temperature $T = 200$ °C did not change. Material creep characteristics $f(\sigma)$, $F_0(\sigma)$, and β have been obtained by the processing of creep curves at unitary loading of samples since the creep data for repeated loading were not known.

Experimental data on cyclic creep of the beam end deflections versus time for consecutive cycles from $k = 2$ to $k = 6$, kindly given to the author by S.A. Nosov, are shown in Fig. 7.6 by points, and the numerals specify cycle numbers. Results of direct computation of the stabilized cycle are shown in Fig. 7.6 by a curve. It was accepted full creep restitution as well as it was observed in the experiment.

The calculation adequately reflects the character and quantity of the cyclic change of the beam deflections with an accuracy that is acceptable in creep design under incomplete information about material characteristics.

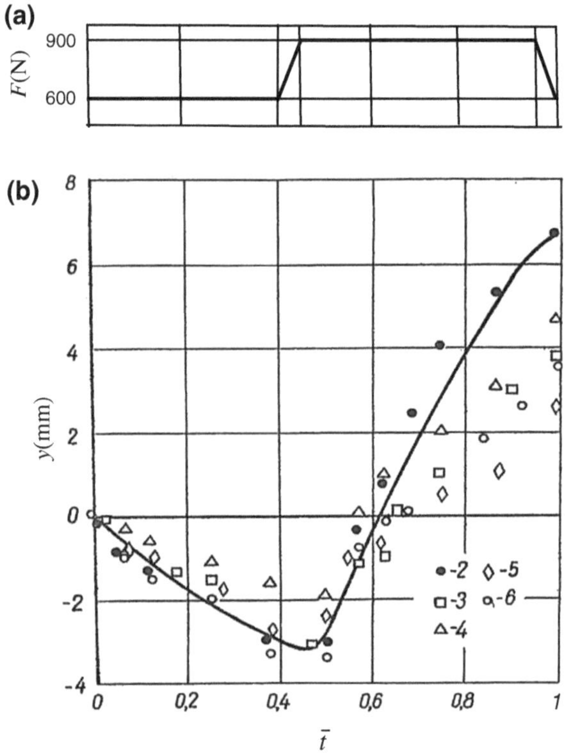

Fig. 7.6 Comparison of calculated and experimental deflection under cyclic bending of a beam

Calculations also showed that stress distribution over the cross sections of the beam in the conditions of cyclic creep considerably differs not only from linear (by elastic design) but also from distribution at constant force values.

7.2 Damage at Unsteady Creep. Thermocyclic Durability

7.2.1 Cyclic Durability Under Unsteady Creep

Unlike "*temporal*" *durability* (a time prior to long-term failure) measured usually in hours, *cyclic durability* is measured in cycle number prior to LCF failure. However, *cyclic durability under creep* depends on both cycle number and cycle duration.

7.2.1.1 Durability at Stepwise Stress Alteration

Under two-level loading, a specimen is usually affected by a stress σ_1 during time $t = t_1$, smaller than its durability $t_{d,1}$. Then, the stress changes to any value $\sigma_2 < \sigma_1$

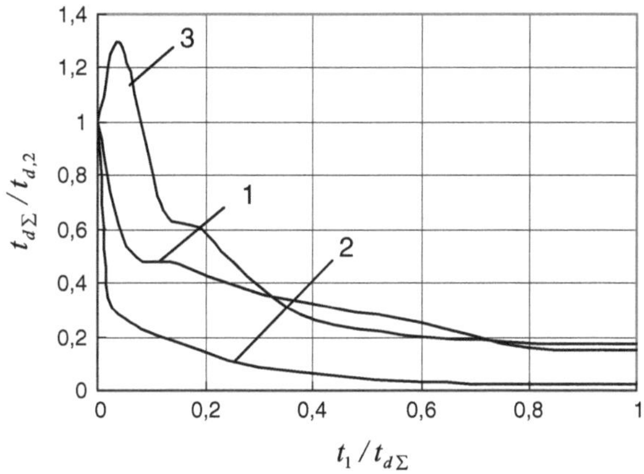

Fig. 7.7 Results of two-step durability tests for some alloys at different ratio $\sigma_1/\sigma_2 > 1$: *1* 1.21; *2* 1.62; *3* 1.38

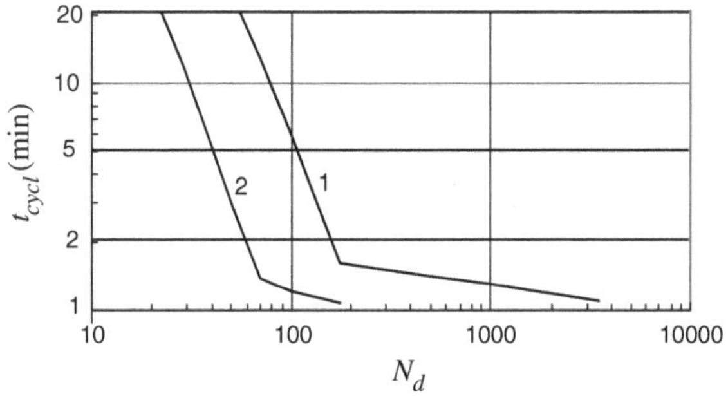

Fig. 7.8 Influence of cycle duration t_{cycl} on cycle number N_d at LCF failure; *1* $T = 750$ °C, *2* $T = 850$ °C

or $\sigma_2 > \sigma_1$ without stress sign altering. At the second stress σ_2, the specimen continues to be tested during some time t_2 before destruction. The total two-level durability is equal to sum $t_{d\Sigma} = t_1 + t_2$.

The durability at constant value σ_2 is $t_{d,2}$.

In Fig. 7.7, typical results of two-level tests of some heat-resistant alloys at the stress ratio $\sigma_1/\sigma_2 > 1$ plotted due to Sizova's data in Birger and Shorr (1975) are presented. The diagram reflects the dependence of relative change of total durability $t_{d\Sigma}/t_{d,2}$ on relative duration $t_1/t_{d\Sigma}$ of the first (in this case—more stressed) step.

In most cases, the initial overload at $\sigma_1/\sigma_2 > 1$ reduces total durability (curves 1 and 2). Moreover, the influence of an initial overload is shown to be intense at a small duration. However, at short-term overloads ($t_1/t_{d\Sigma} \ll 1$), an increase is sometimes observed (curve 3). This can be connected with material hardening at the raised stresses for some temperature.

Experiments show that initial overload influences material behavior at repeated stress changes in the same manner. The sensitivity of a material to repeated loading can considerably change with temperature.

7.2.1.2 Durability at a Variable Temperature

Generally, the design durability at variable stresses and temperatures is defined by the summary damage accumulation from the time-dependent long-term durability and from low cycle fatigue, depending on the number of cycles.

At a small number of cycles with long exposures under constant stresses and temperatures, temperature change decreases the long-term durability qualitatively like the stepped or cyclic loadings described above. But in some cases, when temperature variation leads to a significant change of material structure, special tests will be needed.

Tests of materials for thermal fatigue at a saw tooth change of temperature, when damage is affected only by thermal stresses, present other extreme case. In such tests, the total time of stressed state action at the maximum temperature is not long and (at not too high temperature) the influence of creep and its associated long-term damage are weaker (see Chap. 4).

Thermal fatigue tests are usually carried out using a trapezoidal cycle with exposure $t_{\exp}$ at the maximum temperature. The total duration of a cycle t_{cycl} can exceed several times this exposure, but the operating time at lower temperatures does not render appreciable influence on thermal fatigue. Under increase in exposure at the maximum cycle temperature, the role of creep and operating time becomes more notable.

In Fig. 7.8 (Dul'nev's data in Birger and Shorr 1975), the influence of the duration of a temperature cycle t_{cycl} on the number of cycles before destruction N_d of a heat-resistant alloy for two values of the maximum temperature $T_{\max}$ is shown. At small exposure times (in this case, $t_{\exp} = 10 - 12$ s at $t_{cycl} = 1.5 - 2$ min), the cyclic durability sharply decreases, but the further increase in duration has less influence. It is connected with the increasing rate of a material damage during the first moments of creep occurrence. With rise in temperature T_{max}, the exposure influence increases.

At moderate temperatures, the *total time* before destruction t_d at constant cycle duration t_{cycl} grows proportionally to the number of cycles prior to destruction N_d, i.e., $t_d = t_{cycl} N_d$. But the size N_d is itself affected by the cycle duration, i.e., $N_d = f(t_{cycl})$. Therefore, the time before destruction becomes a nonlinear function of the cycle duration $t_d = t_{cycl}N(t_{cycl})$. This function varies for different materials and test temperatures.

In particular, dependence $t_d = f(t_{cycl})$ can look similar to that presented in Fig. 7.9 (Dul'nev's data in Birger and Shorr 1975).

Initial increase in exposure at short cycle duration (in this case, at $t_{cycl} = (1-3)$ min) strongly decreases low cyclic durability, which reaches a minimum at some value $t_{cycl} \approx 5$ min. With further increase of t_{cycl}, the time t_d grows, as at usual LCF.

The presence of a minimum time before destruction (as a regime of the danger zone) is necessary to consider estimating of machine cyclic durability. On the other hand, this phenomenon allows selecting a trial regime under which the cyclic durability of a structure can be confirmed during the shortest time.

7.2.2 Damage Summation Under Creep

7.2.2.1 Linear Summation of Long-Term Damage

A time prior to failure (durability) at fixed stresses and temperatures is directly defined by experimental creep diagrams or their approximate introducing, as described in Chap. 6.

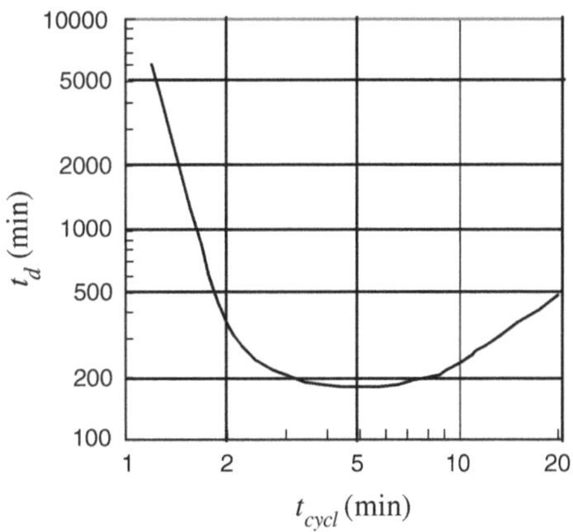

Fig. 7.9 Example of total cycle durability t_d dependence on unit cycle duration t_{cycl}

For estimating a time prior to failure at varying stresses and temperatures, the principle of damage summation presented in Chap. 5 is used. Depending on what is accepted as a criterion of damage under creep, various approaches are under consideration.

Under two-stage tensile loading, a constant stress σ_1 of the first stage might induce a long-term failure at the moment $t_{d,1}$ when strain reaches the value $\varepsilon^c_{d,1}(\sigma_1)$. A value of damage $D_1(\bar{t}_1)$ at a current operating time t_1 will be a function of the relative time $\bar{t}_1 = t_1/t_{d,1}$. Under sequential operation at the second regime with the stress σ_2, the specimen gains additional damage $\Delta D_{1,2}$, and the rupture will occur at a moment $t_{1,\Sigma} = t_1 + t_{1,2}$, when $D_{1\Sigma} = D_1 + \Delta D_{1,2} = 1$. If the operation starts from the regime at stress σ_2, damage during time t_2 will be $D_2 = D(\bar{t}_2)$ where $\bar{t}_2 = t_2/t_{d2}$. Under sequential operation at the regime with the stress σ_1, the rupture will occur at an instant $t_{2,\Sigma} = t_2 + t_{2,1}$ when $D_{2\Sigma} = D_2 + \Delta D_{2,1} = 1$.

Generally, both of the net times to failure $t_{1\Sigma}$ and $t_{2\Sigma}$, and additional damages $\Delta D_{1,2}$, $\Delta D_{2,1}$, calculated accordingly by durations t_{d2} or t_{d1} at the second step, will be different.

In many publications, the principle *of linear damage summation* is used, according to which the sum of damages is considered approximately independent on the order of the regime alternation. In this case, $\Delta D_{1,2} = D(\bar{t}_2), \Delta D_{2,1} = D(\bar{t}_1)$ and

$$D_\Sigma(\bar{t}_1 + \bar{t}_2) = D(\bar{t}_1) + D(\bar{t}_2). \tag{7.16}$$

This means that the damage $D(\bar{t})$ is a linear function of a relative duration $\bar{t}$ under appropriate stresses, i.e., $D(\bar{t}) \approx \bar{t}$ and $D_\Sigma = \bar{t}_1 + \bar{t}_2$.

The damage occurs at $t_1 + t_2 = t_{d\Sigma}$ when it should be

$$D_\Sigma = D(\bar{t}_1) + D(\bar{t}_2) = \bar{t}_1 + \bar{t}_2 = 1. \tag{7.17}$$

Here, relative time $\bar{t}$ appears as some structural parameter, and damage—as a linear function of time.

Presented above in Fig. 7.7, the effects of two-stage trials can be used for estimating the reliability of the linear damage summation.

In Fig. 7.10a, b, adopted values $\bar{t}_1$ for the examined specimens of various materials (curves 1, 2, 3) are stated on the axis of abscissas. On the axis of ordinates of Fig. 7.10a, the values of the relative operating time prior to damage at the second regime $\bar{t}_{d,2} = (t_{d\Sigma} - t_1)/t_{d,2}$ are displayed for each of the specimen. The totals of the relative durations $\bar{t}_1 + \bar{t}_{d,2}$ versus time $\bar{t}_1$ are presented in Fig. 7.10b.

If the totals are $\bar{t}_1 + \bar{t}_{d,2} = 1$, the magnitudes $\bar{t}_{d,2} = 1 - \bar{t}_1$ would meet the oblique straight line displayed in Fig. 7.10a by a bold line. Aberrations of experimental points from this line specify that the relative durability does not quite mirror the time process of a material damage.

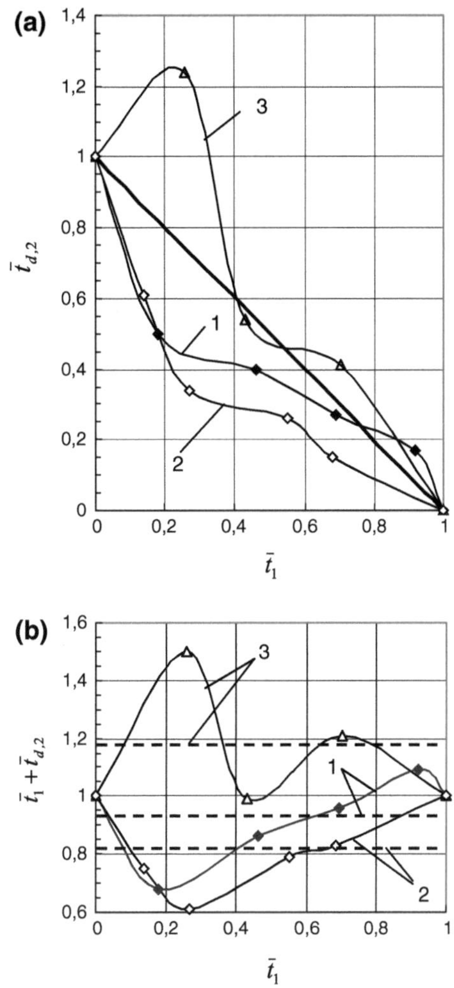

Fig. 7.10 Damage curves for two-step loading of some alloys

A simple approach to correct the formula of linear damage summation at a two-stage stress alteration consists of averaging test data shown in Fig. 7.10b by dashed lines and substitution of Eq. (7.17) by the "equivalent" damage condition

$$\bar{t}_1 + \bar{t}_2 = a_d, \tag{7.18}$$

where $a_d = (\bar{t}_1 + \bar{t}_{d,2})_m$ is a medial observational magnitude of the total of the relative durability at the failure.

A magnitude a_d can be as more than unity (for hardening materials) and as less than unity (for the materials tended to brittleness), e.g., in Fig. 7.10b, it lies within $0.8 \leq a_d \leq 1.2$. Due to Kablov and Golubovsky (1998), parameter a_d can vary within $0.3 \leq a_d \leq 1.8$.

Accounting for a factor a_d, the complete time to failure at a two-stage loading is approximately calculated by the formula

$$t_{d,\Sigma} = t_1 + \left(a_d - \frac{t_1}{t_{d,1}}\right) t_{d,2}. \qquad (7.19)$$

7.2.2.2 Multi-step and Smoothly Varying Loadings

Experience shows that the magnitude of coefficient a_d is basically affected by material structure and test temperature, while the number of cycles and character of the stress change (provided that the stress sign does not vary) have a weaker influence. This allows an operation's durability at complicated multi-stage cycles at sign-constant stress alteration to be evaluated by the formula

$$\sum_{k=1}^{n_d} \bar{t}_k = a_d, \bar{t}_k = \frac{t_k}{t_{d,k}} \qquad (7.20)$$

where magnitude a_d can be found from experiments at simple (two-stage) cycles of stress change or even at a unique two-stage loading.

If a stress sign during a cycle varies, it is necessary to take into account that damage accumulation at compression occurs with lower rate than at extension. Therefore, damage summation within half-cycles with compression and tensile stresses must be carried out separately.

At smoothly varying stresses and temperatures, the damage D is augmented for small time step dt on $dD = dt/t_d(t)$ where $t_d(t)$ is a time prior to failure at current stress and temperature values considered stationary. The damage time $t_{d\Sigma}$ is defined from the equation

$$\int_0^{t_{d\Sigma}} \frac{dt}{t_d(t)} = a_d. \qquad (7.21)$$

At the combined stress states, a step wisely varying magnitude of time to failure $t_{d,k}$ in Eq. (7.20) or smoothly varying time $t_d(t)$ in Eq. (7.21) can be calculated using the criteria presented in Chap. 6.

At multi-regime operation, by analogy to relative damage, the concept about relative accumulated plastic strain D^c_Σ can be introduced. Its increment during a current cycle is $\Delta D^c_k = \Delta\bar{\varepsilon}^c_k = \Delta\varepsilon^c_k/\varepsilon^c_{d,k}$. Here, $\Delta\varepsilon^c_k$ is the increase in creep strain at a regime k with stress σ_k and temperature T_k during the time t_k, and $\varepsilon^c_{d,k}$ is the creep strain at the moment of the specimen failure at constant values σ_k and T_k. Then, the approximate estimation of creep strain at the moment of long-term failure at complicated multi-stage cycles of stress and temperature alteration can be obtained using the equation

$$D_{\Sigma}^{c} = \sum_{k=1}^{n_d} \Delta\bar{\varepsilon}_k^c = a_c, \tag{7.22}$$

where magnitude a_c can be found by measuring a change of creep strain in tests.

7.2.3 Thermocyclic Durability

7.2.3.1 Low Cyclic Fatigue at Thermal Loading

Because of great number of factors influencing the resistance of materials to sign-varying loading and variable heating, the most reliable data can be obtained by direct tests of specimens under the programs imitating a simultaneous change of stresses and temperatures expected in operation. However, for a preliminary estimation of thermocyclic durability, it is possible to use some characteristics leaning against the generalization of specimen tests under typical conditions of heating and loading.

In Chap. 5, it has been shown that at constant elevated temperatures, the general approximate dependences of low cyclic fatigue (LCF) are valid if their experimental parameters are taken as temperature dependent.

At thermocyclic tests of rather plastic materials on a sawtooth cycle $T_{\min} \rightarrow T_{\max} \rightarrow T_{\min}$ (without exposure at the maximum temperature), the maximum cycle temperature T_{maz} becomes decisive. Test results can be approximately presented by the dependence

$$\Delta\varepsilon = \left(\ln\frac{1}{1-\varphi} + \frac{\sigma_B}{E}\right) f(N_d), \tag{7.23}$$

connecting a number N_d of thermocycles prior to destruction with the range of full strain $\Delta\varepsilon$ where material characteristics σ_B, E, ϕ correspond to temperature T_{maz}.

Rather accurate description of experiments for a number of heat-resistant materials is given by the formula (Dul'nev and Kotov 1980)

$$\Delta\varepsilon N^m = A\left[\left(\ln\frac{1}{1-\varphi}\right)^{0.6} + \frac{\sigma_B}{E}\right], \tag{7.24}$$

which is similar to Coffin's (1954) and Manson's (1966) formulas.

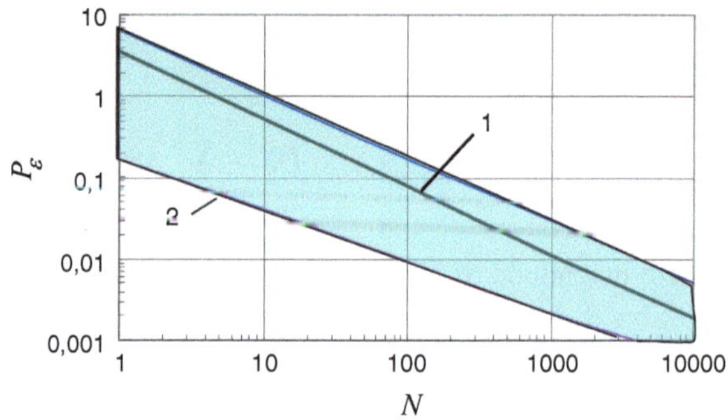

Fig. 7.11 Area of experimental values of deformation parameter P_ε versus cycle number N_d

Factors m and A are obtained by processing and generalizing experimental data of thermocyclic tests in a range of required temperatures for a corresponding class of materials.

In Fig. 7.11, the experimental values area of parameter

$$P_\varepsilon = \Delta\varepsilon\left[\left(\ln\frac{1}{1-\varphi}\right)^{0.6} + \frac{\sigma_B}{E}\right]^{-1} \tag{7.25}$$

in double logarithmic coordinates is presented as dependence on the number N_d for some of the heat-resistant alloys at temperatures $T_{maz} = 700 - 900\,°\mathrm{C}$ (Dulnev's data in Birger and Shorr 1975). Line 1 corresponds to the average, and a line 2—to minimum cyclic durability, for which the factors in Eq. (7.24) have accordingly the values $m \approx 0.7,\ A \approx 3.5$ and $m \approx 0.65,\ A \approx 0.7$.

Due to the tests, the factor m characterizing an inclination of a line $\lg P_\varepsilon = f(\lg N_d)$ changes a little, while the factor A for different materials and temperatures may differ significantly. Therefore, it is recommended to define the value A experimentally at a small number of corresponding thermocycles, for calculating thermocyclic durability at a great number of cycles using the average value $m \approx 0.7$ specified above. For the calculated strain range $\Delta\varepsilon$ and material characteristics σ_B, E, ϕ referring to maximum temperature T_{maz}, one can find under Eq. (7.25) the value P_ε and then the sought quantity

$$N_d = (A/P_\varepsilon)^{1/m}. \tag{7.26}$$

At the combined stress/strain state, equivalent strain ranges in the set of expressions are used.

7.2.3.2 Nonlinear Thermocyclic Damage Summation

Generally, duration prior to failure at thermocyclic loading depends on both the number of cycles and the operating time, especially at temperatures close to the maximum. With increase of exposure at $T_{maz,}$ a number of destruction cycles in comparison with a sawtooth temperature cycles sharply decrease, as shown above in Figs. 7.8, 7.9, and 7.10.

For accounting of the mutual influence of cyclic and long-term static failure, a nonlinear summation of damages can be used (Serensen et al. 1963 and others). Long-term static damage (as a rule, from compression) basically occurs at the maximum temperature T_{maz}. Such damage for a cycle with exposure t_1 at T_{maz} can be estimated as $\tilde{t}_1 = t_1/t_d$ where t_d is a time prior to failure at constant temperature T_{maz}. A low cyclic damage during a unit cycle is estimated as $\tilde{N}_1 = 1/N_d$ where N_d is the number of sawtooth cycles of the considered kind prior to destruction. Nonlinear summation of damages looks like

$$\tilde{t}_1^{\alpha} + \tilde{N}_1^{\beta} = 1 \tag{7.27}$$

where α, β are the experimental factors whose sizes, as a rule, are less than a unit (Fig. 7.11).

For identical cycles with exposures, Eq. (7.27) becomes

$$(\tilde{t}_{\Sigma})^{\alpha} + (\tilde{N}_{\Sigma})^{\beta} = 1, \tag{7.28}$$

where $\tilde{t}_{\Sigma} = Nt/t_d = N\tilde{t}_1$ is a long-term static damage during the total operating time at temperature T_{maz}, and $\tilde{N}_{\Sigma} = N/N_d = N\tilde{N}_1$ is a cyclic damage for N cycles.

Equation (7.28) can be used to find the number N_d of cycles before destruction. At identical values of $\alpha = \beta$, Eq. (7.28) yields

$$N_d = \frac{1}{(\tilde{t}_1^{\alpha} + \tilde{N}_1^{\alpha})^{1/\alpha}}. \tag{7.29}$$

If $\alpha \neq \beta$, the value N_d can be defined from Eq. (7.28) by numerical calculation.

In Fig. 7.12, the averaged experimental dependences between relative quantities $\tilde{t}_{\Sigma} = t/t_d$ and $\tilde{N}_{\Sigma} = N/N_d$ for three heat-resistant alloys within a range of peak temperatures $T_{\max} = (750 - 900)$ °C are plotted (Dul'nev's data in Birger and Balashov 1981). The diagram confirms a nonlinear communication of presented

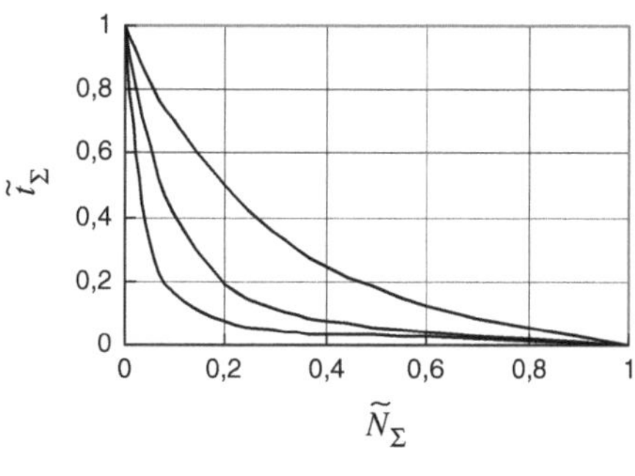

Fig. 7.12 Influence of exposure at peak temperatures on reduction of cyclic durability

parameters and considerable influence at even moderate exposures on cyclic durability reduction.

Generally, damage also occurs in half-cycles of an elastoplastic tension at low temperature. Therefore, as specified above, the summation of damage should be carried out separately for compression and tension half-cycles.

A linear summation of cyclic and long-term damages in most cases is not reliable.

7.2.3.3 Cyclic Creep at Elastoplastic Deformations

In Chap. 4, the problems of low cyclic elastoplastic deformation at constant moderate temperature without accounting for creep have been considered. Tests show that at high temperatures, the influence of cyclic creep can also become significant.

At regimes with rapid increase and reduction of stress and long operating endurance, effects of the plastic flow arising under stress changes and of creep developing over time differ distinctly enough. At relatively slow loading, plastic strains increase simultaneously with creep; therefore, they are difficult for discriminating.

In both cases, the time prior to destruction will depend on the number of cycles loading and accumulation of a long-term damage.

As an example, we will consider a pattern of stress and strain changes at identical thermal cycles in a rod rigidly fixed at the ends. In this case, a total strain ε consisting of the sum

$$\varepsilon = \sigma/E + \alpha_T(T - T_{\min}) + \varepsilon_p + \varepsilon_c = 0 \tag{7.30}$$

is zero during all cycle.

From a periodicity condition, it follows that the inelastic strain component $\varepsilon_p + \varepsilon_c$ should not change for a cycle.

In Fig. 7.13a, dependence of stress σ on the mechanical part of the strain

$$\varepsilon_\sigma = \varepsilon - \alpha_T(T - T_{\min}) = \sigma/E + \varepsilon_p + \varepsilon_c$$

at various portions of a cycle with exposure at the maximum temperature is displayed.

A stabilized thermal cycle is taken identical to considered above in analytical solution (see Sect. 7.1.3.3). The characteristic points of the cycle are marked by the same numerals as in Fig. 7.5. Inelastic strain is assumed to be changed due to creep only during operation at temperature $T_{\max}$ and plastic strain to be increased only near $T_{\min}$. An elasticity modulus is taken a constant.

Reduction (relaxation) of compressing stress at a site 1–2 is caused by creep at constant maximal temperature. At a site 2–3 where temperature falls, unloading at

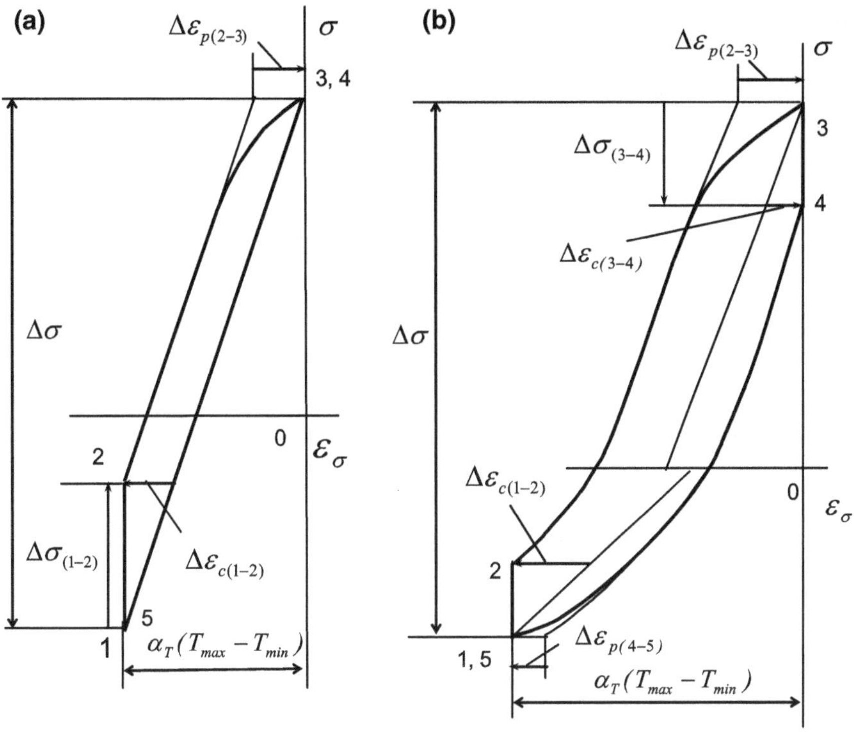

Fig. 7.13 Deformation diagrams for cycles with creep and plastic strains

compression passes into elastoplastic tension with plastic strain increment $\Delta\varepsilon_{p(2-3)}$, and at a site 4–5 where temperature rises, unloading passes into elastic compression. A total of plastic and creep strain increments is $\Delta\varepsilon_{c(1-2)} + \Delta\varepsilon_{p(2-3)} = 0$.

Figure 7.13b corresponds to the more general case when creep increments $\Delta\varepsilon_{c(1-2)}$ and $\Delta\varepsilon_{c(3-4)}$ leading to stress relaxation $\Delta\sigma_{(1-2)}$, $\Delta\sigma_{(3-4)}$, and increase of plastic strain $\Delta\varepsilon_{p(2-3)}$ and $\Delta\varepsilon_{p(4-5)}$ can occur at any temperature. A temperature reduction of the elasticity modulus curving the elastic strain lines is also taken into account. In this case, a total of plastic and creep strain increments is $\Delta\varepsilon_{c(1-2)} + \Delta\varepsilon_{p(2-3)} + \Delta\varepsilon_{c(3-4)} + \Delta\varepsilon_{p(4-5)} = 0$.

To reliably estimate the thermocyclic durability at the joint action of creep and plasticity, it is necessary to reproduce experimentally all cyclic features at trials of samples. This task is extremely difficult.

In the absence of such experimental data, an estimate of the cycle number prior to destruction can be obtained using a full calculated stress range $\Delta\sigma = \sigma^{(3)} - \sigma^{(1)}$, as shown in Fig. 7.13, and the LCF material curve referring to the maximum temperature of a cycle. Some approaches to cyclic creep calculation were considered by Temis (1989).

References

Birger IA (1956) Some mathematical methods for solution of engineering problems. Oborongiz, Moscow
Birger IA, Shorr BF (eds) (1975) Thermal strength of machine parts. Mashinostroenie, Moscow (in Russian)
Birger IA, Balashov BF (eds) (1981) Structural strength of materials and parts of gas-turbine engines. Mashinostroenie, Moscow (in Russian)
Coffin LF (1954) A study of the effects of cyclic thermal stresses on a ductile metal. Trans ASME Ser A 76(6):931–950
Dul'nev RA, Kotov PI (1980) Thermal fatigue of metals. Mashinostroenie, Moscow (in Russian)
Evans RW, Wilshire B (1985) Creep of metals and alloys. Cariton House Terrace, London
Greenwood JN (1949) The influence of vibration on the creep of lead. Proc ASTM 49:834–856
Kablov EN, Golubovsky ER (1998) High-temperature strength of nickel alloys. Mashinostroenie, Moscow (in Russian)
Kennedy AJ (1962) Processes of creep and fatigue in metals. Oliver and Boyd, London
Manson SS (1966) Thermal stress and low-cycle fatigue. McGraw-Hill Company, New York
Rabotnov YN (1966) Creep of construction elements. Physmatgiz, Moscow. Rabotnov YN (1969) Creep Problems in Structure Members (trans). Amsterdam (in Russian)
Serensen SV, Kogaev VP, Shneyderovich RM (1963) Load-ability and strength calculation of machine structures, 1st (3rd edn, 1975) edn. Mashinostroenie, Moscow (in Russian)
Shorr BF (1958) Effect of nonuniform heating under creep conditions on the change in a stress state. Dokl Akad Nauk SSSR 123(5):809–812 (in Russian)
Shorr BF (1964) Cycle creep of nonuniformly heated cylinders. In: Thermal stresses in construction elements, vol 4. Naukova Dumka, Kiev (in Russian)
Shorr BF (1966) Periodical processes and adaptivity at creep. In: Strength and dynamics of aviation engines, 4. Mashinostroenie, Moscow, pp 188–194 (in Russian)
Shorr BF (1970a) Unstable creep design. In: Thermal stresses in construction elements, 9. Naukova Dumka, Kiev, pp 165–173 (in Russian)
Shorr BF (1970b) Cycle creep design of beams using a modified theory of heritable influence (in Russian). In: Thermal stresses in construction elements, 10. Naukova Dumka, Kiev, pp 152–159
Shorr BF, Dul'nev RA (1968) Cycle creep. In: Fridman JaB (ed). Materials strength and deformation in nonuniform physical fields, vol II. Atomizdat, Moscow, pp 34–96 (in Russian)
Shorr BF, Nafikov RM (1963) Cycle creep designs. In: Creep and durability. Sib Otd. AN SSSR, Novosibisk (in Russian)
Temis YM (1989) Plasticity and creep of turbo engines components at cyclic loading. In: Strength and dynamics problems in aviaengines building, CIAM Trudy, 1237(4). Moscow, pp 32–50 (in Russian)

Chapter 8
Creep and Durability at Non-stationary Loading and Heating

Abstract In this chapter, the most validated experimental data for creep and durability of structural materials subjected to unsteady loading and heating are presented. The effect of heredity and adaptability on the material deformation hardening under uni- and multi-axial stress states is demonstrated. A vector theory of stabilized anisotropic-ray hardening (the SARH theory) of a material considering non-isothermal creep under arbitrary loadings, including non-proportional ones, is formulated. The theory is based on the experimentally justified concept of directed and stabilized nature of creep deformation hardening. It develops the results of previous publications on the subject. The design model reveals prominent features of non-steady creep, such as increase in rate of strain development after repeated loadings, spatial variation of the strain rate evolution at rotation of the stress tensor with or without unloading, subsequent stabilizing of the creep rate along a new direction, complete or partial creep inhibition under unloading, and aftereffect. A quasi-steady version of the theory takes into consideration the overall effect of rapid variation of operating regimes on creep evolution.

Detailed reviews of experimental and theoretical researches on creep and durability at variable loading and heating can be found in many works, e.g., Johnson (1960), Kachanov (1960), Shorr and Dul'nev (1964, 1968), Rabotnov (1966), Odqvist (1974), Birger and Shorr (1975), Dul'nev and Kotov (1980), Evans and Wilshire (1985), Temis (2005), and Chaboche (2008).

8.1 Creep at Variable Stresses and Temperatures

8.1.1 Experimental Backgrounds

8.1.1.1 Stepwise Stress Change

"Stepwise" change is understood as a transition from one level of stress to another, while the transition time is essentially smaller than the operation time at constant

B.F. Shorr, *Thermal Integrity in Mechanics and Engineering*,
Foundations of Engineering Mechanics, DOI 10.1007/978-3-662-46968-2_8

stress of each level of loading. Initially, we shall assume that creep is developed without plastic strains.

Stresses of Identical Sign

An elementary test on non-steady creep at constant temperature (Odqvist 1959; Namestnikov and Khvostunkov 1960, and others) comprises of using two-level loading with constant sign: In the beginning, a stress σ_1 operates during the time t_1, then the stress quickly accepts another value $\sigma_2 \neq \sigma_1$: greater, see Fig. 8.1a, or smaller, Fig. 8.1b.

The results of such a test, studied by Rabotnov (1966), are presented in Fig. 8.2. Light points concern experimental data for creep strain ε_c versus time t at $\sigma_1 = 120$ MPa, $\sigma_2 = 160$ MPa, and dark points—to opposite process with $\sigma_1 = 160$ MPa, $\sigma_2 = 120$ MPa. The time t_1 = 24 h, when the stresses have changed

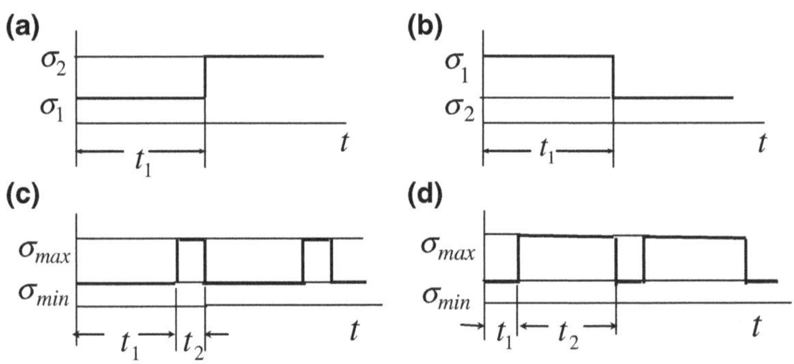

Fig. 8.1 Sketch of stepwise two-leveled creep processes

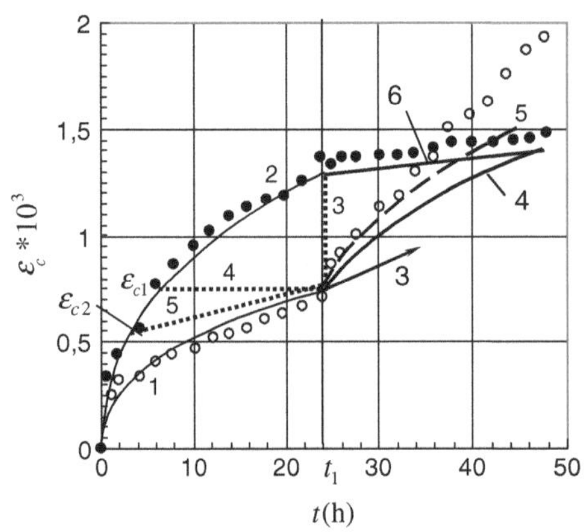

Fig. 8.2 Comparison of experimental and calculated strain curves at two-step creep processes

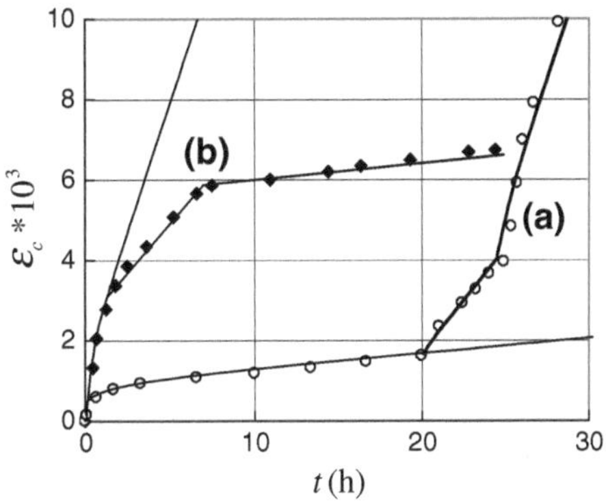

Fig. 8.3 Three-step creep process

their values, is identical. Thin lines at a site from $t = 0$ to $t = t_1$ refer to curves that approximate creep at corresponding constant stresses.

As an example of creep behavior at several steps of stress change, Sizova's (1969) experimental data for the three-stage process of creep for a heat-resistant alloy at a temperature of 900 °C are presented in Fig. 8.3. Light points (a) meet a consecutive stepped stress increase from 270 to 340 MPa and then to 400 MPa, and dark points (b)—stepped stress reduction from 400 to 340 MPa and then to 270 MPa. The creep curves at constant stresses $\sigma = 270$ MPa and $\sigma = 400$ MPa are shown by thin lines.

Other curves in Figs. 8.2 and 8.3 referring to calculated results will be described later.

The presented test data, as with many others, show that each stress increase after sufficiently long work at the previous step leads to a repeating of the initial unsteady stage of the creep, accompanied by the increased creep rate. If the stress decreases, a process is usually decelerated, sometimes to a full temporary stop. In both cases, the creep rate is restored after a while to the value corresponding to a new stress level at the given temperature.

In some tests, the stresses are repeatedly changed by programs $\sigma_{min} \rightarrow \sigma_{max} \rightarrow \sigma_{min}$ shown above in Fig. 8.1c, d with relatively long times t_1 and t_2 of exposure at each stress level, especially at σ_{max}.

The creep accelerates most strongly in the "zero-to-tension" stress cycle $0 \rightarrow \sigma_{max} \rightarrow 0$ at the minimum stress $\sigma_{min} = 0$.

In Fig. 8.4a, some results of Johnson's (1960) experiences are schematically shown. The creep strain ε_c is displayed versus a "regime" operating time $t_{2\Sigma} = \sum t_2$ under which the stress σ_{max} has acted. Creep at repeated change of stress accrues much faster than at the constant stress σ_{max} shown by curve 1.

Curves 2 and 3 concern the processes shown in Fig. 8.1c ($t_2/t_1 \gg 1$) and Fig. 8.1d ($t_2/t_1 \gg 1$), respectively. From Figs. 8.1c, d and 8.4a, it follows that, at reference to the same total time $t_{2\Sigma}$, stepped creep acceleration increases with a reduction of the ratio t_2/t_1 since in this case the repeated creep at the stress σ_{max} will essentially occur corresponding to the conditions of the initial site of unsteady creep where its rate is greatest (curve 2).

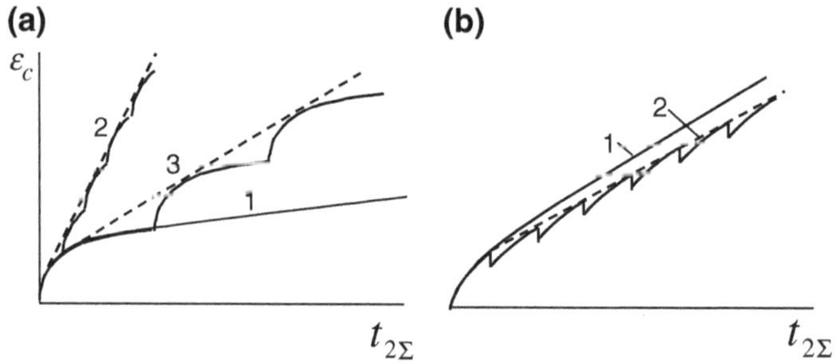

Fig. 8.4 Repeated creep curves under stress alternation $0 \to \sigma_{max} \to 0$

At greater values of ratio t_2/t_1 (curve 3), the average speed of creep at each loading step decreases, and also the effect of repeatability weakens. The increase of repeated stepwise creep rate is especially intensive for a long-time work at moderate stresses and temperatures when the creep rate of the unsteady stage considerably exceeds the rate of the established one. At high levels of stresses and temperatures when creep proceeds with continuously increasing rate, the specified effect is shown poorly.

At constant parameters of loading steps, i.e., at σ_{max} = const, t_1 = const and t_2 = const and a material stable state, repeating unsteady sites of creep curves gradually obtain identical character.

Envelope curves for this process, shown in Fig. 8.4a by dashed lines 2 and 3, are like the usual curve 1 of creep at σ_{max} = const, but with an essentially greater rate of strain development.

Unloading with Rest

In a number of experiments, after unloading and "rest" without cooling, an effect of strain "restitution"(recovery) with partial reduction of residual creep strain was observed (a full restitution to the initial non-deformed state is more typical for non-metallic materials). Creep restitution and an increase in creep rate at repeated loading, operating in the opposite direction, can lead to the delay of visible total rate of strain development as shown in Fig. 8.4b by a dashed curve 2 (in relation to curve 1 for σ_{max} = const) and was observed in Ivanova's (1958) experiences.

Within process sites σ_{min} = 0, there is no creep, but the duration of the "rest" in the unloaded condition may affect further behavior of a material. At very short-term unloading, a material structure has no time to "be restored" and the creep rate at repeated loadings accelerates slower. In this case, the time after unloading (the rest time) happens as a structural parameter.

If minimum stress is no zero ($\sigma_{min} > 0$), the time t_1 of the regime σ_{min} has to be included in the total operating time $t_\Sigma = t_1 + t_2$.

At the transition to lower stress, some additional fall of creep rate $\dot{\varepsilon}_c$ and its sequential restoration to the established level are probable as shown in Fig. 8.5, according to Carreker et al. (1949).

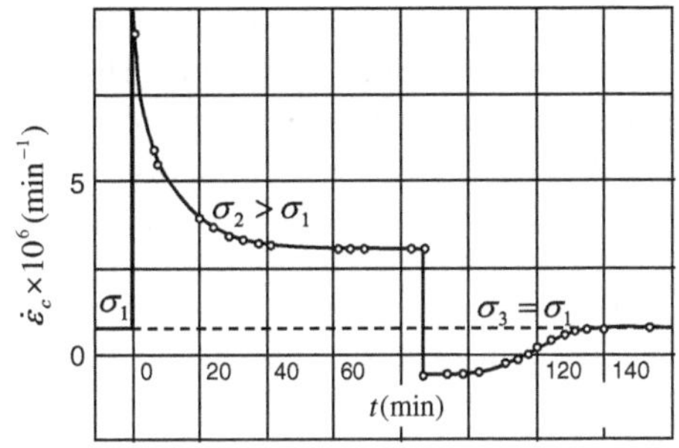

Fig. 8.5 Creep rate following the stress alteration $\sigma_{min} \rightarrow \sigma_{max} \rightarrow \sigma_{min}$

Creep of Combined Stress State

There is limited information available about the creep of combined stress states with a stepwise change.

Namestnikov (1963) describes results of tests for a simultaneous proportionally stepped increasing and decreasing of normal and tangential stresses at tension and torsion of cylindrical samples. The creep curves practically did not differ from uniaxial tension shown in Fig. 8.2.

In other tests by Namestnikov (1957) and also Johnson et al. (1958), studying non-proportional loading, the samples were extended during some time, and then exposed to torsion (and vise versa), while the stress intensity remains the same. No effect of hardening from a preliminary tension on the subsequent torsion (and on the contrary) was noted.

Sign-varying Stresses

Stepwise creep investigation with alteration of stress signs were carried out by Katz (1957), Kennedy (1965), Getzov (1969), Sizova (1969), also Sizova in Birger and Shorr (1975), and others. There are some complications, arising during tension–compression tests with sign-varying stresses. For example, sometimes tests should be carried out with specimen being moved from one testing machine to another using intermediate cooling. Similar simpler tests can be achieved at reverse torsion (Namestnikov 1960).

8.1.1.2 Stepwise Temperature Change

For many machine parts operating in conditions of creep development, periodic change of loading and heating, accompanied by periodic alteration of stresses and temperature fields, is rather typical. During specimen testing, the simultaneous change of stresses and temperatures are reproduced; however, in practice, these fields do not always change simultaneously. Keeping stress strictly constant in a specimen during an entire test is rather difficult due to the varying thermal stresses caused by non-stationary heating. Thus, some creep tests at changing temperatures may be conditionally considered to reflect only temperature changes.

At constant stress and cyclic two-level temperature change $T_{\min} \leftrightarrow T_{\max}$, the creep rate, as a rule, increases in comparison with test at $T_{\min}$, and yet sometimes in relation to test at average temperature $T_m = 0.5(T_{\min} + T_{\max})$.

Temperature change that may occur in the course of tests affects the deformation and strength material characteristics rather variously. In particular, some experiences show that after specimens have been cooled without unloading, the creep will develop after repeated heating to a former temperature as if a continuous temperature was maintained (Malinin 1952). On the other hand, upon transitioning from an elevated temperature to a higher one, unsteady creep can occur anew and creep rate grows, as well as with an increase of stress.

Some experimental results for sign-varying cyclic creep have been presented in Figs. 7.1–7.4 of Chap. 7.

8.1.2 Simplest Approaches to Creep Design

Tensile trials of double-step (or, more seldom, three-step) creep at constant temperature, schematically shown in Fig. 8.1, are widely used as the simplest control condition for accurately testing theoretical approaches to creep analysis.

Numerical results are compared with the test data, of which an example has been presented above in Figs. 8.2 and 8.3.

8.1.2.1 Aging Theory

This theory (see Sect. 6.2.1.2), in which an operating time t is taken as a single structural parameter determining a creep process, is the simplest creep theory. Due to this approach, the creep rate after transition at the moment t_1 to the stress $\sigma_2 > \sigma_1$ (see Fig. 8.2) should be found as a point on curve 2 corresponding the same time moment t_1 (a dotted vertical line 3) and creep strain should follow arrow 3 in Fig. 8.2. This result sharply disperses from experimental data depicted by light dots.

Therefore, the use of this theory even at stepwise creep, not considering more complicated processes, is unacceptable.

8.1.2.2 Flow Theory

As in the yielding theory of plasticity, a flow theory in creep also finds application, but preferably for smoothly varying loads and temperatures since it is based on the correlation $\dot{\varepsilon}_c = f(\sigma, T, t)$, see Sect. 6.2.1.2.

At rapid stress change from σ_1 to σ_2, creep rate increases just as in the aging theory and has the similar deficiencies.

Duration $\Delta t = t - t_*$ after a specified moment (e.g., after moment of full unloading) can sometimes mirror the real temporal properties of a material shown during some time period.

8.1.2.3 Theory of Hardening

A more widespread and rather simple method of the creep design is an approach based on the theory of hardening (see Sect. 6.2.1.2). According to this theory, a unique structural parameter of creep process is a current strain ε_c reached to the moment of consideration (Davenport 1938; Rabotnov 1948, and others).

In this case, for describing of stepwise transition at the time moment t_1 from the smaller stress σ_1 to the higher σ_2, one should find a point on the curve σ_2 corresponding to the value of the creep strain ε_{c1} that has been reached earlier at the stress σ_1 (dotted horizontal line 4 in Fig. 8.2).

The further process of creep will be assumed proceeding along a curve that is equidistant to the continuation of curve $\sigma_2 = \text{const}$ from the point ε_{c1}, as shown in Fig. 8.2 by bold line 4.

The continuation of the creep process during the transition from higher to low stress (bold line 6 in Fig. 8.2) must be similarly defined; however, this is sometimes impossible since the value of ε_{c1} becomes too large.

From Fig. 8.2, one can see that the comparison of computed and experimental curves referring to two-step sign-constant creep shows good qualitative and satisfactory quantitative conformity, at least near to a point of transition from one regime to another. Using such an approach, the equations of creep rate can be presented in any form convenient for computation, e.g., as shown by Eqs. (6.11a, b) of Chap. 6.

Work of Plastic Strain

Even better describing of the transition from one stress level to another can be obtained if the *work* of the previous plastic strain W_c is used as the structural creep parameter of hardening. Under transition from smaller stress σ_1 to greater σ_2, this work for the given example is equal to $W_{c1} = \sigma_1 \varepsilon_{c1}$. A point 5 in Fig. 8.2 at curve $\sigma_2 = \text{const}$, in which the creep strain is $\varepsilon_{c2} = \varepsilon_{c1}\sigma_1/\sigma_2$, corresponds to the same work W_{c1}. Further calculation must be continued, beginning from this point along the curve, equidistant to curve $\sigma_2 = \text{const}$ as it has been shown by dashed line 5.

Under transition from greater stress σ_2 to smaller σ_1, it is necessary to find a point on curve $\sigma_1 = \text{const}$ where $\varepsilon_{c1} = \varepsilon_{c2}\sigma_2/\sigma_1$, (if such point exists) and continue the computation as above.

The simple hardening theory with structural parameter presented both in creep strain, and in work of plastic deformation, rather adequately describes creep processes at constant-sign loading within the unsteady creep area. But this theory also does not reflect a number of substantial effects of unsteady creep: A significant rate increase at stepwise stress transition for sign-alternating loading, rate changes at stepwise stress loading for the stabilized area within which creep has constant rate for each of the fixed stress, and others.

The advance theory of creep hardening taking into account the mentioned and other practically important effects is described below.

8.2 Model of Creep-Stabilized Anisotropic-Ray Hardening

In works on solid mechanics, it is acceptable to name processes of a plastic yielding and creep connected with the development of appropriate inelastic strains by the common term "hardening." So as not to disturb this tradition, we name the approach considered below by the analogous term, as in Chap. 5, *the model of creep stabilized anisotropic hardening* (SAH).

A part of the equations of the SAH creep and analogous equations of the SAH plasticity theories are alike. However, to facilitate independent usage of Chaps. 8 (on creep) and 5 (on plasticity), the author has preferred to briefly repeat some formulations.

8.2.1 Empirical and Theoretical Backgrounds

The general backgrounds of the SAH theory regarding the inherently isotropic material have been stated in application to the theory of plasticity in Sect. 5.3.3 of Chap. 5. Those statements remain valid with reference to a creep problem with certain modification and additions.

8.2.1.1 Deformation Anisotropy in Creep

Anisotropic Hardening

Inherently isotropic materials become anisotropic over the course of development of inelastic deformation affected by the directional changes of material structure. These changes may be both reversible and irreversible, depending on accumulated strains or regime temperature and duration.

As a result, a creep rate induced by the loads of the same magnitude occurs differently in various (including opposite) directions.

The structural parameters that most closely reflect specified changes in material are creep strain, work of creep strain, and some associated quantities. Approaches to theories of creep based on the assumption of *anisotropic hardening* were justified by many researchers (see e.g., Malinin and Khadjinsky 1969, 1972).

Axes of Anisotropy

The quantitative material characteristics (such as elasticity moduli, dependencies of the stationary creep rate and the unsteady creep strain "jumps" on stresses, and also strength and durability limits) have isotropic character, i.e., under repeated loading along any direction their magnitudes have identical values.

Other characteristics of a material behavior (the rate of unsteady creep and its alteration under changes of loading regimes) are linked with deformation hardening of a material in a certain direction and have therefore anisotropic character. Current direction of the hardening process rate affected by the current direction of the stress vector $\boldsymbol{\sigma}_i$ has been named in Chap. 5 "*an axis CH of the current anisotropic hardening.*" However, a rate of the unsteady creep magnitude depends also on the axis CH veering in relation to the *former* direction of the unsteady strain vector $\boldsymbol{\varepsilon}_i^*$. We have named the latter direction "*an axis FH of the former anisotropic hardening.*" An analogous axis in theory of plasticity has been affected by the direction of the unsteady part of the micro-stresses.

Isotropic and Anisotropic Parts of Creep

In order to justify the theoretical approach to creep analysis at arbitrary loadings, we will start with the premises founded on the effects of restricted experiments (partially described above) and some logical deductions, considering the above-mentioned reasons.

- The stationary part of creep has isotropic character and develops under the same stress over any direction with the equal rate. The vector of a stationary creep rate is defined uniquely in magnitude and direction by a current stress vector, i.e., does not depend on the previous loading history.
- The non-stationary part of creep has anisotropic character and at an identical magnitude of a stress vector develops in different directions with various rates depending on previous loading history. The *direction* of the rate vector $\mathbf{v}_i^* = \dot{\varepsilon}_i^*$ is determined by the current direction of the stress vector CH, coinciding with it under active loading. The conditions for determine a *magnitude* of the vector $v_i^* = \left|\mathbf{v}_i^*\right|$ will be considered below.

8.2.1.2 Heredity Influence on Material Behavior

Creep behavior of a material depends not only on a strain quantity referring to the current moment, but also on the peculiarity of previous processes of stress–strain state alteration. Such dependence is usually named *heredity of material properties.*

The influence of heredity on material behavior was considered, to some extent, in a number of procedures with reference to linearly viscous-elastic polymers.

Hereditary influence on metal creep was studied by Rozovsky (1951), Arutjunjan (1952), Rabotnov (1966), and some others.

The use of accumulated creep strain ε_c or the corresponding work W_c as the structural parameters in the theory of hardening only particularly reflects this influence. Experiments show that the hereditary influence of preceding sign constant loadings and headings gradually fades, following which the creep precedes as stabilized with constant rate up to the third stage of creep. At repeated stress and temperature alterations, the non-stabilized creep occurs again each time, and its rate can depend on a number of factors.

Various approaches to describe material behavior both at transient and steady regimes were discussed in Malinin (1952, 1975), Malinin and Khadjinsky (1969, 1972), Murakami and Ohno (1982), Chaboche (1989), Lemaitre and Chaboche (1990, 2004), Temis (2005), and others. Some approaches to modeling unsteady creep behavior were considered by Shorr (1964, 1966, 1970a, b) and also in Birger and Shorr (1975). Its substantive provisions are then added and explicated, taking into account further theoretical and experimental results published within the last years.

A theoretical approach more completely reflective of the fundamental characteristics of metal creep behavior but at the same time adjustable to practical application is described below.

8.2.1.3 Steady and Unsteady Rates of Creep

Steady Rate of Creep

To reflect as far as possible the specified material properties, a total rate of creep strain v for any moment of time is presented as the sum of steady v_{st} and unsteady v_* parts. At a primary uniaxial loading

$$v(t) = v_{st}(t) + v_*(t). \tag{8.1a}$$

At multi-axial loading, the analogous equality is written in terms of strain rate intensities

$$v_i(t) = v_{st,i}(t) + v_{*i}(t). \tag{8.1b}$$

In the creep hardening theory, a concept "time" acts in two functions discussed in Chap. 1: As the index of the process moment—in the equations defining varying values of deformation parameters, i.e., $v(\sigma) = f[\sigma(t)]$, and as a physical quantity—in the equations reflecting the duration of the process, i.e., $\varepsilon(t) = \varepsilon_{in} + f(t - t_{in})$. However, in the latter case, creep parameters referring to the stable state of a material are affected not by entire process time, but only by its duration between changes of loading direction.

In Eq. (8.1a, b) and further the script (c) at creep strain is absent if there will be no necessity in it. Dependence of material characteristics and creep course on temperature is assumed to be apparent and is not marked explicitly.

Integration of Eq. (8.1a) at $\sigma = \text{const}$ yields

$$\varepsilon(t) = v_{st}t + \varepsilon_*(t). \tag{8.2}$$

The steady part v_{st} corresponds to the minimal strain rate of the secondary creep stage and the unsteady part $v_*(t)$—to the varying excess of total rate at the primary creep stage.

The value of steady creep ε_{st} does not change after unloading and cooling and begins again changing with new growth of the stresses. Its rate $v_{st}(\sigma)$ depends only on current values of stress and temperature.

Unsteady Fading Rate of Creep

After rapid stress increase (as a "jump") from zero to a quantity σ, the unsteady rate v_* also rapidly increases from zero to $v_*(\sigma)$ and then at $\sigma = \text{const}$ gradually fades over time to zero. The fading of unsteady creep rate leads to the process stabilization with the rate $v(\sigma, t) \to v_{st}(\sigma)$, constant in magnitude and direction. This effect mirrors material "*adaptability*" to permanently operating factors (as well as in case of plastic yielding).

Stabilization bears an asymptotic character; a finish magnitude of unsteady creep strain at which there comes stabilization we designate as $\varepsilon_{*\infty}$. In common case, this strain depends on creep process. At the primary uniaxial loading

$$\varepsilon_{*\infty} = (\text{sign}\sigma)\varepsilon_{\lim}(\sigma), \tag{8.3}$$

where $\varepsilon_{\lim}(\sigma)$ is a material characteristic limiting the area of unsteady creep. It depends on stress and temperature and coincides with the value that has been considered in Chap. 6 as "inherent" value $\varepsilon_{c0}(\sigma)$ for stepped representation of a creep diagram.

The quantities of the strains $\varepsilon_{\lim}(\sigma_1) = \varepsilon_{0.1}$ and $\varepsilon_{\lim}(\sigma_2) = \varepsilon_{0.2}$ can be found, prolonging rectilinear sites of creep curves up to the intersection with the axis of ordinates, as shown below in Fig. 8.12 for stresses σ_1 and σ_2. Making this procedure for some stresses, we obtain a positive function $\varepsilon_{\lim} = f(\sigma)$. For real operating domains, the values of strains $\varepsilon_{\lim}$ usually increase with an increase of stresses σ, however, at very high temperatures and stresses the signs of $\Delta\varepsilon_{\lim}$ and $\Delta\sigma$ do not always coincide. Since the role of non-stabilized creep at those conditions becomes uncertain, we propose taking $\Delta\varepsilon_{\lim} = 0$ if $\Delta\varepsilon_{\lim} < 0$ at $\Delta\sigma > 0$.

Generally, the finish unsteady strain $\varepsilon_{*\infty}$ depends on both the current stress and the process prehistory, whose influence can be described by one or several *inner variables* q_{var}.

Direction of the steady strain rate v_{st} always coincides with stress direction. As of the rate of unsteady strain v_*, its direction after unloading may be opposite to stress due to phenomenon of restitution.

The allocation of the unsteady part of creep strain in the absence of restitution was considered by Odqvist (1953). Malinin (1975) additionally proposed to divide the unsteady creep strain into two parts: reversible and irreversible.

8.2.1.4 Two Phases of Creep Step-by-Step Computation

Operating conditions, in which rapid changes of regimes in any time instant alternate with long-time operation at almost invariable stresses and temperatures, have the basic practical value. For computation target, we shall regard each step of any creep process during a time Δt also as a sequence of two phases: (i) rapid change ("jump") of stresses $\Delta\sigma_i$ and temperatures ΔT with creep strain intensity $\varepsilon_i = \text{const}$ and (ii) creep strain evolution over time $\Delta\varepsilon_i = f(t)$ under new stresses $\langle\sigma_i\rangle = \sigma_i + \Delta\sigma_i$ and temperatures $\langle T\rangle = T + \Delta T$ invariable during this step.

We assume that stress-dependent material characteristics $\varepsilon_{\lim}(\sigma)$ and others vary at these jumps *simultaneously* with the stress vector $\boldsymbol{\sigma}_i$ and temperature alteration. This allows a magnitude and direction of the "after-jump" creep rate vector $\mathbf{v}_i$ to be calculated. Then the creep gradually changes during given step time Δt under new constant values of the stresses, temperatures, and creep parameters.

In common case, the stress and temperature jumps using for step-by-step computation of any creep process are assumed to be small. If they are considerable at actual change of an operating regime, the regime ought to be divided into several parts.

8.2.2 *Unsteady Uniaxial Creep*

8.2.2.1 Governing Parameters of the Creep Process

It is useful to present a current value of creep strain ε as a sum of the steady strain ε_{st} and two parts of unsteady strain: A "transmitting" strain ε_{*0} referred to current position of the center of the unsteady strain area and "stabilized" part ε_*

$$\varepsilon = \varepsilon_{st} + \varepsilon_{*0} + \varepsilon_*. \tag{8.4}$$

For the primary loading, $\varepsilon_{*0} = 0$ and $\varepsilon = \varepsilon_{st} + \varepsilon_*$.

An absolute value of stabilized strain ε_* varies within the range $0 \le |\varepsilon_*| \le \varepsilon_{\lim}$. The quantity $\varepsilon_{\lim} = f[\sigma(t)]$ is a half of maximum possible change ("jump") of the unsteady strain for current calculation step.

At the beginning of the first step when $\sigma = 0$, the quantities are $\varepsilon_{*0} = \varepsilon_{\lim} = \varepsilon_* = 0$. In common case, a current jump of the loading step starts from nonzero quantities $\sigma(t)$, $\varepsilon_{\lim}(t)$, and $\varepsilon_{*0}(t)$.

Furthermore, we shall write $\varepsilon_{\lim}$ instead of $\varepsilon_{\lim}[|\sigma(t)|]$ and ε_{*0} instead of $\varepsilon_{*0}(t)$. Process parameters at the termination of the first phase are marked as $\langle\varepsilon_{*0}\rangle = \varepsilon_{*0} + \Delta\varepsilon_{*0}$, $\langle\varepsilon_{\lim}\rangle$, etc.

8.2.2.2 Phase of Creep Parameters Alteration

Let us consider the stress σ and unsteady strain parts ε_* and ε_{*0} at the beginning of the current step to be known from the proceeding calculation steps, and a value of stress increment $\Delta\sigma$ to be set.

A new value of the limiting strain corresponding to new stress value $\langle\sigma\rangle = \sigma + \Delta\sigma$ is $\langle\varepsilon_{\lim}\rangle$. Then, there are $\langle\varepsilon_*\rangle = \varepsilon_* + \Delta\varepsilon_*$, $\langle\varepsilon_{*0}\rangle = \varepsilon_{*0} + \Delta\varepsilon_{*0}$, $\varepsilon_{*\infty} = \varepsilon_{*0} + (\operatorname{sign}\sigma)\langle\varepsilon_{\lim}\rangle$, and $\Delta\varepsilon_{\lim} = \langle\varepsilon_{\lim}\rangle - \varepsilon_{\lim}$.

During the first calculation phase, a total unsteady creep strain does not change, i.e.,

$$\Delta\varepsilon_{*0} + \Delta\varepsilon_* = 0. \tag{8.5}$$

To satisfy experimental data, increments of the unsteady creep parameters $\Delta\varepsilon_{*0}$ and $\Delta\varepsilon_*$, and also a new value of stabilized strain $\langle\varepsilon_*\rangle$ might be determined by the following relations:

(a) In case of the creep process occurring *without stress sign varying*, when

$$\sigma\langle\sigma\rangle \geq 0 : \tag{8.6}$$

(a1) for *active* loading, when relations $\sigma\Delta\sigma \geq 0$ and $\Delta\varepsilon_{\lim} \geq 0$ are valid, both transmitting and stabilized strains do not change,

$$\Delta\varepsilon_{*0} = \Delta\varepsilon_* = 0, \tag{8.7}$$

$\langle\varepsilon_*\rangle = \varepsilon_*$ and $\langle\varepsilon_{*0}\rangle = \varepsilon_{*0}$,

(a2) for *unloading,* when inequalities $\sigma\Delta\sigma < 0$ and $\Delta\varepsilon_{\lim} < 0$ are valid, the change of these unsteady strain parts may be calculated as

$$\Delta\varepsilon_{*0} = -\Delta\varepsilon_* = -(1-\zeta)\varepsilon_*\overline{\Delta\varepsilon_{\lim}}, \tag{8.8}$$

where $\overline{\Delta\varepsilon_{\lim}} = \Delta\varepsilon_{\lim}/\varepsilon_{\lim} < 0$.

The ratio $0 < |\overline{\Delta\varepsilon_{\lim}}| \leq 1$ shows a relative change of unsteady strain jump at partial unloading. There is $\overline{\Delta\varepsilon_{\lim}} = -1$ for full unloading to $\langle\sigma\rangle = 0$. In this case, the limit strain is zero, $\langle\varepsilon_{\lim}\rangle = 0$.

A material restitution factor ζ varies within $0<\zeta\leq 1$. At full restitution, it is $\zeta=1$, then we have $\Delta\varepsilon_{*0}=\Delta\varepsilon_*=0$ that leads to $\langle\varepsilon_*\rangle=\varepsilon_*$. After full unloading without restitution, $\zeta=0$, $\Delta\varepsilon_*=-\varepsilon_*$ and the unsteady part of the strain becomes zero, $\langle\varepsilon_*\rangle=0$.

(b) The creep process *with varying stress sign*,

$$\sigma\langle\sigma\rangle<0. \tag{8.9}$$

The calculation step can be presented as a sequence of a full unloading, for which Eq. (8.8) is valid, and new active loading in the opposite direction described by Eq. (8.7).

8.2.2.3 Phase of Creep Unsteady Strain Evolution

Due to the idea that the fading rate of the unsteady creep strain v_* should be proportional to the difference between the limiting and running strain values, we assume

$$v_*(t)=\dot{\varepsilon}_*(t)=\beta\{[\mathrm{sign}\sigma(t)]\varepsilon_{\mathrm{lim}}(t)-\varepsilon_*(t)\}. \tag{8.10a}$$

At the beginning of the second calculation phase, Eq. (8.10a) yields

$$\langle v_*\rangle=\beta[(\mathrm{sign}\langle\sigma\rangle)\langle\varepsilon_{\mathrm{lim}}\rangle-\langle\varepsilon_*\rangle], \tag{8.10b}$$

where unsteady creep parameters $\langle\varepsilon_{\mathrm{lim}}\rangle$ and $\langle\varepsilon_*\rangle$, calculated due to previous Sect. 8.2.2.2, are the initial parameters which allows the solution of Eq. (8.10a) to be obtained.

Within some stress and temperature ranges, the factors of creep stabilization β and restitution ζ may be averaged.

Under active loading, the rate $\langle v_*\rangle$ can vary from zero to $2\beta\langle\varepsilon_{\mathrm{lim}}\rangle$. After full unloading, $\langle v_*\rangle=-\beta\langle\varepsilon_*\rangle$.

Solution of Eq. (8.10a) for constant value of the stress $\langle\sigma\rangle>0$, i.e., for $\langle\varepsilon_{\mathrm{lim}}\rangle=\mathrm{const}$, with initial condition $\varepsilon_*(t)=\varepsilon_{*in}$ at $t=t_{in}$ yields

$$\varepsilon_*(t)=\varepsilon_{*in}+(\langle\varepsilon_{\mathrm{lim}}\rangle-\varepsilon_{*in})\{1-\exp[-\beta(t-t_{in})]\}. \tag{8.11}$$

In the course of time, at $\beta(t-t_{in})\geq 4.5$, the value of stabilized part of the unsteady strain aspires to the limiting value $\langle\varepsilon_{\mathrm{lim}}\rangle$ for any initial strain ε_{*in}.

During a small step Δt, a total creep strain changes for a quantity

$$\Delta\varepsilon\approx(\langle v_{st}\rangle+\langle v_*\rangle)\Delta t+\Delta\varepsilon_{*0}. \tag{8.12}$$

For usual creep test when a constant stress σ affects a specimen practically right after loading at the initial time $t_{in}=0$, Eq. (8.11) yields

$$\varepsilon_*(t) = \varepsilon_{\lim}(\sigma)[1 - \exp(-\beta t)], \tag{8.13}$$

and the initial creep rate is $v_*(0) = \beta\varepsilon_{\lim}(\sigma)$.

Knowing a test value $\varepsilon_{\lim}(\sigma)$, we find from Eq. (8.13)

$$\beta = \frac{1}{t}\ln\frac{1}{1-\kappa(t)}, \tag{8.14a}$$

where

$$\kappa(t) = \frac{\varepsilon_*(t)}{\varepsilon_{*\infty}}. \tag{8.14b}$$

Just as in a formally analogous task of the theory of plasticity, see Eqs. (5.24)–(5.25), the averaged quantity of the factor β can be calculated using values of time instants t and corresponding values of the experimental curve $\varepsilon_*(t)$ in some points within the range $0 < \varepsilon_*(t) < 1$. The factor β can also be found as $\beta = v_*(0)/\varepsilon_{\lim}$; however, accurate estimation of the rate $v_*(0)$ from a creep diagram is problematical.

8.2.2.4 Calculating Procedure

In Fig. 8.6, calculating procedures due to SAH theory for two typical situations are demonstrated: (a) the first loading and then for repeated loading to the previous stress after unloading, and (b) the transition to the opposite stress of the same level. Curves in Fig. 8.6c, d refer to the creep without restitution, $\zeta = 0$, in Fig. 8.6e, f—to the full restitution, $\zeta = 1$. Bold lines refer to unsteady creep total strain $\varepsilon_{*0} + \varepsilon_*$, thin lines correspond to strain $\varepsilon_{*\infty} = \varepsilon_{*0} + \varepsilon_{\lim}$, dashed ones—to the position of the point ε_{*0}. Little circles indicate positions of the points $\varepsilon_{\lim} = 0$ for moments t_* and t_{**}, where $\sigma = 0$. Arrows show the direction of stabilized strain ε_* evolution during the second phase of the calculating procedure.

Comparison of total creep strain ε calculated using the described approach with empirical data (Samarin and Sorokin 1969) is displayed in Fig. 8.7. Specimens of a heat-resistant alloy were tested at constant temperature and repeated loadings with unloadings and intermediate rest (Fig. 8.7a). Averaged experimental data are shown Fig. 8.7b by bold curves. The blacked out fields show a scatter of calculated results obtained in view of the scatter of primary experimental creep curves due to the tests of just over 30 specimens.

In Fig. 8.8, Namestnikov's (1960) test results are presented by bold curves. A tubular specimen was initially affected by a constant torque for duration of 50 h (curve 1), then after unloading it was repeatedly loaded by the same torque but in the opposite direction. This is shown by curve 2 where the absolute values of the deference $\gamma(t) - \gamma(50)$ are displayed versus time t—50 h. Calculated results obtained by $\beta = 3.33\ \ 1/h$ and $\zeta = 0.833$ are shown by light circles and quadratics.

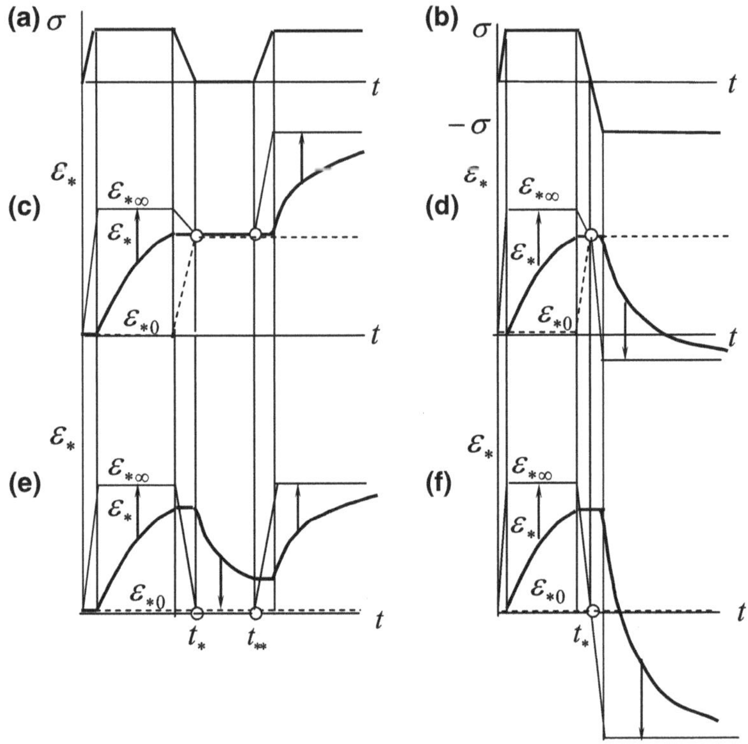

Fig. 8.6 Examples of calculations due to SAH theory: **a** repeated loading after unloading, **b** transition to opposite stress

The presented computational results rather adequately reflect different peculiarities in unsteady creep processes.

8.2.3 Unsteady Multi-axial Creep Under Proportional Loading

Further, for simplification, some subscripts will be transposed in superscripts. The dependence of the materials properties on temperature is always assumed.

8.2.3.1 Unsteady Multi-axial Creep

Above in Sect. 8.2.1, a concept of two axes CH and FH of anisotropic hardening has been introduced. Under sign-constant proportional loading, directions of both axes coincide. Under sign-varying proportional loading, the stress and strain vectors

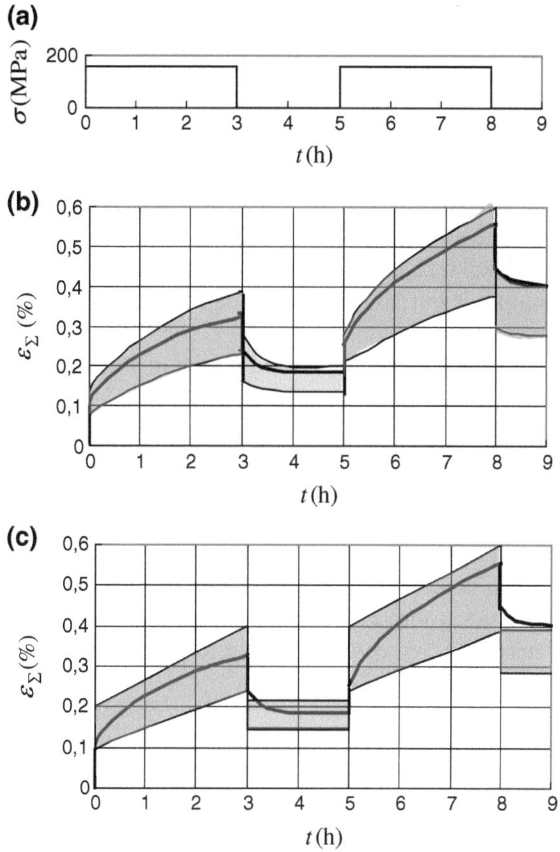

Fig. 8.7 Comparison of calculated and experimental results of non-stationary creep under repeated step-wise loading; **a** stress versus time, **b** theory of stabilized anisotropic hardening, $\zeta = 0.5$; **c** quasi-stabilized approach (see Sect. 8.2.5.2)

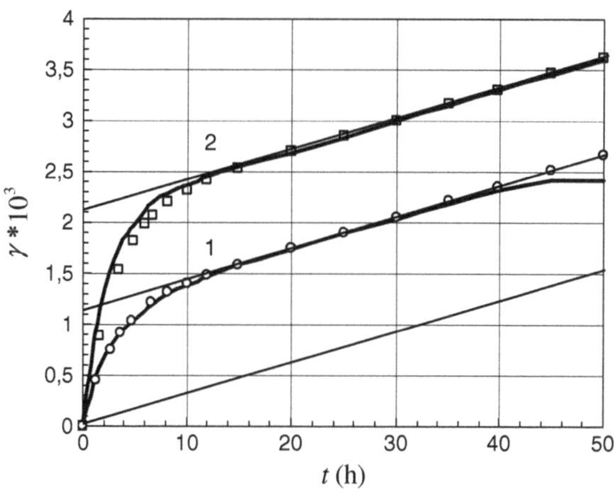

Fig. 8.8 Comparison of experimental data and computational results for sign-varying loading in torsion

are collinear, projections of all vectors onto the CH axis coincide in magnitude with their intensities but may be differ in sign.

Theory of the steady multi-axial creep has been considered in Chap. 6. As in the theory of plasticity, experimental data on multi-axial creep under proportional loading allow accepting the correlations between the stress and strain intensities in a form similar to the case of uniaxial tension/compression. Analogous correlations for the unsteady sign-varying creep deal with projections of the corresponding vectors onto the CH axis.

Material characteristics—a rate of the steady creep strain v_i^{st} and a value of the limiting stabilized strain $\varepsilon_{\lim}$—are assumed to be given as the functions of the stress intensity σ_i.

Equation (8.4) presented in a form of strain parts projections onto the CH axis is

$$\hat{\varepsilon}_i = \hat{\varepsilon}_i^{st} + \hat{\varepsilon}_{i0}^{*} + \hat{\varepsilon}_i^{*}. \tag{8.15}$$

When the stresses alter the sign, i.e., vector $\boldsymbol{\sigma}_i$ turns on 180°, the CH axis also turns after it. This moment the FH axis do not change, so as the strain vectors $\boldsymbol{\varepsilon}_i$ and all its parts do not vary while their projections $\hat{\varepsilon}_i$ and also the others alter their signs.

The stress projection $\hat{\sigma}_i = 3s_{ij}s_{ij}/2\sigma_i$, numerically equal to the stress intensity σ_i, is always positive. Projections of vectors $\boldsymbol{\Delta\sigma}_i$ and $\boldsymbol{\varepsilon}_i$, denoted by $\Delta\hat{\sigma}_i$ and $\hat{\varepsilon}_i$, are determined, respectively, by the formulas

$$\Delta\hat{\sigma}_i = \frac{3\Delta s_{ij}s_{ij}}{2\sigma_i} \tag{8.16a}$$

and

$$\hat{\varepsilon}_i = \frac{\varepsilon_{ij}s_{ij}}{\sigma_i}. \tag{8.16b}$$

Analogous relations are valid for the rates of the creep strain v_i and its parts v_i^{st} and v_i^{*}, e.g.,

$$\hat{v}_i^{*} = \frac{v_{ij}^{*}s_{ij}}{\sigma_i}. \tag{8.17}$$

For the first step of loading, which starts from the unstressed state $\sigma_i = 0$ with strain $\varepsilon_i^{*} = 0$ and terminates at $\langle\sigma\rangle_i = \Delta\sigma_i$ and $\langle\varepsilon_i^{*}\rangle = \Delta\varepsilon_i^{*}$, initial direction of the CH axis is defined by the direction of the stress increment $\Delta\sigma_i$.

8.2.3.2 Reversible Directions of Loading

In order to analyze the proportional loading, when the direction of the vector $\boldsymbol{\sigma}_i$ alternates, it is useful to pay attention to the following:

While at the beginning of the current step, we take the stress projection $\hat{\sigma}_i \geq 0$, its increment $\Delta\hat{\sigma}_i$ defined by Eq. (8.16a), and also a sum $\langle\hat{\sigma}_i\rangle = \hat{\sigma}_i + \Delta\hat{\sigma}_i$ may be of any sign. Therefore, conditions (8.6) and (8.9) may be written in the forms $\langle\hat{\sigma}_i\rangle \geq 0$ and $\langle\hat{\sigma}_i\rangle < 0$, respectively. However, direction of the CH axis in the last case is assumed to be changed during the first phase of creep development and the projection $\langle\hat{\sigma}_i\rangle$ again becomes positive.

For full unloading to $\langle\hat{\sigma}_i\rangle = 0$, Eq. (8.8) yields

$$\Delta\hat{\varepsilon}_{i0}^* = -\Delta\hat{\varepsilon}_i^* = (1-\zeta)\hat{\varepsilon}_{*i}. \tag{8.18}$$

and a new projection of the stabilized strain for the first phase termination is$\langle\hat{\varepsilon}_i^*\rangle = \zeta\hat{\varepsilon}_i^*$.

Equations (8.10b) and (8.12) written in terms of strain intensities yield, respectively,

$$\langle\hat{v}_i^*\rangle = \beta(\langle\varepsilon_{\lim}\rangle - \langle\hat{\varepsilon}_i^*\rangle) \tag{8.19}$$

and

$$\Delta\hat{\varepsilon}_i \approx (\langle\hat{v}_i^{st}\rangle + \langle\hat{v}_i^*\rangle)\Delta t + \Delta\hat{\varepsilon}_{i0}^*. \tag{8.20}$$

In the second phase of the step, all strain components alter simultaneously and can be determined by the equations similar to Eq. (6.44a), e.g.,

$$\langle\Delta\varepsilon_{ij}\rangle = \frac{3}{2}\frac{\langle\Delta\varepsilon_i\rangle}{\langle\sigma_i\rangle}\langle s_{ij}\rangle. \tag{8.21}$$

Namestnikov's (1963) tests had confirmed that material behavior at multi-axial stepwise proportional stressing demonstrated the same creep development as at uniaxial loading/unloading.

8.2.4 Unsteady Creep Under Non-proportional Loading

8.2.4.1 Various Cases of Non-proportional Loading

Non-proportional loading can be most simply carried out by unloading of a specimen to $\sigma_i = 0$ and subsequent loading by another system of forces (for example, extension followed by torsion). The stress axis CH turns at the same moment when the forces change and then it remains constant. The unsteady strain axis FH is turning gradually in a new direction during the process of creep evolution.

Another extreme case corresponds to the gradual change of correlations between the components of a stress vector so that its magnitude (the stress intensity) remains constant while directions of the anisotropy axes are continuously altering.

In general case, the stress vector can change during each step both in direction and magnitude. The creep process might be different depending on step durations and value of the restitution parameter ζ.

The specified creep cases can be considered starting from the uniform positions, answering the concepts of the theory of stabilized anisotropic-ray hardening (SARH), analogous to the SARH theory of thermal plasticity considered in Chap. 5.

8.2.4.2 General Case of the Stress Vector Varying

Creep Parameters at the First Phase Termination

If components S_{ij} of the stress vector have the increments Δs_{ij}, their new values become

$$\langle s_{ij} \rangle = s_{ij} + \Delta s_{ij}, \tag{8.22}$$

a new quantity of the stress intensity is

$$\langle \sigma_i \rangle = (3 \langle s_{ij} \rangle \langle s_{ij} \rangle / 2)^{0.5}, \tag{8.23}$$

and its increment is $\Delta \sigma_i = \langle \sigma_i \rangle - \sigma_i$.

When a new value of the stress intensity has been determined, a rate of stationary creep $v_i^{st} = \varphi_v(\langle \sigma_i \rangle, \langle T \rangle)$ and a limiting unsteady part of creep $\langle \varepsilon_{\lim} \rangle = \varphi_\varepsilon(\langle \sigma_i \rangle, \langle T \rangle)$, and its increment $\Delta \varepsilon_{\lim} = \langle \varepsilon_{\lim} \rangle - \varepsilon_{\lim}$ can be obtained.

An angle ϕ_* between the current direction of an axis of anisotropy FH and new direction of the stress axis CH is defined from the equation

$$\phi_* = \arccos \frac{\varepsilon_{ij}^* \langle s_{ij}^* \rangle}{\varepsilon_i^* \langle \sigma_i^* \rangle}. \tag{8.24a}$$

Due to experimental supervision specified above in Sect. 8.1, unsteady creep rate after rotation of the stress axis can be defined by projecting a former stabilized part of unsteady creep vector onto a new direction of the stress axis CH, i.e., by a quantity

$$\hat{\varepsilon}_i^* = \varepsilon_i^* \cos \phi_* = \frac{\varepsilon_{ij}^* \langle s_{ij} \rangle}{\langle \sigma_i \rangle}. \tag{8.24b}$$

We assume that conditions of loading and unloading described above in Sects. 8.2.2–8.2.3 remain valid if they should be related to projections of the corresponding vectors onto the axis CH.

Some special cases are considered briefly below.

Chart of unsteady creep parameters

Plane chart indicating the changes of unsteady creep parameters $\boldsymbol{\varepsilon}_*$, ε_{lim}, and value of a creep rate $\langle v_i^* \rangle$ after the stress vector rotations with *constant vector magnitude*, for a material without restitution is presented in Fig. 8.9.

It is accepted that before the stress vector rotation, the creep strain has attained the limited quantity $\varepsilon_i^* = \varepsilon_{\text{lim}}$. Position of point ε_i^* on the axis I before rotation is indicated by a large black square, position of point ε_{*0}—indicated by black circle. Parameters after rotation are marked by angle brackets.

If the stress vector turns through an angle ϕ'_*, a position of the point ε_i^* does not change and a new position of the point ε_{*0} is defined by the projection $\langle \hat{\varepsilon}_{*0} \rangle'$ on a new direction II, as shown by the dashed arrow; then, a distance $\varepsilon_i^* - \langle \hat{\varepsilon}_{*0} \rangle$ reduces. At $\sigma_i = \text{const}$, the value $\langle \hat{\varepsilon}_{\text{lim}} \rangle = \varepsilon_{\text{lim}}$ does not also change and a rate $\langle v_i^* \rangle = \beta(\langle \hat{\varepsilon}_{\text{lim}} \rangle - \langle \hat{\varepsilon}_i^* \rangle)$, where $\langle \hat{\varepsilon}_i^* \rangle = \varepsilon_i^*$, arises in the direction of the vector $\boldsymbol{\sigma}_i$ as shown by the arrow.

At $\phi_* = 90°$, projection of the site $\langle \varepsilon_i^* \rangle - \langle \hat{\varepsilon}_{*0} \rangle$ is zero, the rate of non-steady creep grows and becomes equal in magnitude to the value in the beginning of the process, but now in a new direction. For further magnification of a rotation angle (axis III), the projection $\langle \hat{\varepsilon}_i^* \rangle$ changes the sign and the rate v_i^* continues to grow, doubling to 180°.

Envelope of the ends of unsteady creep rate vectors, displayed by the dotted lines, mirrors not only the properties of a material, but also a loading process, including its previous history.

8.2.4.3 Sign-Constant Creep

Active Loading

The first phase of the step. In Fig. 8.10, referred to case $\varepsilon_{i0}^* = 0$, we can see a case when a turn angle varies within the range $0 \leq \phi_* \leq 90°$, initial stabilized unsteady

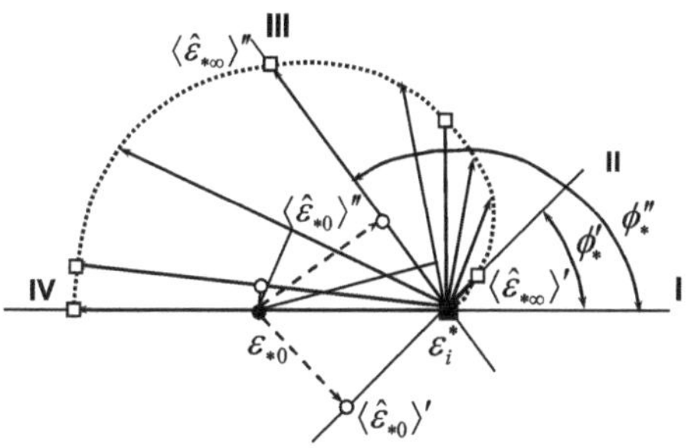

Fig. 8.9 Chart of unsteady creep parameters altering after stress vector rotation without change of magnitude for $\zeta = 0$

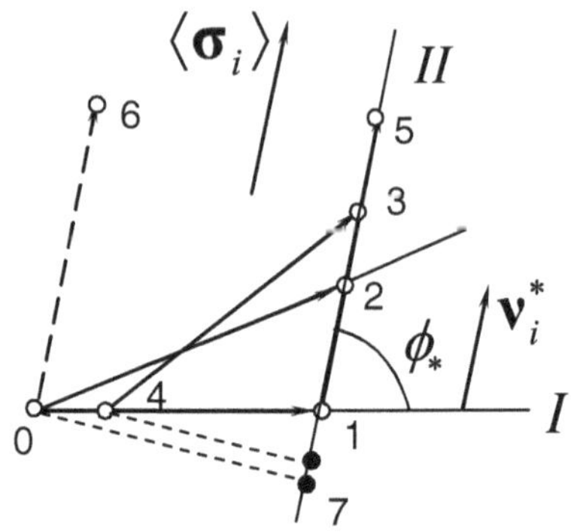

Fig. 8.10 Development of sign-constant unsteady creep

strain ε_i^* (line 0–1 along the direction I of the FH axis) is less than the limit value $\varepsilon_{\lim}$ and projection of the strain ε_i^* onto new direction II of the CH axis (site 7–1) is non-negative, $\hat{\varepsilon}_i^* \geq 0$.

Due to Eq. (8.7), for active loading when $\Delta\sigma_i \geq 0$ and $\Delta\varepsilon_{\lim} \geq 0$ all parts of unsteady strains projections do not change,

$$\Delta\hat{\varepsilon}_{i0}^* = \Delta\hat{\varepsilon}_i^* = 0. \tag{8.25}$$

During *the second phase,* unsteady creep increases due to Eqs. (8.19) and (8.20) along the CH axis II.

Limiting Rotation of Unsteady Strain Vector

Let the vector $\langle \boldsymbol{\sigma}_i \rangle$ be kept any time after rotation by a constant direction φ_* and magnitude σ_i. While a projection $\hat{\varepsilon}_i^*$ for the angles $\varphi_* > 0$ is less than $\varepsilon_{\lim}$, growth of unsteady strain along the axis II and a magnitude of the unsteady strain vector $\langle \boldsymbol{\varepsilon}_i^* \rangle = \boldsymbol{\varepsilon}_i^* + \Delta\boldsymbol{\varepsilon}_i^*$ will be continued during the following steps. Thus, the vector $\langle \boldsymbol{\varepsilon}_i^* \rangle$ will gradually rotate until the equality $\varepsilon_i^* = \varepsilon_{\lim}$ is reached as it is shown by line 0–2. However, unsteady creep along direction II will proceed since the inequality $\hat{\varepsilon}_i^* < \varepsilon_{\lim}$ remains valid (see line 7–2). Further, the strain vector $\boldsymbol{\varepsilon}_i^*$ will turn keeping the constant size $\varepsilon_i^* = \varepsilon_{\lim}$ (line 4–3). Transient process stops only when the strain vector $\langle \boldsymbol{\varepsilon}_i^* \rangle$ takes the direction of the stress vector $\langle \boldsymbol{\sigma}_i \rangle$ (lines 1–5 or 0–6).

Analogous problem has been discussed in Sect. 5.4.1.2 of Chap. 5. In case of creep, rotation of the vector $\langle \boldsymbol{\varepsilon}_i^* \rangle$ must be accompanied by increase of the vector $\boldsymbol{\varepsilon}_{i0}^*$ on additional quantity $\Delta\boldsymbol{\varepsilon}_{i0}^* = (1-\zeta)(\Delta\boldsymbol{\varepsilon}_i^*)_{ad}$. A magnitude of the vector $(\Delta\boldsymbol{\varepsilon}_i^*)_{ad}$ is equal to the site Δ_{0-4} shown in Fig. 8.10. This site can be obtained due to geometrical relationships from the set of equations

$$\begin{aligned} \Delta_{0-4} &= \varepsilon_i^* - \langle \varepsilon_{\lim} \rangle \cos \phi_{**} + \Delta\varepsilon_i^* \cos \phi_* \\ \langle \varepsilon_{\lim} \rangle \sin \phi_{**} &= \Delta\varepsilon_i^* \sin \phi_* . \end{aligned} \tag{8.26a}$$

A new value of the unsteady strain vector is

$$\langle \boldsymbol{\varepsilon}_i^* \rangle = \boldsymbol{\varepsilon}_i^* + \Delta\boldsymbol{\varepsilon}_i^* - (\Delta\boldsymbol{\varepsilon}_i^*)_{ad}. \tag{8.26b}$$

A final position of the center of unsteady strain area is $\varepsilon_{i0}^* = (1 - \zeta)\varepsilon_i^*$ meeting point 1 for lack of restitution ($\zeta = 0$) or remains at point 0 under full restitution ($\zeta = 1$).

The pattern does not change for stress σ_i increasing, but thus the size $\langle \varepsilon_{\lim} \rangle$ also increases.

Rotation of the Stress with Unloading

Under unloading the projections $\hat{\varepsilon}_{i0}^*$ and $\hat{\varepsilon}_i^*$ change during the first phase according to Eq. (8.8). In case of full unloading, the magnitude of the vector $\boldsymbol{\varepsilon}_i^*$ and its projection $\hat{\varepsilon}_i^*$ decrease during the second phase to zero. A center of unsteady strain area ε_{i0}^* allocates position at a point of the line 0–1 depending on a value of factor ζ and projection $\hat{\varepsilon}_{i0}^*$ allocates a corresponding position at a point of the line 7–1.

8.2.4.4 Sign-Varying Unsteady Creep

Under turn of the stress vector through an angle within the range $90 < \phi_* \leq 180°$, the projection of stabilized unsteady strain part onto new direction II of the stress vector is negative, $\langle \hat{\varepsilon}_i^* \rangle < 0$. In this case, evolution of the intensity of the unsteady strain ε_i^* renews after stress vector turn with larger rate, as it follows from Eq. (8.19) and the chart presented above in Fig. 8.9.

Further, the creep process occurs analogously aforementioned one accordingly to correlations of vectors' projections on the stress axis CH.

Solution procedure for the SARH model of creep is described below in Sect. 8.2.6.

8.2.5 Special Problems of Unsteady Creep

8.2.5.1 Calculation Within the Entire Unsteady Stage

In some cases, the sign-varying creep development occurs at each step entirely within an unsteady area. Therefore, it is of interest to study material behavior based only on the primary stage of the creep diagram. To this goal, parameter β can be

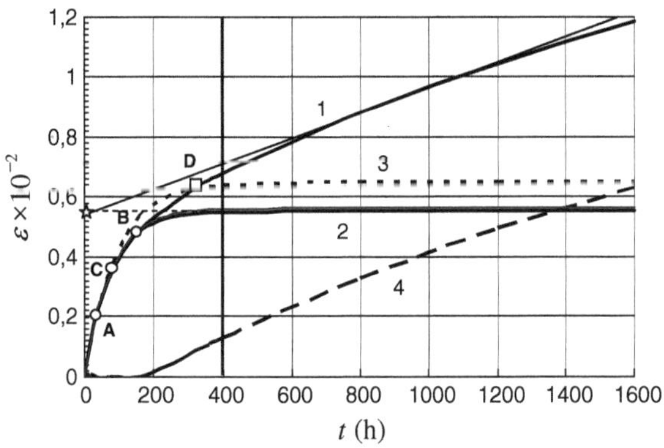

Fig. 8.11 Processing of a creep diagram for calculation entirely within an unsteady stage

calculated from experimental values of ε_* at two points of the creep diagram. In the beginning of this stage, a steady part of the strain is very small in comparison with the unsteady one. Thus, we take $\varepsilon_* \approx \varepsilon$. Marking $\bar{t} = t_1/t_2 < 1$, $\bar{\varepsilon} = \varepsilon_1/\varepsilon_2 < 1$, and $x = \beta t_2$, one can obtain the equation in x

$$1 - \bar{\varepsilon} + \exp(-x)\{\bar{\varepsilon} - \exp[x(1 - \bar{t})] = 0, \qquad (8.27)$$

which is easily numerically calculated. Then, we find $\beta = x/t_2$ and $\varepsilon_{\text{lim}} = \varepsilon_2/[1 - \exp(-x)]$.

In Fig. 8.11, a creep diagram obtained by Davis (1943) is presented by curve 1. Assuming that only its initial part until 400 h would be known, we take two points A and B, find $\beta = 0.013\ \text{h}^{-1}$ and $\varepsilon_{\text{lim}} = 0.55 \times 10^{-2}$, and build a fading function $\varepsilon_*(t)$ shown by curve 2. Having defined the parameter β on the base of any other point C underlying within the initial part of the creep diagram, we arrive practically at the same curve 2. If a calculation badly represents the test data (when it is based, e.g., on point D), its results must not be used (curve 3).

The obtained value ε_{lim} is very close to the quantity yielded from the full experimental diagram shown by asterisk for the moment t = 0. Line 4 corresponds to the difference $\varepsilon_{st} \approx \varepsilon - \varepsilon_*$, which allows rough estimation of a rate of steady creep.

An approach to study the cyclic creep occurring entirely within the unsteady area without calculation of limit value ε_{lim} was considered by Murakami and Ohno (1982).

8.2.5.2 Quasi-stabilized Creep

For a process of long-term duration, the effects specified above can be approximately taken into account using a model of quasi-stabilized creep as a generalization of a stepped representation of creep diagrams (see Chap. 7).

8.2.5.3 Final Effect Induced by Non-stationary Creep

In some cases, it is possible to present the working conditions in the form of n consecutive regimes $k = 1, 2, \ldots, n$ with constant stress σ_k and temperature T_k.

At major duration of the regimes, interest is represented not by delicacies of transient strain development but a *finite effect* from such transitions. This effect, consisting in rather rapid additional creep strain modification at the change of a regime, can be described by assumption of "instantaneous" unsteady creep strain ε_* transition from one limiting value $(\varepsilon_{\lim})_{k-1}$ to another value $(\varepsilon_{\lim})_k$ under sudden change of stress from σ_{k-1} to stress σ_k.

For simplicity, we shall write ε_{*k} instead of $(\varepsilon_{\lim})_k$. Then an unsteady strain "jump" is defined as $\Delta\varepsilon_{*k} = \varepsilon_{*k} - \varepsilon_{*(k-1)}$. The total creep strain during n quasi-stabilized regimes is obtained as a sum

$$\varepsilon_\Sigma(t_n) = \sum_{k=1}^{k=n} [\Delta\varepsilon_{*k} + v_k(t_k - t_{k-1})], \tag{8.28}$$

where $v_k = (\operatorname{sign}\sigma_k) v_{st}(|\sigma_k|, T_k)$ is a steady creep rate at the regime k.

The idea of using a "jump," or limit function $\varepsilon_{\lim}(\sigma)$, for modeling the final effect of unsteady creep was delivered by Odqvist (1953, 1974) and applied to non-isothermal problems in Birger and Shorr (1975). The use of this idea for describing arbitrary creep processes demands a special approach to determine the jumps $\Delta\varepsilon_{*k}$ in dependence on loading history and material heredity properties.

8.2.5.4 Example of Quasi-stabilized Creep

An example of a quasi-stabilized approach for multi-step stress sign-constant process is shown in Fig. 8.12 where two stress levels σ_1 and σ_2 are allocated. Curves of creep for constant stress values σ_1 and σ_2 are shown by the thin lines in Fig. 8.12b. Temperatures at these regimes are assumed to be identical.

The bold broken line refers to $\zeta = 0$, and the dashed line refers to $\zeta = 1$.

The return to the former stress under creep after complete or incomplete unloading renews at an exaggerated rate.

Comparison of total strain ε_Σ calculated using the formulas of quasi-stabilized creep with empirical results (Samarin and Sorokin 1969) presented in Fig. 8.7a has been displayed in Fig. 8.7c. Calculated pattern is less accurate than in Fig. 8.7b but it is acceptable for practical use.

8.2.5.5 Creep Computation of Machine Parts

In the presented correlations of creep theory, the stresses and their changes on operation regimes were assumed to be known. The general approach to creep computing of machine parts consists of the following:

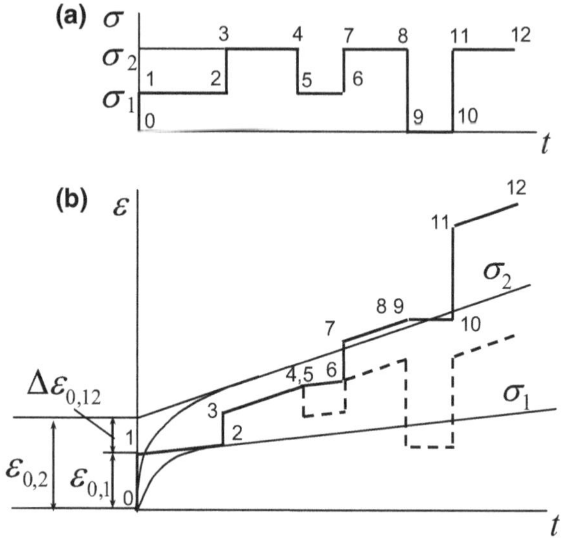

Fig. 8.12 Multi-step creep of sign-constant loading; **a** stress versus time; **b** quasi-stabilized approach

For each regime, the calculations of thermal plastic stresses in a detail (see Chap. 5) are fulfilled. The "jumps" of the unsteady creep parameters for the first phase can be established by the methods specified above in this chapter. The obtained parameters are used as reference values for computing creep evolution during the second (time dependent) phase of each regime.

8.2.5.6 Modification of the Constitutive Equations

Equation (8.8) may be modified using acceptable experimental data. For example, if it will be established for a material that doubling of speed of a creep unsteady part occurs at turn angle $\phi = 180°$ at any value of the restitution factor ζ, it is enough to accept a value of the projection $\hat{\varepsilon}_{*0}$ be zero if its sign changes.

8.2.6 Analysis of Creep Problems Using Finite Element Method

8.2.6.1 General

As it has been stated above in Sect. 5.4.3, the reader is supposed be familiar with the main principles of FEM analysis (see, i.e., Zienkiewicz 1967; Zienkiewicz and Taylor 2000) and constructions of the universal software codes, delighted in special managements.

In theory of FEM, a tensor (matrix-column) **v** and a matrix **K** are usually designated as $\{v\}$ and $[K]$. However, describing a calculation procedure for an element, we shall use for vector an index notation.

For creep problem, a term $\{\Delta\varepsilon_p\}$ in Eq. (5.61) and in what follows must be replaced by $\{\Delta\varepsilon_c\}$ where $\{\varepsilon_c\}$ is a creep part of the total strain tensor $\{\varepsilon\}$ Thus, Eq. (5.61) takes a form

$$\{\Delta\sigma\} = [D](\{\Delta\varepsilon\} - \{\Delta\varepsilon_c\} - \{\Delta\varepsilon_T\} - \{\Delta\varepsilon_T^e\} \tag{8.29}$$

The increments of the total and creep strains and also stresses during a current step are determined from the FEM solution of the entire structure by successive approximations accounting the convergence conditions (see Chap. 5 and also Temis 1980, 2005). Below, procedures concerning calculation of the creep increment during a step are only considered.

8.2.6.2 Solution Procedure for the SARH Model of Creep

For the creep problem analyzing with use of the SARH model, the following material characteristics, along with elasticity parameters, must be given: an intensity of the steady creep rate v_i^{st}, a limit value of creep unsteady part $\varepsilon_{\lim}^c$, a parameter of creep stabilization β_c, and also a factor of strain restitution ζ. Generally, specified characteristics depend on stress intensity σ_i and temperature T.

Thereafter, the index (c) is omitted (if there is no necessity to use it).

Before the beginning of a current step, i.e., for a time instant t, the following parameters are assumed to be known from the proceeding calculations or initial conditions: tensors components of stresses σ_{ij}, total creep strains ε_{ij}, creep strain steady part ε_{ij}^{st}, and also unsteady stabilized part ε_{ij}^*. Initial values of all specified parameters are zero.

Computation process includes two phases. Creep parameters referred to the phases terminations are parenthesized: for the first phase in angle brackets $\langle\,\rangle$ and for the second—in double ones $\langle\langle\,\rangle\rangle$.

Step Beginning

At the step beginning, all material characteristics may be recalculated for new temperature $\langle T\rangle$ that yields $v_i^{st}(\langle T\rangle)$, $\varepsilon_{\lim}(\langle T\rangle)$, $\langle\beta\rangle$, and $\langle\zeta\rangle$. Then, the following procedures are executed:

Values of initial step parameters of stress/strain state are calculated, namely:

- components of the stress deviator

$$s_{ij} = \sigma_{ij} - \delta_{ij}\sigma_{ii}/3;$$

- stress intensity

$$\sigma_i = (3s_{ij}s_{ij}/2)^{0.5};$$

- and intensity of the stabilized part of unsteady creep

$$\varepsilon_i^* = (2\varepsilon_{ij}^*\varepsilon_{ij}^*/3)^{0.5}.$$

Iteration Procedures

The subsequent procedures concern each of the FEM iterations for the creep problem solution. During entire step, the external loads and temperatures are assumed remaining constant.

New Parameters after Input of Stress Increment Using components $\Delta\sigma_{ij}$ of the stress tensor increment, which have to be set or obtained in the previous calculation step or iteration, we define:

- new components of the stress vector

$$\langle\sigma_{ij}\rangle = \sigma_{ij} + \Delta\sigma_{ij},$$

- components of the stress deviator increment

$$\Delta s_{ij} = \Delta\sigma_{ij} - \delta_{ij}\Delta\sigma_{ii}/3,$$

- new values of the stress deviator components

$$\langle s_{ij}\rangle = s_{ij} + \Delta s_{ij},$$

- new value of the stress intensity

$$\langle\sigma_i\rangle = (3\langle s_{ij}\rangle\langle s_{ij}\rangle/2)^{0.5}$$

  and its change for the given step

$$\Delta\sigma_i = \langle\sigma_i\rangle - \sigma_i;$$

- new limit value of the stabilized part of unsteady creep

$$\langle\varepsilon_{\lim}\rangle = \varphi_\varepsilon(\langle\sigma_i\rangle, \langle T\rangle)$$

and its change for the given step

$$\Delta\varepsilon_{\lim} = \langle\varepsilon_{\lim}\rangle - \varepsilon_{\lim};$$

- new rate of the steady creep intensity $\langle v_i^{st}\rangle = \varphi_v(\langle\sigma_i\rangle, \langle T\rangle)$.

First phase of the step. New values of creep process parameters are calculated for the conditions that all components of the stain vectors remain invariable during this phase, i.e., $\Delta\varepsilon_{ij} = 0$, $\Delta\varepsilon_{ij}^{st} = 0$ and $\varepsilon_{ij0}^* + \Delta\varepsilon_{ij}^* = \Delta\varepsilon_{ij} - \Delta\varepsilon_{ij}^{st} = 0$; hence $\Delta\varepsilon_{ij0}^* = -\Delta\varepsilon_{ij}^*$.

We define:

- A cosine of the angle between directions of the vectors $\boldsymbol{\varepsilon}_i^*$ and $\langle\boldsymbol{\sigma}_i\rangle$

$$\cos\phi_* = \frac{\varepsilon_{ij}^*\langle s_{ij}\rangle}{\varepsilon_i^*\langle\sigma_i\rangle}.$$

- A projection of the vector $\boldsymbol{\varepsilon}_i^*$ onto new direction of the stress axis

$$\hat{\varepsilon}_i^* = \varepsilon_i^* \cos\phi_*.$$

Two situations are possible:

- *Loading continue or a stationary stress state*: $\Delta\sigma_i \geq 0$, i.e., $\Delta\varepsilon_{\lim} \geq 0$. A change of the vector $\boldsymbol{\varepsilon}_i^*$ projection is accepted to be zero, $\Delta\hat{\varepsilon}_i^* = 0$.
- *Unloading*: $\Delta\sigma_i < 0$, i.e., $\Delta\varepsilon_{\lim} < 0$. Then

$$\Delta\hat{\varepsilon}_i^* = (1-\zeta)\hat{\varepsilon}_i^*\overline{\Delta\varepsilon_{\lim}},$$

where $\overline{\Delta\varepsilon_{\lim}} = \Delta\varepsilon_{\lim}/\langle\varepsilon_{\lim}\rangle < 0$.

Increments of the stabilized unsteady creep components are

$$\Delta\hat{\varepsilon}_{ij}^* = \frac{3\Delta\hat{\varepsilon}_i^*\langle s_{ij}\rangle}{2\langle\sigma_i\rangle}.$$

New value of the vector $\boldsymbol{\varepsilon}_i^*$ projection at the end of the first phase is

$$\langle\hat{\varepsilon}_i^*\rangle = \hat{\varepsilon}_i^* + \Delta\hat{\varepsilon}_i^*.$$

Both conditions i and ii can be united accepting

$$\langle\hat{\varepsilon}_i^*\rangle = \hat{\varepsilon}_i^*(1-q),$$

where

$$q = \begin{cases} 0 & \text{at } \Delta\sigma_i \geq 0 \\ -(1-\zeta)\overline{\Delta\varepsilon_{\lim}} & \text{at } \Delta\sigma_i < 0 \end{cases}.$$

Second phase of the step. Creep development during a time step Δt is calculated for the conditions that the stress and creep parameters determined for the first phase remain invariable during the second phase.

Rate intensity of the stabilized part of the unsteady creep at the beginning of the second phase is

$$v_i^* = \beta[\langle\varepsilon_{\lim}\rangle - \langle\hat{\varepsilon}_i^*\rangle].$$

Intensity increment of the stabilized part of the unsteady creep is

$$\langle\Delta\varepsilon_i^*\rangle = \frac{v_i^*}{\beta}[1 - \exp(-\beta\,\Delta t)],$$

that yields for very small steps $\langle\Delta\varepsilon_i^*\rangle \approx v_i^*\Delta t$.

Components of the vector $\langle\boldsymbol{\Delta\varepsilon}_i^*\rangle \| \langle\boldsymbol{\sigma}_i\rangle$ are

$$\langle\Delta\varepsilon_{ij}^*\rangle = \langle\Delta\varepsilon_i^*\rangle\frac{3\langle s_{ij}\rangle}{2\langle\sigma_i\rangle}.$$

and its magnitude is $\langle\varepsilon_i^*\rangle = (2\langle\varepsilon_{ij}^*\rangle\langle\varepsilon_{ij}^*\rangle/3)^{0.5}$.

Increment of a steady strain part during the second phase is

$$\langle\langle\Delta\varepsilon_{ij}^{st}\rangle\rangle = \langle v_i^{st}\rangle\frac{3\langle s_{ij}\rangle}{2\langle\sigma_i\rangle}\Delta t.$$

The final values of the components of the creep vector's increment for the current iteration are

$$\langle\langle\Delta\varepsilon_{ij}\rangle\rangle = \langle\langle\Delta\varepsilon_{ij}^{st}\rangle\rangle + \langle\Delta\varepsilon_{ij}^*\rangle.$$

Transition to the next iteration Calculated increments of the creep strain $\langle\langle\Delta\varepsilon_{ij}\rangle\rangle$ for each element are entered into corresponding equations of FEM code and the given iteration is carried out. New values of the stress increment vector $\Delta\sigma_{ij}$ allow a following iteration of a current step to be executed.

Step termination Upon completion of the iterations, we calculate values of the additional increment of unsteady stabilized strain.

If a calculated value of the vector's magnitude is $\langle \varepsilon_i^* \rangle > \langle \varepsilon_{\lim} \rangle$, then the additional increment of unsteady stabilized strain is

$$(\Delta\varepsilon_i^*)_{ad} = \varepsilon_i^* - \langle \varepsilon_{\lim} \rangle \cos\phi_{**} + \langle \Delta\varepsilon_i^* \rangle \cos\phi_*$$

where

$$\phi_{**} = \arcsin\phi_* \frac{\langle \Delta\varepsilon_i^* \rangle}{\langle \varepsilon_{\lim} \rangle}.$$

Components of the corresponding vector are

$$(\Delta\varepsilon_{ij}^*)_{ad} = (\Delta\varepsilon_i^*)_{ad} \frac{\varepsilon_{ij}^*}{\varepsilon_i^*}.$$

New component values of the total creep strain tensor and its parts are

$$\begin{aligned}
\langle\langle \varepsilon_{ij} \rangle\rangle &= \varepsilon_{ij} + \langle\langle \Delta\varepsilon_{ij}^{st} \rangle\rangle + \langle \Delta\varepsilon_{ij}^* \rangle \\
\langle\langle \varepsilon_{ij}^* \rangle\rangle &= (1-q)\varepsilon_{ij}^* + \langle \Delta\varepsilon_{ij}^* \rangle - (1-\zeta)(\Delta\varepsilon_{ij}^*)_{ad} \\
\langle\langle \varepsilon_{ij}^{st} \rangle\rangle &= \varepsilon_{ij}^{st} + \langle\langle \Delta\varepsilon_{ij}^{st} \rangle\rangle.
\end{aligned}$$

These quantities and the value of the last iteration of the stress σ_{ij} are taken as new components for the next step.

References

Arutjunjan NK (1952) Some problems of theory of creep. Gostekhizdat, Moscow (in Russian)

Birger IA, Shorr BF (eds) (1975) Thermal strength of machine parts. Mashinostroenie, Moscow (in Russian)

Carreker RP, Leschen JG, Lubahn JD (1949) Transient plastic deformation. Trans Amer Soc Eng 180:139–146

Chaboche JL (1989) Constitutive equations for cyclic plasticity and cyclic viscoplasticity. Int J Plast 5:247–302

Chaboche JL (2008) A review of some plasticity and viscoplasticity constitutive equations. Int J Plast 24:1642–1693

Davenport CC (1938) Correlation of creep and relaxation properties of copper. J Appl Mech 60:56

Davis EA (1943) Creep and relaxation of oxygen free copper. J Appl Mech 10(2)

Dul'nev RA, Kotov PI (1980) Thermal fatigue of metals. Mashinostroenie, Moscow (in Russian)

Evans RW, Wilshire B (1985) Creep of metals and alloys. Cariton House Terrace, London

Getzov LB (1969) To problem of cyclic creep and relaxation of high-resistant alloys. J Strength Prob 5:34–38 (in Russian)

Ivanova GM (1958) Creep of the alloy EI437B at variable temperatures. Izv. AN SSSR<OTN, Mech Mach Build 4:98–99 (in Russian)

Johnson AE (1960) Complex stress creep of metals. Metall Rev 5(20):447–506

Johnson AE, Henderson J, Mathur VD (1958) Creep under changing complex creep systems, Engineer, V. 206
Kachanov LM (1960) Theory of creep. Fizmatgiz, Moscow (in Russia)
Katz SN (1957) Creep and failure of tubes affected by inner pressure. Izv. AN SSSR ONT Mech Mashinostroenie 10:86–89 (in Russian)
Kennedy AJ (1965) Creep and fatigue in metals. Metallurgiya, Moscow
Lemaitre J, Chaboche JL (1990) Mechanics of solid materials. Cambrodge University Press, Cambridge
Lemaitre J, Chaboche JL (2004) Mecanique des materiaux solides, 2nd edn. Dunod, Paris
Malinin NN (1952) Design on creep. In: Ponomarev SD (ed) Foundations of contemporary methods of strength design in machinery, Mashgiz, Moscow (in Russian)
Malinin NN (1975) Applied theory of plasticity and creep, 2nd edn. Mashinostroenie, Moscow (in Russian)
Malinin NN, Khadjinsky GM (1969) Formulation of the creep theory with anisotropic hardening. Izv. AN SSSR MTT 3:148–152 (in Russian)
Malinin NN, Khadjinsky GM (1972) Theory of creep with anisotropic hardening. Int J Mech Sci 14(4):235–246
Murakami S, Ohno N (1982) A constitutive equation of creep based on the concept of a creep-hardening surface. Int J Solids Struct 18(7):597–609
Namestnikov VS (1957) On creep under variable loads at complex stress state. Izv. Akad. Nayk SSSR OTN 10 (in Russian)
Namestnikov VS (1960) Direct and opposite torsion under creep conditions cJourn. PMTF 1:121–122
Namestnikov VS (1963) Combined stress creep under changing loads. In: Proceedings of the joint international conference on creep
Namestnikov VS, Khvostunkov AA (1960) Creep of duralumin at constant and variable loads. J PMTF 4:90–95 (in Russian)
Odqvist FKG (1953) Influence of primary creep on stresses in structural parts. Trans R Inst Technol 66
Odqvist FKG (1959) Engineering theories of creep "Mechanics" 2
Odqvist FKG (1974) Mathematical theory of creep and creep rupture, 2nd edn. Clarendon Press (1st edn., 1966)
Rabotnov YuN (1948) Designing machine parts for creep. Ixv. Akad. Nauk SSSR, OTN. 6 (in Russian)
Rabotnov YN (1966) Creep of construction elements. Physmatgiz, Moscow; Rabotnov YN (1969) Creep problems in structure members (trans). North Holland, Amsterdam
Rozovsky MI (1951) Creep and durable damage of materials. J Theor Phys v.XXI(11) (in Russian)
Samarin YP, Sorokin OV (1969) On formulation of stochastic equations for creep. In: Rapport at conference of Kuybyshev Polytechnic Institute, Kuybyshev, vol 31 (in Russian)
Shorr BF (1964) Cycle creep of nonuniformly heated cylinders. In: Thermal stresses in construction elements, vol 4. Naukova Dumka, Kiev (in Russian)
Shorr BF (1966) Periodical processes and adaptability at creep. In: Strength and dynamics of aviation engines, vol 4. Mashinostroenie, Moscow, pp 188–194 (in Russian)
Shorr BF (1970a) Unstable creep design. In: Thermal stresses in construction elements, vol 9. Naukova Dumka, Kiev, pp 165–173 (in Russian)
Shorr BF (1970b) Cycle creep design of beams using a modified theory of heritable influence. In: Thermal stresses in construction elements, vol 10. Naukova Dumka, Kiev, pp 152–159 (in Russian)
Shorr BF, Dul'nev RA (1964) Investigations of thermal stresses and creep at varying temperatures. J Zavodskaya Laboratoriya 3:340–347 (Review in Russian)
Shorr BF, Dul'nev RA (1968) Cycle creep. In: Fridman JB (ed) Materials strength and deformation in nonuniform physical fields, vol II. Atomizdat, Moscow, pp 34–96 (in Russian)

Sizova RN (1969) Some regularities of non-stabilized creep and their influence on a stress state. Thermal strength of materials and construction elements. Naukova Dumka, Kiev, pp 35–41 (in Russian)

Temis YM (1980) Method of successive approximations with error correction in geometrically nonlinear elastic problems (in Russia). In: Proceeding of applied problems of strength and plasticity. Algoriphmic solution procedures in elasticity and plasticity problems. Izd. Gor'kovskogo Gos. Universiteta. Go'kiy, V.16:3–10

Temis YM (2005) Plasticity and creep modeling of turbo engines constructive materials. In: Proceedings of 49th international conference AAI, Part 2, MAMI, Moscow, pp 25–76 (in Russian)

Zienkiewicz OC (1967) The finite element method in structural and continuum mechanics. McGraw-Hill, London

Zienkiewicz OC, Taylor RL (2000) The finite element method, 5th edn. Butterworth-Heinemann, Oxford

Chapter 9
Estimation of Design Thermal Integrity

Abstract Working capacity of machine parts in various possible operation conditions must be maintained by introduction of strength and durability margins (safety factors). These margins are necessary in view of possible random industrial deviations, unforeseen adverse combinations of loads, temperatures, and operating time at some regimes, and many other reasons. General principles for safety factor formation require analyzing the most probable ways for deviation of stresses, temperatures, and work duration or any other specific parameter, over its design value—on separateness or in aggregate with other ones. The possibility of creating a design of "equal strength" at non-isothermal loading is illustrated by the optimization of a turbomachine blade model non-uniformly heated along its length. The benefits of introducing a "weak link" that reaches destruction under overload before the entire system are discussed. It is shown that "equivalent" trials replicating the lifetime of the system can significantly accelerate the verification of the most stressed machine parts. The trials ensure a machine functioning with the same safety factors as under the work conditions, but during smaller duration. The cyclic durability margins for non-isothermal cyclic fatigue, taking into account influence of exposure at the maximum cycle temperature and asymmetric loading, are considered. Along with the evaluation of local strength and durability margins for the most stressed elements of a structure, computation methods, on a bearing ability of the structure "in whole," are stated. Use of the determined safety factors is shown to be principally necessary for the reliable probabilistic estimation of details' low-cycle fatigue (LCF). For this purpose, rational methods for generation of probabilistic-statistical strength and durability margins of machine parts, using results of the limited scope of sampling tests, are proposed.

B.F. Shorr, *Thermal Integrity in Mechanics and Engineering*,
Foundations of Engineering Mechanics, DOI 10.1007/978-3-662-46968-2_9

9.1 General Principles of Thermal Integrity Assurance

9.1.1 Margins of Thermal Integrity

9.1.1.1 Role of Safety Factors

The necessary requirement for a design's reliable behavior during a demanded work time (in-service life, resource) is the maintenance of structural integrity. A strength analysis (calculations of integrity) should consider not only the expected design operating conditions and material properties, but also the inevitable random deviations from them. Therefore, the integrity estimation should have a probabilistic character based on experimental statistical data accumulated in preceded researches.

Various aspects of such an approach were considered by Serensen et al. (1963), Bolotin (1965), Shorr et al. (1972), Birger and Shorr (1975), and others. When using a probabilistic approach, it is still necessary to keep a certain determined distance between the operation and destruction parameters of a system. This distance is determined by the safety factors.

In a broad sense, the safety factors (*strength margins*) for a system Q are understood as the margins on any parameter influencing the working ability of the entire structure or its separate units. The loss of working ability can occur both after the first loading, and as a result of action of long-term or repeated loading and heating.

Some margins take a *common character* and concern the factors affecting a structure in operating and failure conditions as a whole, for example, temperature and air pressure upon an input of the engine and frequency of rotation. Common margins are usually determined experimentally by trials of a structure both in nominal and in more severe conditions, sometimes down to destruction. Some failure parameters can be approximately calculated at designing (e.g., the rotation frequency of disk failure). The development of computer technology has allowed increasingly more reliable results to be obtained for the estimation of common strength margins.

Margins of local strength concern the conditions of material damage in a danger point of a system and can refer to stresses, strains, temperatures, work duration, and the number of cycles. Sometimes, a margin of displacement is of interest. During designing, the values of the destruction parameters are determined on the basis of experimental–computational methods, as it has been stated in the previous chapters. A margin may refer to failure conditions or greatest admissible quantity of the considered parameter.

The working ability of a detail at any *k*th operating regime is defined, as a rule, not by a single, but several state parameters Q_k. To estimate the safety of the regime, it is necessary to examine the most probable ways of a deviation of each parameter from its nominal value (set or calculated). The reasons for scatter of material properties can occur due to metallurgical and technological deviations or special operating conditions (in particular, corrosion), and also a rise in temperature connected with the deterioration of cooling. The discrepancies of calculation, which are caused, for example, by the uncertainty of the boundary conditions or unexpected operating conditions, could increase the actual stresses.

When the immediate reason for a deviation affects one of the parameters, others could remain constant or also vary due to the physical connection to each other (for example, durability to stresses). Thus, the safety factors can appear variously depending on a probable path of parameters deviation. The ratio

$$n_k = \frac{Q_{k,d}}{Q_{k,a}}, \tag{9.1}$$

where $Q_{k,d}$ and $Q_{k,a}$ are, respectively, a limit and active operating quantity of a given parameter under constant values of other parameters, may be called "a particular safety factor."

The limit parameter's quantity can be defined as the failure of a structure and also such a quantity above which the normal functioning of the structure during the assigned resource becomes impossible.

An approach of defining the particular safety factors has been offered by Birger (1948) with reference to fatigue strength margins.

A calculated safety factor n_k must be larger than unity, which is necessary for ensuring design reliability. The minimally admissible size $[n_k]$ is established by the operating experience of similar designs. Therefore, the condition of strength looks like

$$n_k \geq [n_k], \tag{9.2}$$

where

$$[n_k] > 1. \tag{9.3}$$

Equation (9.1), strictly speaking, means that parameters $Q_{k,a}$ and $Q_{k,d}$ can be defined identically. But the quantity $Q_{k,d}$ is mostly established experimentally, and the quantity $Q_{k,a}$ is quite often calculated. Then, the safety factor n_k has conditional character, and the value $[n_k]$ should be based on an experience of application of Eq. (9.1) in the ways in which parameters $Q_{k,a}$ and $Q_{k,d}$ have been defined.

The concept of "constant" values of other parameters also requires a certain improvement that will be shown below.

9.1.1.2 Margins at Stationary Loading

A mechanical state of a machine part, working for a long time at a regime k under slightly varying loads and stationary temperatures, is characterized by stress tensor σ_k, temperature T_k, and also duration t_k. A common approach to the estimation of safety factors will be illustrated for a case of 1D extension of a body at uniform temperature and constant stress. The index of mode k is omitted here.

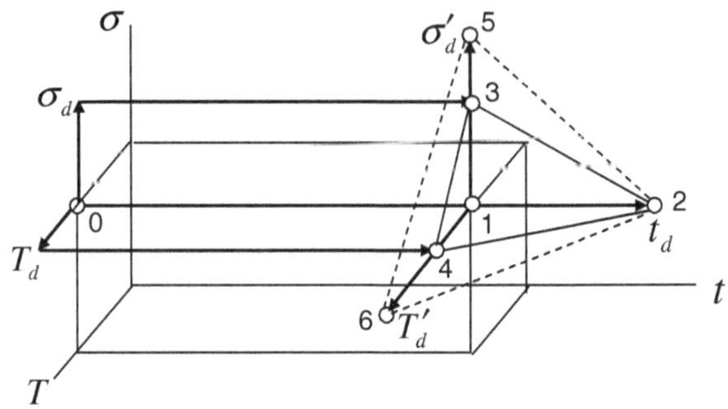

Fig. 9.1 Strength margins at different ways of deviation from nominal regime

In Fig. 9.1, representing a material mechanical state in coordinates σ, T, and t, a point 1 corresponds to a state with designed parameters σ_{calc}, T_{calc} at the nominated resource t. Probable ways of specified parameter deviations from nominal values are conditionally shown by arrows. These lead to various possible conditions of failure, and accordingly to the various safety factors, available at the designed regime.

The arrow 1–2 corresponds to time deviation when *a durability margin*

$$n_t = \frac{t_d}{t} \tag{9.4}$$

specifies how long (during what time t_d) a detail could still work under the same stress and temperature before destruction.

The deviation along arrows 0–1–3 corresponds to the case when for any reason (for example, because of incorrectly appointed loads) the actual stress σ_d during all operating time t has exceeded the design value σ_{calc}. The admissibility of such a deviation is defined *by a margin of long-term strength* which is calculated as a ratio of the limit of long durability in tension σ_d to the stress σ_{calc}

$$n_\sigma = \frac{\sigma_d}{\sigma_{calc}} \tag{9.5}$$

under the nominal operating time t and temperature T_{calc}.

The deviation along arrows 0–1–4 characterizes the possible influence of a structure overheating during the entire operating time t because of insufficient cooling when the stress remained constant. A quantity

$$n_T = \frac{T_d}{T_{calc}} \tag{9.6a}$$

characterizes *a temperature margin*, which is sometimes more convenient to define as a difference of temperatures

$$\tilde{n}_T = T_d - T_{calc}. \tag{9.6b}$$

Sometimes, another character of deviations can also occur. The arrow 1–5 demonstrate a possible short-term overload at the end of the nominal operating time, connected, for example, with over-speeding of a turbine rotor due to a fall of resistance torque in the system. The ratio

$$n'_\sigma = \frac{\sigma'_d}{\sigma_{calc}} \tag{9.7}$$

along arrow 1–5 specifies a margin of short-term strength in the case when the previous operation has run during the time t at nominal stress and temperature. Obviously, there is $\sigma'_d > \sigma_d$ and accordingly $n'_\sigma > n_\sigma$.

Finally, the arrow 1–6 shows a similar case for a short-term overheat at the end of the set operating time. The temperature margin of the overheat

$$n'_T = \frac{T'_d}{T_{calc}} \tag{9.8}$$

exceeds a margin due to Eq. (9.6a).

Under elastic deformations, the margin in strains

$$n_\varepsilon = \frac{\varepsilon_d}{\varepsilon_{calc}}, \tag{9.9}$$

is identical to the margin of long-term strength in stresses n_σ defined by Eq. (9.5).

Under plastic deformations and creep, the specified margins become different, and in some cases, for example, by estimation of destruction danger of few hardening materials and at intensive creep, the margin in strains becomes more informative.

At moderate temperatures and stresses when creep can be neglected and the time before destruction at a constant stress is very great, the high-cycle fatigue is directed by amplitude σ_a and mean σ_m stresses of a cycle. In these conditions, the margins in total stresses $\sigma = \sigma_m + \sigma_a$ or only in their variable part σ_a are necessary to be defined depending on the most probable way of process deviation (Birger et al. 1993).

The time t_d prior to the failure of a material is connected with stress $\sigma_{d/t}$ and temperature $T_{d/t}$ operated during this time by an experimental dependence, which in certain limits of parameters can be approximated by analytical expressions

$$f(\sigma_{d/t}, T_{d/t}, t_d) = 0. \tag{9.10}$$

In this case, the safety factors n_σ, n_T and also n_t can be expressed one through another, and Eq. (9.10) generates a *surface of long-term destruction*, of which the borders are conditionally shown in Fig. 9.1 by lines 2–3–4. Analogously to this, borders of the surface of short-term overload and overheat at the end of a resource are conditionally presented by dashed lines 2–5–6.

The same concerns margins in strains.

Specified "surfaces" are defined not only by properties of a material, but also depend on a process of stress and temperatures altering during the nominal operation time t, and further to the increased value t_d. The detail can be considered as strong only if all specified margins satisfy the *strength and durability conditions.*

In some cases, the working ability of details is defined by the size of safe displacements.

9.1.1.3 Margins at Arbitrary Processes

In the case of variable stresses and temperatures, a similar approach to the definition of thermal integrity margins is applicable by replacing in Fig. 9.1 the strength parameters with their ranges in each cycle of loading. *A cyclic durability margin*

$$n_N = \frac{N_d}{N} \tag{9.11}$$

is used to this goal instead of margin in time n_t.

Here, N is the operating number of loading cycles, and N_d is a number of identical cycles before destruction for constants stress $\Delta\sigma$ or full strain $\Delta\varepsilon$ and temperature ΔT ranges.

Analogously, the margins in stress and temperature ranges at other possible deviations from the nominal state can be calculated.

In statically indeterminate problems, the cyclic durability margin is usually defined at constant strain ranges.

At a combined stress–strain state, the margins must be calculated using those components or strength criteria which define failure conditions (see previous chapters).

With reference to sufficient plastic materials, the long-term durability margins are usually defined using equivalent stresses

$$n_{eqv,\sigma} = \frac{\sigma_{eqv,d}}{\sigma_{eqv,calc}} \tag{9.12a}$$

or maximal tangential stresses

$$n_\tau = \frac{\tau_{\max,d}}{\tau_{\max,calc}}. \tag{9.12b}$$

For brittle materials, the margins are defined by the first principle (the greatest tensile) stress.

The margins in strain components, durability, and temperature are calculated using the same reasoning.

For 1D extension of an evenly heated body, the safety factors for all points of a body are identical; therefore, a condition $n_t \leq 1$ concerning any point specifies failure of the entire body. For non-uniform fields of stresses (strains) and temperatures, calculation defines "a field" of one or another safety factor from where the points with the least quantity of the given margin must be found.

The margins *of "local"* durability appear varied in different points. Therefore at a moment when the safety factor in the "weakest" point decreases to $n_t = 1$, the other parts of a detail can still keep functioning. The opportunity of a structure to continue operating after damage in its separate points or parts is named "*construction load-carrying (bearing) capacity.*"

It is obvious that durability estimation on the basis of the weakest point gives a greater guarantee of reliability, while calculation of bearing capacity reveals additional design spares which in some cases can be useful.

9.1.2 Non-isothermal-Stressed Structures of Equal Strength

9.1.2.1 Optimization of a Non-isothermal-Stressed State

The general principle of thermo-stressed state optimization of a machine part implies that an optimum combination of fields of operating and failure parameters during all work ought to be found.

In many cases, the temperature field and its change in time are considered known. Then optimization concerns only mechanical parameters.

To align the local safety factors, it is necessary to choose the geometrical forms most corresponding to the character of acting loads and temperatures. The same purpose pursues the reduction of stress concentration at holes, fillets, and other places of sharp change of geometry.

More generally, the optimization of a heat-stressed state of a design should also cover a temperature field. This notably concerns cooled machine parts. However, the simultaneous solving of thermal and mechanical problems of real details is rather complicated, especially in non-stationary statement. Therefore, such optimization is usually carried out using consecutive steps. As computing capabilities grow, optimization techniques also improve.

The following example from Birger et al. (1969) illustrates the elementary approach to optimization of a heat-stressed member.

9.1.2.2 Rotating Non-uniformly Heated Rod

A rotating rod of variable cross section, non-uniformly heated along its length, approximately models a blade of axial turbomachine. Rational distribution of a blade's cross-sectional area, $A(r)$, along the radius, r, of a rotating wheel should be calculated so that in all cross sections of the blade, the margins of long-term strength referring to the temperature field $T(r)$

$$n_\sigma(r) - \frac{\sigma_d[T(r)]}{\sigma_{calc}(r)} \tag{9.13}$$

should satisfy Eq. (9.10), namely

$$n_\sigma(r) \geq [n_\sigma]. \tag{9.14}$$

For a reduction of the wheel's weight, it is desirable to provide Eq. (9.14) as an equality $n_\sigma(r) = [n]$. However, the value of cross-sectional area should not decrease with reduction of the cross-sectional radius because of the maintenance of the dynamic strength.

In Fig. 9.2a, the basic chart of calculation is shown. Curve 1 corresponds to stress distribution in a rod of cross section $A(r) = \text{const}$ rotating with angular speed ω

$$\sigma_{\omega 0}(r) = 0.5\rho\omega^2(R^2 - r^2), \tag{9.15}$$

where ρ is the density of the material, and R is the outer radius of the wheel.

Curve 2 shows admissible values of the stress

$$[\sigma_{calc}(r)] = \frac{\sigma_d[T(r)]}{[n_\sigma]}. \tag{9.16a}$$

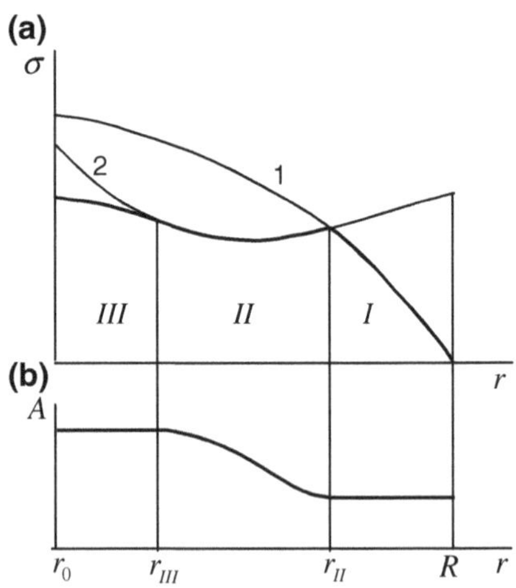

Fig. 9.2 Design scheme of uniform-strength turbine blade

Within site I at the free end of the rod $r = R$ where operating stresses are small, the inequality (9.14) is satisfied for a rod of constant cross-sectional area, since $\sigma_{\omega 0}(r) < [\sigma_{calc}(r)]$. This allows setting a value $A(r) = A_R$ for cross-sectional areas located at $r_{II} \leq r \leq R$ where the size A_R is chosen from constructive reasons.

Within site II where the stresses in a rod of constant cross-sectional area are $\sigma_{\omega 0}(r) > [\sigma_{calc}(r)]$, the stress $\sigma_{\omega}(r)$ varies providing equality

$$\sigma_{\omega}(r) = [\sigma_{calc}(r)]. \tag{9.16b}$$

Distribution of the area along radius $A(r)$ can be obtained by integrating the differential equation of a rotating rod

$$\frac{d}{dr}(\sigma A) + \rho\omega^2 rA = 0$$

under the condition of Eq. (9.16b).

It yields

$$A(r) = A_R \frac{\sigma_{calc}(r_{II})}{[\sigma_{calc}(r)]} \exp\left[\rho\omega^2 \int_r^{r_{II}} \frac{r_1 dr_1}{[\sigma_{calc}(r_1)]}\right]. \tag{9.17a}$$

A function $T(r)$, simulating a turbine blade temperature field, decreases the closer it is to the point where the blade is fixed in the disk; therefore, the limit of long-term durability and admissible stresses will increase. The area calculated by Eq. (9.16a) can start decreasing with radius reduction that is undesirable. Therefore, constant or slightly increasing areas should be supported by a reduction of the radius within site III. The value of radius r_{III} is defined by condition

$$\frac{d\sigma_{\omega 0}}{dr} = \frac{d[\sigma_{calc}(r)]}{dr}, \tag{9.17b}$$

from which the zero derivative $dA/dr = 0$ can be found numerically or graphically.

The calculated theoretical laws of optimum stress $\sigma_{\omega}(r)$ and areas of cross-sectional $A(r)$ distributions, shown in Fig. 9.2a, b, respectively, can serve as a useful start for the subsequent designing.

9.1.2.3 Equal Strength and a "Weak" Link

Alongside alignment of the safety factor field within a separate detail, it is necessary to analyze a problem of margin distribution between details. To this goal, various probable deviations of operating regimes must be examined in view of the interaction of parts in the entire system, and also considering operational and economic requirements. It is usually more useful to accept smaller admissible durability for the hottest members than for the colder ones, replacing the first in repairs.

Destructions of various parts of the structure can lead to differing consequences and unequally influencing its reliability as a whole. To protect the most vital parts against failure, a fortiori weak links (as well as "safety device" in electric circuits) are entered. In case of a seldom but possible severe deviation of a regime from normal, a weak link would collapse first and lead to cessation or restriction of structure operation without dangerous destruction.

9.2 Margins of Local Strength

9.2.1 Long-Term Durability

9.2.1.1 One-Regime Operation

Due to Eq. (6.16), an experimental dependence of the long-term strength limit σ_d on time t at constant temperature T is displayed in double logarithmic coordinates by a direct line $A_1 - A_2$ (Fig. 9.3), described by the equality

$$\sigma_{d1}^m t_1 = \sigma_{d2}^m t_2 \tag{9.18}$$

at $m > 1$.

Margins in stress $n_\sigma = \sigma_{d1}/\sigma$ and in time $n_t = t_2/t$ at any point $A(\sigma, t)$ are connected between themselves by the equation

$$n_t = n_\sigma^m. \tag{9.19}$$

The exponent m for heat resisting materials usually lies within the limits of $m = 4-15$, so $n_t \gg n_\sigma$. For example, at $n_\sigma = 1.4$ and $m = 7$, the margin in time is $n_t > 10$.

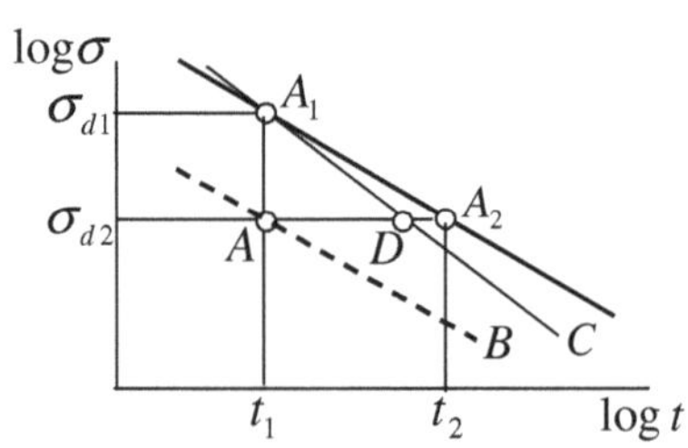

Fig. 9.3 Calculation of durability margins at different temperatures

If a direct line $A - B$ is drawn through a point A (Fig. 9.3) parallel to the line of destruction $A_1 - A_2$, then in any other point situated on this straight line the margins n_t, n_σ will have the same values.

With a rise in temperature, the exponent m mostly decreases. Then, the line of destruction $A_1 - C$ falls more abruptly, and a smaller margin n_t determined by a piece $A - D$ will correspond to the same margin n_σ.

Margins of thermal strength, including the temperature margin n_T, can be calculated by means of the Larsen-Miller's parameter $P_L = T(C + \log t)$, see Chap. 3. In Fig. 9.4, a typical dependence between logarithms of long-term strength limit $\log \sigma_d$ and "destruction" value of parameter $P_{Ld} = T_d(C + \log t_d)$ is shown. Let, point A refer to operating stress σ, temperature T, and duration t. Having prolonged the line $\sigma = \text{const}$ up to crossing the curve P_{Ld} at point B, we find the piece AB equal to

$$AB = P_{Ld} - P_L = T_d(C + \log t_d) - T(C + \log t), \tag{9.20}$$

and a margin expressed in parameter P_L

$$n_P = \frac{P_{Ld}}{P_L} = n_T \frac{C + \log t_d}{C + \log t}, \tag{9.21}$$

where the temperature margin n_T is determined by Eq. (9.6a).

If there is a deviation of the regime in temperature during the same time $t_d = t$, Eq. (9.21) yields the equality of margins $n_T = n_P$.

A deviation in time for constant temperature $T_d = T$ results in correlation $\log n_t = (P_{Ld} - P_L)/T$.

Having prolonged the line $P_L = \text{const}$ up to crossing a curve of destruction at point C, we find that $\log n_\sigma = AC$.

From Fig. 9.4, it is visible that if a curve $\log \sigma = f(P_L)$ is displaced equidistantly along the vertical axis to point A, a line of constant long-term strength margins n_σ will be shown by a dashed line. If the similar displacement of specified curve CB

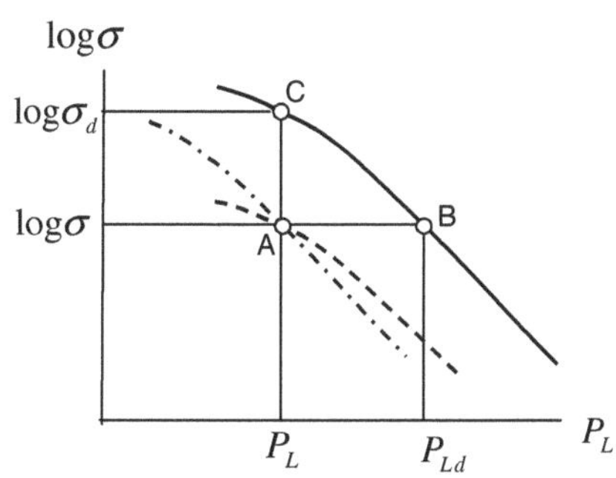

Fig. 9.4 Calculation of durability margins using Larson–Miller's parameter P_L

will be horizontal (seen by dot-and-dashed line), we come for identical duration to a line of constant margins in temperature n_T.

As an approximate estimation of thermal strength, the margins under combined stressed states can be obtained using the same formulas as those at uniaxial extension, substituting characteristics that follow from the corresponding theory of strength (see Chap. 3).

9.2.1.2 Multi-regime Operation

According to the concept on the accumulation of material damage D (see Chaps. 3 and 6), its value to the moment of destruction becomes $D_d = 1$, and the long-term durability margin at a regime k is

$$n_{D,k} = \frac{D_d}{D_k} = \frac{1}{D_k}. \tag{9.22}$$

At a great number of various regimes, a principle of linear summation of damages may be used independently of regimes order. A total damage during several regimes N will be equal to

$$D_\Sigma = \sum_{k=1}^{N} D_k, \tag{9.23}$$

which yields $n_{D,\Sigma} = 1/D_\Sigma$ or

$$\frac{1}{n_{D,\Sigma}} = \sum_{k=1}^{N} \frac{1}{n_{D,k}}. \tag{9.24}$$

It is convenient to use a *reduced* regime (a regime of *reduction).* Damage D_{eqv} at this regime is supposed equal to total damage at all regimes D_Σ. A *reduced margin of long-term durability* is

$$n_{D,red} = n_{D,\Sigma}. \tag{9.25}$$

The general deviation of any state parameter at a multi-regime operation consists of deviations of the same parameter at each regime and also depends on their number. Correct accounting all these effects is difficult, and the deviations usually refer to the reduced regime.

In addition to a one-regime operation, reduced margins can be expressed in various ways.

Reduced Margin of Long-term Durability

According to Eqs. (9.4) and (9.24), the reduced margin in durability is

$$\frac{1}{n_{t,red}} = \sum_{k=1}^{N} \frac{1}{n_{t,k}}, \tag{9.26}$$

where a particular margin of the regime k is

$$n_{t,k} = \frac{t_{d,k}}{t_k}. \tag{9.27}$$

Having chosen any regime with duration t_1, stress σ_1 and temperature T_1 as a regime of reduction, i.e., putting $\sigma_{red} = \sigma_1$ and $T_{red} = T_1$, we can find *a reduced operating time* t_{red}, resulting in the margin $n_{t,red}$:

$$t_{red} = \frac{t_{d,red}(\sigma_{red}, T_{red})}{n_{t,red}}. \tag{9.28}$$

The reduced operating time always exceeds a real time of the reference regime, $t_{red} > t_1$. On the other hand, in view of Eq. (9.27), we find

$$\frac{t_{red}}{t_\Sigma} = t_{d,1} \sum_{k=1}^{N} \frac{1}{n_{t,k}} = \sum_{k=1}^{N} t_k \frac{t_{d,1}}{t_{d,k}}, \tag{9.29}$$

where the common operating duration t_Σ is

$$t_\Sigma = \sum_{k=1}^{N} t_k, \tag{9.30}$$

If the most sever regime, for which all ratios $t_{d,1}/t_{d,k} < 1$, is chosen as the reduced one, then the reduced time t_{red} will be less than the common duration t_Σ.

The use of such reduced regime allows one to carry out the accelerated equivalent tests for long-time durability of machine parts. If possible, load conditions at the reduced regime of a trial can be taken heavier than at a real maximal operating regime with minimal time $t_{d,1}$, which results in further shortening of test duration.

Reduced Margin of Long-Term Strength

Due to Eq. (9.19), an equality $n_{t,k} = n_{\sigma,k}^{m_k}$ is valid for each regime k, where

$$n_{\sigma,k} = \frac{\sigma_d(T_k, t_k)}{\sigma_k}. \quad (9.31)$$

Analogously, for a reduced regime a reduced margin of long-term strength is

$$n_{\sigma,red} = \frac{\sigma_d(T_{red}, t_{red})}{\sigma_{red}}. \quad (9.32)$$

Equations (9.32) and (9.27) yield

$$\frac{1}{n_{\sigma,red}} = \sqrt[m_{red}]{\sum_{k=1}^{N} \frac{1}{n_{\sigma,k}^{m_k}}}. \quad (9.33)$$

At moderate temperatures, margins of long-term durability cease to be informative; therefore, it is more convenient to evaluate durability by margins of long-term strength.

Reduced Stress

Sometimes a reduced stress σ_{red} is used for the estimation of design durability. It follows from Eq. (9.18) that at $m \approx \text{const}$

$$\sigma_{d,red}^m t_{red} = \sigma_{d,k}^m t_k \quad (9.34)$$

which in connection with Eq. (9.33) leads to

$$\sigma_{red} = \sqrt[m]{\sum_{k=1}^{N} \sigma_k^m \frac{t_k}{t_{red}}}. \quad (9.35)$$

A quantity of σ_{red} depends on duration t_{red}, e.g., on choice of the reduction regime.

Smoothly Varying Stress and Temperature

When stress and temperature are any functions of time, then damage increases in $dD = dt/t_d[\sigma(t), T(t)]$ during a time interval dt. A total damage over time t is

$$D(t) = \int_0^t \frac{dt_1}{t_d(t_1)}, \tag{9.36}$$

and a margin of long-time durability to the moment t is

$$n_t(t) = D(t)^{-1} = \left[\int_0^t \frac{dt_1}{t_d(t_1)}\right]^{-1}. \tag{9.37}$$

The condition of long-time strength exhaustion is $n_t(t_d) = 1$. Formulas for other margins can be obtained in a similar way.

The summation of damage and the equivalent margin can also be calculated with the use of Larsen-Miller's parameter (Shorr et al. 1970).

9.2.2 *Thermocyclic Fatigue*

Failures examined above concern low frequency loading at significant stress and temperature ranges within the number of cycles $N_d = 10^2 - 10^5$ corresponding to the area of low-cycle fatigue (LCF).

Methods of estimation of thermocyclic fatigue safety factors are based, basically, on the same principles as calculations of long-term durability. But, more complex characters of thermocyclic fatigue processes demand the use of increased values of admissible margins.

9.2.2.1 Cyclic Durability at Constant Elevated Temperature

Cyclic durability is estimated by a *margin of cyclic durability* due to Eq. (9.11) in which cycle numbers N and N_d refer to operating temperature T and identical stress $\Delta\sigma$ or strain $\Delta\varepsilon$ ranges.

Test results for elevated temperature can be represented in the form as shown in Chap. 4

$$\Delta\varepsilon^{m_\varepsilon(T)} N_d = C_\varepsilon(T). \tag{9.38}$$

The margin n_N is connected with a margin in variable strains $n_{\Delta\varepsilon}$ by correlation

$$n_{\Delta\varepsilon} = \frac{\Delta\varepsilon_d}{\Delta\varepsilon} = n_N^{-m_\varepsilon(T)}. \tag{9.39}$$

Analogous formulas can be written for the relation of N_d and stress ranges and margins $n_{\Delta\sigma} = \Delta\sigma_d/\Delta\sigma$. Generally, margins $n_{\Delta\sigma}$ and $n_{\Delta\varepsilon}$ are not equal.

9.2.2.2 Cyclic Durability at Alternating Temperature

In Chap. 5, it has been shown that temperature alternating at a constant stress leads, even in the elastic area, to the change of strains. On the contrary, temperature alternating at a constant strain leads to the change of stresses. This effect does not allow carrying out trials at temperature alternating “in the pure state.” Therefore, results of cyclic tests at temperature range ΔT are represented in the form of dependences of the thermocycle number $N_{d/\Delta T}$ prior to failure and under certain conditions of stress or strain changes (see Chap. 7). According to this, the thermocycle margins are expressed in the form

$$n_{N/\Delta T} = \frac{N_{d/\Delta T}}{N_{\Delta T}}. \tag{9.40}$$

where $N_{\Delta T}$ is a thermocycle number at operation conditions.

In a number of cases, cyclic changes of stresses and temperatures with corresponding ranges $\Delta\sigma$ and ΔT occur synchronously. At the moment when the stress reaches a maximal absolute value σ_{max}, the temperature also becomes maximal T_{max}. The usual thermocycle tests under a sawtooth cycle $T_{min} \rightarrow T_{max} \rightarrow T_{min}$ (without exposure at maximal temperature) and constant strain range $\Delta\varepsilon$, described in Chap. 7, are carried out in such a way. In this case, a thermocycle number $N_{d/\Delta T}$ before destruction depends on strain range $\Delta\varepsilon$ and maximal cycle temperature T_{max} that allows using Eq. (9.11) with parameters corresponding to temperature $T = T_{max}$.

As the majority of relations for thermocycle durability and long-term strength are identical, the margins of thermocyclic fatigue can be calculated by replacing time t in formulas of Sect. 9.2.1 for number of cycles N.

At multi-regime thermocycles, a reduced number of cycles before destruction or reduced margins of cyclic durability can be used. To consider the rapid accumulation of damage at the stepped transition from one cyclic process to another, it is useful to introduce a factor $a \leq 1$ in the formulas of linear summation of damages.

Equation (9.27) in this case becomes

$$\frac{1}{n_{N,red}} = \frac{1}{a}\sum_{k=1}^{N} \frac{1}{n_{N,k}}. \tag{9.41}$$

A quantity of factor a for the given material depends on temperature and strain ranges and to a lesser extent on the concrete character of thermocycles.

9.2.3 Combined Loadings

9.2.3.1 Thermocyclic Loadings with Exposures

In real conditions, thermocyclic loadings are usually accompanied by exposures at some parts of a cycle. When stress and temperature achieve maximal values at the same moment of time, an exposure under the maximal cycle's temperature plays a decisive role, while exposure at the minimal temperature does not influence durability. But a total (calendar) operating time, naturally, depends on the total working duration.

Such a process is represented in a form of symmetric alternation of stresses in a range $\Delta\sigma$ (i.e., with amplitude $\sigma_a = 0.5\Delta\sigma$) and exposure at each cycle during time t_{dur} at the temperature $T_{\max}$ and stress σ_a. Influence of the duration of transitive cycle parts near to the maximal temperature can be accounted by calculating the reduced time. A material is stressed by σ_a at temperature $T_{\max}$ over N cycles during the total time

$$t_\Sigma = t_{dur}N. \tag{9.42}$$

Dependence $\sigma_{a,d} = f(N, t_\Sigma)$ of failure stress amplitude $\sigma_{a,d}$ on cycle number N and time t_Σ under constant temperature $T_{\max}$ can be represented by a surface of "long-term thermocyclic fatigue failure," shown in Fig. 9.5. Special cases of loading are displayed by schemes I and II, directed along the corresponding axes. The painted sections N = const, t_Σ = const, and also σ_a = const show dependences between each pair of these parameters. Analyzing the specified dependences, it is necessary to keep in mind the time t_Σ to the cycle number N connection.

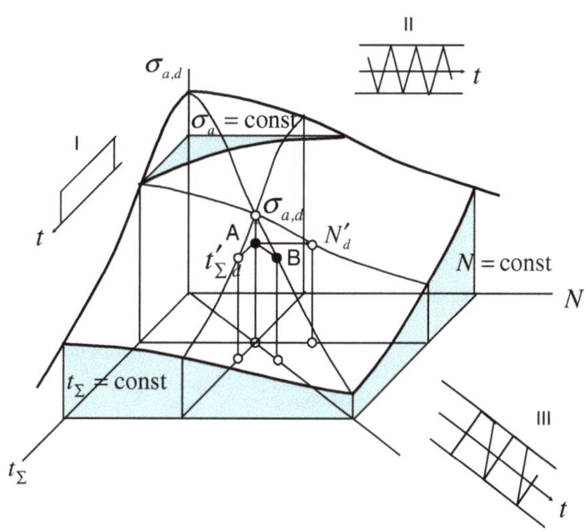

Fig. 9.5 Surface of long-term thermocyclic fatigue failure at $T_{\max}$ = const

For any regime (black point A), specific failure values $\sigma_{a,d}$, N'_d, and also $t'_{\Sigma,d}$ are displayed using light points. The corresponding margins are given as follows:

$$n_\sigma = \frac{\sigma_{a,d}}{\sigma_u}; \quad n_N = \frac{N'_d}{N}; \quad n_t = \frac{t'_{\Sigma,d}}{t}. \tag{9.43}$$

A case, when total loading time t_Σ increases proportionally to the number of cycles N under constant quantities of amplitude σ_a, maximal temperature $T_{\max}$, and exposure t_{dur}, is of great importance for many machine parts. In Fig. 9.5, a ray III with number N_d of destruction cycles and total durability $t_{\Sigma,d} = t_{dur}N_d$ corresponds to such a process (black point B). The margin of thermocyclic durability is

$$n_{t,N} = \frac{N_d}{N} = \frac{t_{\Sigma,d}}{t_\Sigma}. \tag{9.44}$$

For a process with small exposures, this expression yields $n_{t,N} \approx n_N$, and in the case of a small number of cycles, it yields $n_{t,N} \approx n_t$. Generally, a quantity $n_{t,N}$ is always less than the minimum of n_N or n_t.

A curve of thermocycle fatigue with amplitude σ_a and exposures at $T_{\max}$, representing a dependence $f(t_{\Sigma d}, N_d)$, can be described by the equation

$$\left(\frac{t_{\Sigma d}}{t_d}\right)^a + \left(\frac{N_d}{N_{0,d}}\right)^b = 1, \tag{9.45}$$

where t_d is a long-term durability at $\sigma = \sigma_a = \text{const}$, and $N_{0,d}$ is the number of thermocycles before destruction without exposure.

Constants $a, b \leq 1$ depend mainly on temperature $T_{\max}$ and can accept values from 1 up to 0.2.

Due to Eqs. (9.44) and (9.45), a margin of *long-term thermocyclic fatigue* $n_{t,N}$ can be found numerically from the nonlinear equation (Dul'nev 1971)

$$\left(\frac{n_{t,N}}{n_{t0}}\right)^a + \left(\frac{n_{t,N}}{n_{N0}}\right)^b = 1, \tag{9.46}$$

where the margins n_{t0} and n_{N0} are defined by Eqs. (9.4) and (9.11), respectively.

The influence of an operating time at transitive sites of a cycle can be accounted for by reduced exposure $t_{dur,eqv}$ referring to $T_{\max}$:

$$t_{dur,red} = t_d(T_{\max}, \sigma_a) \int_0^{t_{cycl}} \frac{dt}{t_d[T(t), \sigma(t)]}. \tag{9.47a}$$

Then, a specific margin of durability at $\sigma_a = \text{const}$ will be

$$n_{t0} = \frac{t_d(T_{\max}, \sigma_a)}{t_{dur,eqv} N}. \tag{9.47b}$$

9.2.3.2 Asymmetric Thermocyclic Loading

At synchronous alternation of stress and temperature, the total stress reaches the maximum absolute value $|\sigma = \sigma_m + \sigma_a|_{\max}$ at $T_{\max}$. Here, σ_m and $\sigma_a = 0.5\Delta\sigma$ are the mean stress and amplitude over the cycle, respectively.

By analogy to conditions of isothermal cyclic fatigue failure, it is possible to accept the following expression as a limit of thermocyclic fatigue at asymmetric loading

$$\sigma_{d,m} = (\sigma_d)_{-1}\left(1 - \frac{\sigma_m}{\sigma_d}\right), \tag{9.48}$$

where $(\sigma_d)_{-1}$ is a thermocyclic fatigue limit at a symmetric cycle, for which $\sigma_{\min}/\sigma_{\max} = -1$, and σ_d is a limit of long-term durability at temperature $T_{\max}$.

A thermocyclic fatigue margin in the stress amplitude will be

$$n_{\sigma,a} = \frac{\sigma_{d,m}}{\sigma_a}. \tag{9.49a}$$

and a margin of thermocycle durability

$$n_{N/\Delta T} = \frac{N_{d/\Delta T}}{N_{\Delta T}} = n_{\sigma,a}^{-m(T_{\max})}. \tag{9.49b}$$

9.3 Bearing Capacity of Mechanical Structures

9.3.1 Material and Structure Destructions

9.3.1.1 Non-isothermal Structure Destruction

At constant temperature, a strength limit is not achieved simultaneously at varies points of a body. The initial damage at any point leads to the stress change in other parts of the body. The stresses in points that were less stressed earlier will grow until the body destructs entirely.

As a rule, local cracks of thermal LCF are formed in zones where the stress concentrates. As the number of cycles grows, so do the sizes of the cracks. The rate of crack development and its quantity, critical for design reliability, is studied by using special methods of failure mechanics.

The moments between the appearance of the first local damage and full structural failure or cessation of its normal function can be notably differ, especially for the destruction of LCF.

At high temperatures, the bearing capacity of a structure is affected by creep development.

Testing of natural details or their models gives the fullest representation on the bearing capacity of a structure. But reproduction of various operational conditions is difficult or in general impossible. There are some ways of bearing capacity approximation when designing the structure.

9.3.1.2 Ideally Plastic Body

A non-isothermal bearing capacity for designs of materials with high plastic properties can be estimated by a simple method based on the theory of the limit balance of ideally plastic bodies.

In Fig. 9.6a, a system of $z = 5$ parallel rods of identical cross-sectional area A is shown. The rods are assumed to be heated up to different temperatures T_k $(k = 1,\ldots, z)$ with corresponding thermal elongation, and in this hot position connected among themselves by a rigid plate.

In this state, the system starts to be extended by gradually increasing the total force $F_\Sigma = \sum_{k=1}^{z} F_k$, so all rods obtain identical strain $\varepsilon_k = \varepsilon_0$. The rod material is assumed to be elastoplastic without hardening. At different temperatures T_k, the rods have different Young's moduli E_k and different yield points in stress $\sigma_{y,k}$ and in strain $\varepsilon_{y,k} = \sigma_{y,k}/E_k$. Deformation diagrams expressed as $F_k(\varepsilon_k) = \sigma_k A = f(\varepsilon_k)$ and dependence $F_\Sigma(\varepsilon_0)$ (in the same scale) are shown in Fig. 9.6b.

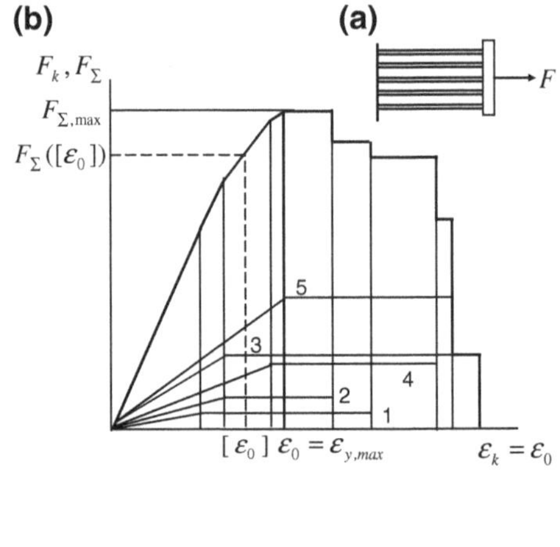

Fig. 9.6 Limit state calculation for a rod structure of a high plastic material

In the elastic state for $\varepsilon_0 \le \varepsilon_{y,k}$, the total force is

$$F_\Sigma(\varepsilon_0) = A\sum_{k=1}^{z} \sigma_k(\varepsilon_0) = A\varepsilon_0 \sum_{k=1}^{z} E_k(T_k). \tag{9.50}$$

The first rod $k = 1$, having the least value $\varepsilon_{y,1}$, will reach its yield point and break at the total force $F_{1\Sigma}$. Afterward, the increase of total force $F_\Sigma(\varepsilon_0)$ is shown by a broken bold line. Bearing capacity of the system will be exhausted when stresses in all rods reach their yield points $\sigma_{y,k}$. At this moment,

$$(F_\Sigma)_{\max} = A\sum_{k=1}^{z} \sigma_{y,k}(T_k). \tag{9.51}$$

If boundary conditions maintain the further gradual increase of the common deformation, total force will reduce as the rods break. However, in view of the bearing capacity of the system, this does not matter.

The margin of bearing capacity of the system referring to the nominal operating force F'_Σ is equal to

$$n_F = \frac{(F_\Sigma)_{\max}}{F'_\Sigma}. \tag{9.52}$$

If a structure working capacity is limited by admissible strain $[\varepsilon_0]$, the bearing capacity margin expressed by strains is

$$n_\varepsilon = \frac{[\varepsilon_0]}{\varepsilon'_0}. \tag{9.53}$$

and expressed by force

$$n'_F = \frac{F_\Sigma([\varepsilon_0])}{F'_\Sigma}. \tag{9.54}$$

Values of the force $F_\Sigma([\varepsilon_0])$ and $[\varepsilon_0]$ are shown in Fig. 9.6b by dashed lines.

9.3.1.3 Hardening Body

For hardening materials, Eqs. (9.50)–(9.51) state a "bottom" evaluation of the bearing capacity. In order to obtain an estimate of the upper bearing capacity of the system, we assume that the stresses in all rods simultaneously reach their strength limits, $\sigma_{B,k}(T_k)$. In the given case, this would be possible if strains at destruction at various temperatures are constant. Then

$$F_B = A \sum_{k=1}^{z} \sigma_{B,k}(T_k). \tag{9.55}$$

Applying this approach to long-term strength, one can insert a limit of durability $\sigma_{d,k}(T_k, t)$ instead of $\sigma_{B,k}(T_k)$ into an analogous equation. Then

$$F_d(t) = A \sum_{k=1}^{z} \sigma_{d,k}(T_k, t). \tag{9.56}$$

A similar approach is used for the estimation of breaking rotation frequency of turbomachine disks (Birger et al. 1993).

Other criteria of definition of a system bearing capacity were proposed in Serensen et al. (1963).

In reality, the deformation curves $\sigma_k(\varepsilon_k)$ and also the ultimate stresses $\sigma_{B,k}(T_k)$ and strains $\varepsilon_{B,k}(T_k)$ at different temperatures are not constant. The assumed pattern of curves, applied to the previous example, is shown in Fig. 9.7.

While neither rod has been collapsed, i.e., $\varepsilon_0 < \varepsilon_{B,k}$, the total force is defined by Eq. (9.51).

At the strain $\varepsilon_0 = \varepsilon_{B,1}$, the weakest rod $k = 1$ breaks. The total resistance force of the system then decreases and becomes less than the outer force $F_{1\Sigma}$ as shown by double line. But due to the condition of the static balance, the other rods will be filled up at increased strain. A similar situation will arise at the destruction of the following rod $k = 2$, and only after the third rod breaks at $\varepsilon_0 = \varepsilon_{B,\max}$ does a further increase of loading become impossible. So the limit state (bearing capacity) of the given system is determined as

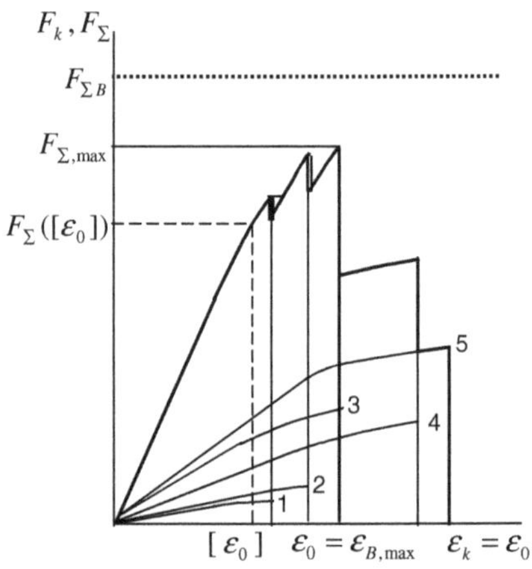

Fig. 9.7 Limit state calculation for a rod structure of a hardening material

$$F_{\Sigma\max}(\varepsilon_{0\max}) = A \sum_{k=z-l}^{z} \sigma_k(\varepsilon_{0\max}), \tag{9.57}$$

where l is the number of rods which have already been broken. In the example given above, $k = 2$ and $\varepsilon_{0\max} = \varepsilon_{B,3}$.

The quantity of $F_{\Sigma\max}$ is lower than the "top" estimation of the ultimate load $F_{\Sigma B}$ calculated by Eq. (9.55) and shown in Fig. 9.7 by dotted line. In some cases, the bearing capacity can be limited by admissible deformation $[\varepsilon_0]$.

9.3.2 Bearing Capacity Margins of Machine Parts

9.3.2.1 Control of Local Damage Parameters

Calculations of the bearing capacity of a structure under short-term static loading with the use of modern FE software packages might consist of the following:

- Two failure criteria should be established:
  (a) Criterion for the durability of a material at the combined stress state; for example, an equivalent stress (see Chaps. 4 and 6) and its limiting value at which local material damage occurs;
  (b) Criterion for the bearing capacity of a part or a whole structure, for example, the permissible stretch of a disk rim or permissible rate of stretch increase.
- Detailed 3D calculation of elastoplastic deformation of a detail is carried out under gradually increasing loads, and a "critical" point in which the criterion of local damage has its maximal value is found.
  If a temperature field alters, one takes into account its influence on the redistribution of stresses and limiting value of durability criterion of the material.
- At each computation step, the current and limiting values of the durability criteria in the selected critical point are compared.

If criterion (a) of a material durability at this point reaches its value at local destruction, the given element can be excluded from the design model, another critical point must be found, and the computation proceeds. The load at which there occurs a full loss of bearing capacity of the structure is estimated according to criterion (b). The bearing capacity margin of a design is defined by the ratio of the value of the destruction load to its nominal size under formulae of the type Eq. (9.54).

For the criterion of bearing capacity, it is possible to use the specific potential energy of deformation U in a zone of stress concentration.

During the process of step-by-step design, the increasing value of specific potential energy of deformation at the most stressed point may be calculated by the equation

$$U(t) = \int_0^t \sigma_{ij} d\varepsilon_{ij} \quad (9.58)$$

and destruction is expected at the moment $t = t_d$ when $U(t_d) = U_d$.

A "critical" quantity of the specific potential energy of deformation U_d is defined by the area under the curve of deformation diagram (see, e.g., Fig. 3.12) referred to the destruction moment

$$U_d = \int_0^{\varepsilon''_d} \sigma_0 d\varepsilon_0. \quad (9.59)$$

The ratio $\bar{U} = U(t)/U_d$ can give a good appreciation about material damage accumulation as it includes not only the shape but also the volume deformations.

Comparison of experimental and numerical results for such an approach is presented in Servetnik (2012).

In Fig. 9.8, a vertical line $\bar{n}_d$ corresponds to the burst rotation speed obtained under a spin test of a turbine disk, where a ratio $\bar{n} = n/n_0$ refers to operational value n_0 for this disk. Critical quantities of the specific potential energy of deformation U_d and calculation dependencies of the energy $U(\bar{n})$ in a zone of maximum stress concentration versus rotation speed $\bar{n}$ are displayed. Calculations were fulfilled with the use of the deformation diagrams obtained for the several specimens cut out from similar disks. Values $U_{d,m}$ and $U_{d,\min}$ and curves $U_m(\bar{n})$ and $U_{\min}(\bar{n})$ were defined using average and minimum material characteristics, respectively.

Calculated values of the burst rotation speed are close to the experimental value $\bar{n}_d$ obtained at the disk spin test

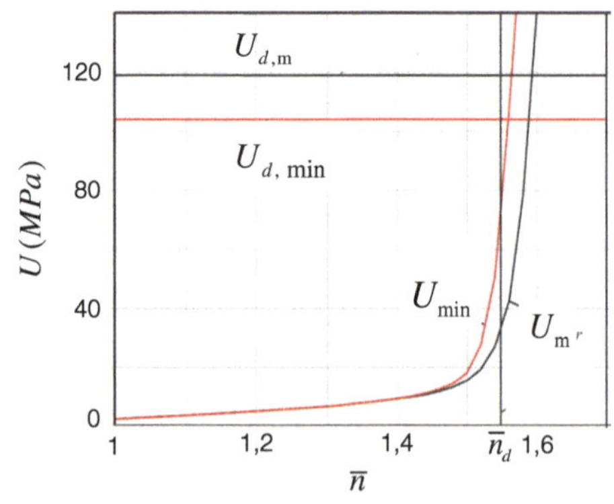

Fig. 9.8 A disk burst speed due to spin test and calculations with using of specific potential energy at the most stressed point (Servetnik 2012)

9.3.2.2 Control of Global Damage Parameters

Not only do local characteristics of the process (stresses, strains, specific deformation work in volume unit) increase with gradual growth of loads, but also the parameters of a whole structure (for example, the deformation work in all volume) increase.

Therefore, an estimation of the limiting condition of a structure should be carried out instead of stress/strain control.

Calculations show that the estimation of the moment of final fracture using various global characteristics is less sensitive to local destruction criteria and leads to the results which do not usually transcend the limits of test results scatter (Karimbaev and Starodubtzev 2010).

9.4 Probabilistic-Statistical Approaches to Strength Estimation

9.4.1 Probability of Failure

9.4.1.1 Casual Character of Thermal Integrity Parameters

As specified in the beginning of this chapter, when estimating design thermal integrity, inevitable casual deviations should be considered. The deviations can concern operating conditions (loads, temperatures, regimes duration) and properties of materials, production and assembly tolerances, approximate nature of solution methods, and an opportunity of influence from other random factors. The scatter of characteristics can have various origins and be presented in different ways.

In the case of controllable productions, small deviations of material mechanical characteristics can be shown at constant stresses and temperatures in the form of *random variables*, slowly varying in time or time independent.

By statistical processing of array test results of many alloys, it has been established that distributions of mechanical characteristics comply with the normal law for the fixed time moments. The temporal characteristics (duration and number of cycles before destruction) usually comply with the logarithmic normal law (usually named the lognormal law).

Small geometrical deviations within the production tolerances might influence the durability in zones of stress concentration and contact strength, and also the operation of the cooling system making changes in temperature fields.

As a rule, mechanical and thermal loads affecting machine parts are *stochastic functions of time*. These functions can usually be approximately presented as *a set of random quantities* allocated for one or several typical operating conditions.

Laws and parameters of the distribution of such random quantities can be obtained by studying the operating experience of similar machines.

The scatter of parameters can take place in three different directions:

- when some nominally identical details are present in a machine or design (e.g., blades of the same driving wheel);
- when an array ("a park") of nominally identical machines is used for operating under conditions that are assumed to be the same; and
- when a machine is used for operation under different nominally identical programs.

In reality, all three types of parameter scatter can arise simultaneously.

Some parameter scatter alters in operation time. Analysis of operational data shows (Velikanova 1990) that mean distribution characteristics mainly change, while parameter dispersions remain rather stable.

9.4.1.2 Theoretical Probability of Failure

Let us be distracted from real features of operation of machine parts and admit that distributions of probability $P(\sigma_a)$ of operating stress σ_a and $P(\sigma_d)$ of breaking stresses σ_d are known. Then, the theoretical probability of destruction P_d can be determined as the probability of a casual event

$$P_d(\sigma_a \geq \sigma_d), \tag{9.60}$$

where σ_a and σ_d are the continuous independent positive random variables.

In Fig. 9.9a, curves of distribution densities $p(\sigma_a)$ and $p(\sigma_d)$, which are assumed to cross each other in some narrow area of values σ, are schematically shown.

The entire probability of an event due to Eq. (9.60) is determined as the sum of all possible combinations of probabilities of products

$$dP(\sigma_d) = p(\sigma_d)d\sigma_d \quad \text{and} \quad P(\sigma_a) = \int_{\sigma_d}^{\infty} p(\sigma_a)d\sigma_a,$$

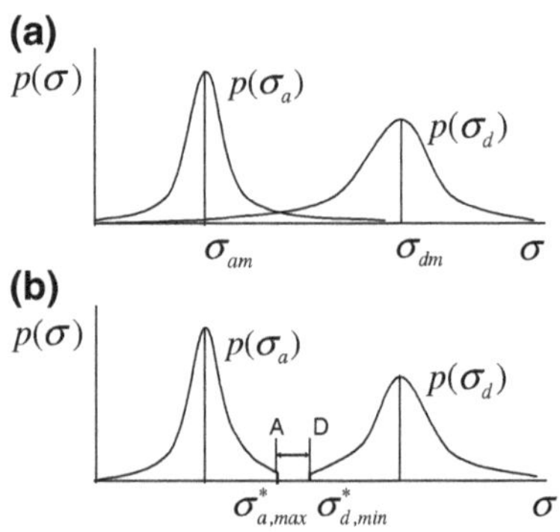

Fig. 9.9 Distribution densities of acting $p(\sigma_a)$ and damaging $p(\sigma_d)$ stresses; **a** curves intercrossed with each other, **b** curves separated by a margin AD

where

$$P_d = \int_0^\infty p(\sigma_d) \left[\int_{\sigma_d}^\infty p(\sigma_a) d\sigma_a \right] d\sigma_d. \quad (9.61)$$

Unfortunately, a direct use of Eq. (9.61) for practical estimation of non-destruction probability $P_{nd} = 1 - P_d$ is impossible. It is connected (besides reasons resulted above) with the following arguments:

- The use of Eq. (9.60) for the definition of the failure probability assumes that random variables σ_a and σ_d are completely comparable characteristics of the same physical phenomena. However, destruction stresses are established on the basis of standard sample tests, while operating stresses are determined by calculations or strain measurements (gauging) on installations of casual copies of working objects. Therefore, the probability of event $\sigma_a \geq \sigma_d$ cannot be completely identified with the probability of real failure.
- Admissible probabilities of destruction $[P_d]$ in machinery are small (10^{-4} to 10^{-5} and below) and defined by "tails" of distributions σ_a and σ_d. A distribution of probability $P(\sigma_d)$ and sometimes also of $P(\sigma_a)$ can only be established experimentally, but within a practically realizable number of measurements, the functions of the distribution based on these measurements cannot be used with a comprehensible degree of trust for calculation of such small probabilities. To find a function $P(\sigma_a)$ by calculation that in some cases is principally possible, it is necessary to take into consideration a great number of casual design factors.
- Rare detail destructions that happen, despite the presence of developed monitoring systems, are mostly defined by unforeseen mistakes of manufacturing, operation, and human factor, i.e., essentially other sources than the reasons of small casual deviations of parameters near to the center of dispersion. Therefore, statistically certain probability of very rare events cannot be estimated even by the significant increase in the number of measurements or calculated variants describing the mentioned deviations.

In view of this, the theoretical probability of destruction calculated by Eq. (9.61) can only be considered as a *conditional quantity* describing design intensity but not reflecting the true probability of destruction. Therefore, the strength condition may be represented in the form of

$$P_d \leq [P_d], \quad (9.62)$$

where $[P_d]$ is an acceptable value which must be established by operating experience.

The size P_d can also be used for comparative estimations of changes introduced into a design from the standpoint of its integrity.

9.4.2 Probabilistic Safety Factors

9.4.2.1 Probabilistic-Statistical Safety Factor

From the previous analysis, it follows that between a maximal value of a random variable σ_a and minimal value of σ_d calculated on the basis of the probability approach, there should always remain some "distance" AD, as shown in Fig. 9.9b. Such a distance can be ensured by a safety factor considered as not being a random variable.

In Birger (1970) and Shorr et al. (1972), it was offered to use *a determined probabilistic-statistical margin* K^* as a ratio of statistically extreme values of destruction $\sigma^*_{d,\min}$ and operating $\sigma^*_{a,\max}$ stresses

$$K^* = \frac{\sigma^*_{d,\min}}{\sigma^*_{a,\max}}. \tag{9.63}$$

Each of the stresses $\sigma^*_{d,\min}$ and $\sigma^*_{a,\max}$ is estimated at admissible distance from the corresponding distribution centers (within areas of "3 sigma") and can be determined, reliably enough, using the distributions σ_a and σ_d.They describe the influence of numerous typical random factors occurring within these areas. As for strength insurance under the affects of less probable factors, it is provided by the margin $K^* > 1$. The necessary value of K^* must be established from the analysis of operating experience and calculations.

Distributions $P(\sigma_a)$ also $P(\sigma_d)$ near to the centers of distributions in the majority of practically important cases are well described by normal law. The casual characteristics, reflecting the influence of time t_d or number of cycles N_d before destruction, are usually described more appropriately by lognormal law of distribution where the argument is $\log t_d$ or $\log N_d$.

Parameters of distribution $P_a = P(\sigma_a)$ are approximately established by results of the limited number z_a of measurements or calculations, and parameters of distribution $P_d = P(\sigma_d)$—by z_d number of measurements.

The sampling mean value of the operating stresses $\bar{\sigma}_a$ is determined as

$$\bar{\sigma}_a = \frac{1}{z_a}\sum_{k=1}^{z_a} \sigma_{ak}, \tag{9.64}$$

the sampling mean square deviation is

$$s_a = \sqrt{\frac{1}{z_a - 1}\sum_{k=1}^{z_a} (\sigma_{ak} - \bar{\sigma}_a)^2}, \tag{9.65a}$$

and the factor of variation is

$$v_a = \frac{s_a}{\bar{\sigma}_a}. \tag{9.65b}$$

Analogous formulae are referred to sampling parameters of distribution P_d.

At the chosen values of probability P_a^* and P_d^* of statistically extreme quantities $\sigma_{a,\max}^*$ and $\sigma_{d,\min}^*$, statistically extreme stresses should satisfy the equations

$$\begin{aligned} P\left[P(\sigma_a \le \sigma_{a,\max}^*) \ge P_a^*\right] &= \gamma, \\ P\left[P(\sigma_d \ge \sigma_{d,\min}^*) \ge P_d^*\right] &= \gamma, \end{aligned} \tag{9.66}$$

where γ is a confidence probability used in the calculations with a view that the probabilistic analysis is based on limited sampling data.

According to the methods of mathematical statistics, solutions of Eqs. (9.66) can be represented in the form of relations

$$\begin{aligned} \sigma_{a,\max}^* &= \bar{\sigma}_a\left[1 + k_t(P_a^*, \gamma, z_a)v_a\right], \\ \sigma_{d,\min}^* &= \bar{\sigma}_d\left[1 - k_t(P_d^*, \gamma, z_d)v_d\right] \end{aligned} \tag{9.67}$$

where k_t are the so-called one-sided tolerant factors depending on probabilities P^*, confidential factor γ, and available volumes of sampling z_a and z_d.

Values of factors $k_t(P^*, \gamma, z)$ for normal distribution are shown in some works on the theory of probability. Obviously, the accuracy of calculation must increase with an increase in the sample volume z_a.

Statistically extreme quantities $\sigma_{a,\max}^*$ and $\sigma_{d,\min}^*$ depend on rational values of border probability of distributions P_a^*, P_d^*, and also γ, which must be chosen in view of concrete features of specific problems.

In Shorr et al. (1972), such an approach has been applied to the estimation of statistically maximal variable stresses $\sigma_{a,\max}^*$ determined by the results of stress measurement of 400 samples of turbine blades in operation conditions.

The analysis has shown that to the specified requirements, it is better to satisfy the values $P^* = 0.995$ and $\gamma = 0.95$ at which the first Eq. (9.67) is approximated precisely enough by the expression

$$\sigma_{a,\max}^* = \bar{\sigma}_a\left\{1 + 3[(z_a + 1)/(z_a - 1)]^2 v_a\right\},$$

which is valid at $z_a > 1$.

For several turbine wheels, the variable resonant stresses are measured at different numbers of blades z_a, and the maximum values of measured $\sigma_{a,meas}$ and forecasted $\sigma_{a,\max}^*$ stresses are found.

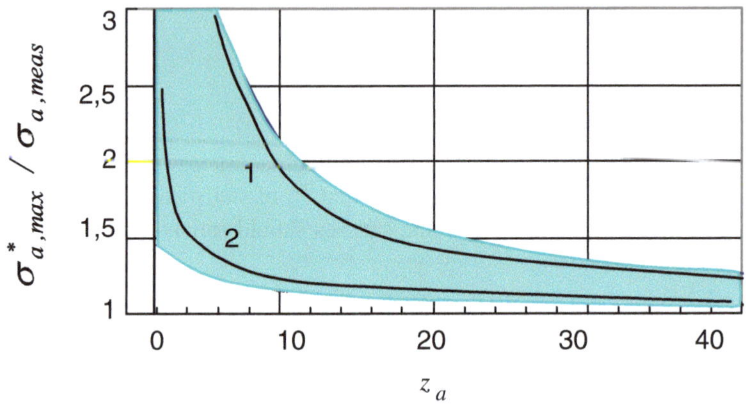

Fig. 9.10 Ratio of forecasted stresses $\sigma^*_{a,\max}$ to maximal measured stresses $\sigma_{a,meas}$ versus number of tested blades

In Fig. 9.10, the area which encloses the values of ratio $\sigma^*_{a,\max}/\sigma_{a,meas}$ versus number z_a is colored, where the stresses $\sigma^*_{a,\max}$ were calculated using the various experimental values of $\bar{\sigma}_a$ and v_a. Lines 1 and 2 correspond to the analogous ratio $\sigma^*_{a,\max}/\sigma_{a,meas}$, but with stresses $\sigma^*_{a,\max}$ calculated by constant factors of variation $v_a = 0.4$ and $v_a = 0.1$, respectively. All experimental data are stacked within the specified area. At small number z_a, the size $\sigma^*_{a,\max}$ can exceed the maximum measured stress two-three times. It specifies that such a small volume of tests is insufficient to directly estimate the possible maximum stresses in view of their scatter. With increase in number z_a, the ratio $\sigma^*_{a,\max}/\sigma_{a,meas}$ decreases, remaining more than unit.

9.4.2.2 Safety Factor as a Random Variable

The ratio of the random variables σ_a and σ_d also represents a random variable—safety factor (margin)

$$K = \frac{\sigma_d}{\sigma_a}, \tag{9.68a}$$

which implies a value $K \leq 1$ at the coming of the event $\sigma_a \geq \sigma_d$.

For a normal distribution, a mean value of margin is expressed through mean stresses $K_m = \sigma_{dm}/\sigma_{am}$ (Fig. 9.9a), and a variation of distribution $P(K)$ can be determined as

$$v = \frac{\sqrt{v_a^2 + K_m^2 v_d^2}}{K_m - 1}, \tag{9.68b}$$

where v_a and v_d are the variations, accordingly, of distributions σ_a and σ_d.

The condition $K > [K]$, where $[K]$ is an admissibly calculated value of safety factor, can be used as a conditional characteristic representing a probabilistic feature of machine member strength.

9.4.3 Probabilistic-Statistical Margin of Durability

9.4.3.1 General Approach

Probabilistic-statistical margins of durability, expressed in time or in a number of loading and heating cycles, can be basically calculated analogously to the approaches of Sect. 9.4.2 replacing the normal distribution parameters to lognormal one.

Probabilistic-statistical characteristics of loadings and temperatures, affecting machine parts in working conditions, can be obtained by calculation, based on the data referred to operation of similar units.

As for long-term trials of machine parts, they are expensive and require long period of time, in view of which the tests are usually carried out with very small number of copies. To account for the casual character of destruction parameters, some special approaches are necessary. At small volumes of a specific test, it is impossible to accurately calculate its mean square deviation s_d and instead of this an averaged statistical value $s_{d,m}$ is used.

Below, for concreteness, a margin of cyclic durability K_N of a machine part, e.g., a turbine disk made of a typical material with the use of standard technology, is discussed.

From the statistical analysis of cyclic test results of various turbine disks, it is established that a distribution of cycle number before destruction N_d is described by the lognormal law. A typical scatter of values N_d connected with the disks' material and manufacturing deviations occupies some range of values from $N_{d,\min}$ to $N_{d,\max}$. The ratio

$$\tilde{N}_d = \frac{N_{d,\max}}{N_{d,\min}} \tag{9.69}$$

might be taken for disks' typical materials used in turbomachines as the quantity $\tilde{N}_d = 4{-}6$, and $\log \tilde{N}_d = 0.60{-}0.78$ where

$$\log \tilde{N}_d = \log N_{d,\max} - \log N_{d,\min}.$$

A real area of realizations of $\log N_d$ distribution keeps within a range of values $\pm 3 s_d$ and $\log \tilde{N}_d \approx 6 s_d$. Assuming $s_d \approx s_{d,m}$ yields

$$s_{d,m} = \frac{\log \tilde{N}_d}{6} \approx 0.1 - 0.13.$$

The average value of destruction cycle number $N_{d,m}$ for the distribution $\log N_d$ is connected with the minimal value of $\log N_{d,\min}$ by dependence

$$\log N_{d,m} = \log N_{d,\min} + 0.5 \log \tilde{N} \approx \log N_{d,\min} + (0.3 - 0.39),$$

or

$$N_{d,m} = N_{d,\min} \sqrt{\tilde{N}_d} \approx (2 - 2.45) N_{d,\min}. \tag{9.70}$$

The quantity $N_{d,\min}$ is the minimal statistically probable value of destruction cycle number for the tested disk's assemble in the given trial conditions.

The maximal acceptable value of cycle number for such a disk's assemble under operating conditions $N_{a,\max}$ is

$$N_{a,\max} = \frac{N_{d,\min}}{K_N^*}, \tag{9.71}$$

where K_N^* is the *statistically probable margin of disk's cyclic durability.*

Necessary cyclic durability of the disk is provided with a condition of durability

$$K_N^* \geq [K_N^*], \tag{9.72}$$

where $[K_N^*]$ is an acceptable value of the durability margin.

Equation (9.72) takes into account that the disk samples might be tested at loadings and temperatures that only roughly simulate possible scatter of service conditions.

9.4.3.2 Using the Sampling Tests

To estimate the statistically probable value $N_{d,\min}$ for an investigated disk with the necessary margin, cyclic trials of several copies, z_{test}, of such disk are carried out: Each of them is tested until $N^{(k)}$. This group represents a casual sample of the disk's universe corresponding to the sample of unrestrictedly large size. Tests of a part of disks $N_d^{(z_1)}$ can come to an end with LCF crack detection that can be qualified as the beginning of destruction. The tests of the other part of disks $N^{(z_2)}$ could be interrupted for the different reasons with undetected LCF cracks. Since the value of a minimal statistically probable destruction $N_{d,\min}$ must be less than the minimal value $N_{d,\min}^{(z_1)}$ obtained by the trial, results of tests interrupted at $N^{(k)} < N_{d,\min}^{(z_1)}$ may be excluded as not informative. The disks with undetected LCF cracks tested until the cycles $N^{(k)} > N_{d,\min}^{(z_1)}$ can be assumed to fail at the cycle number $N^{(k)} + 1$, the next after checked.

So, the informative sample of the tested disks z may be less than z_{test}.

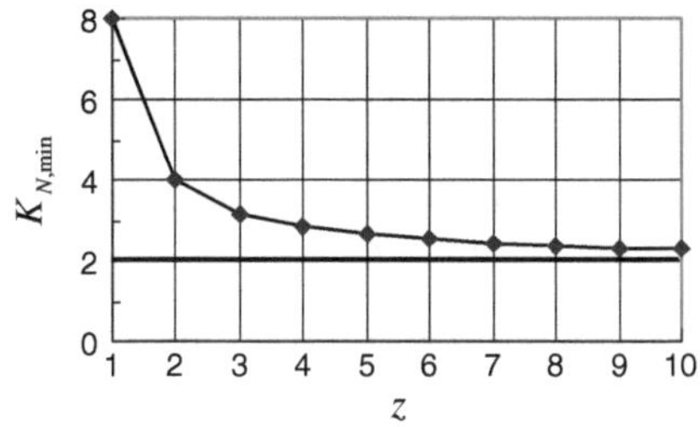

Fig. 9.11 Ratio of recommended margin $K_{N,\min}$ versus quantity of tested disks z_d

Unlike a sampling mean square deviation $s_d^{(z)}$, a sampling mean value $N_{d,m}^{(z)}$ may be found more accurate on the base of a small number of tested disks.

Known methods and tables of the statistical analysis for the lognormal law of distribution allow statistically probable sizes of $N_{d,m}$ and $N_{d,\min}$ to be found. This is done by use of the averaged value of mean square deviation $s_{d,m}$, the sampling values $N_{d,\min}^{(z)}$ or $N_{d,m}^{(z)}$ estimated from specific tests, and a set value of a confidence probability γ, see Nozhnitsky et al. (2006). A curve in Fig. 9.11 displays the ratio $K_{N,\min}(z) = N_{d,\min}^{(z)}/N_{d,\min}$ versus specimen number z within the range $1 \leq z \leq 8$. The failure probability was taken as $P^* = 1.35 \times 10^{-3}$ that corresponds to the quintile $3s_d$ and confidence probability $\gamma = 0.95$ for $\log \tilde{N} = 0, 13$.

A value of minimum statistically probable value of destruction cycle number is

$$N_{d,\min} = N_{d,\min}^{(z)}/K_{N,\min}(z). \tag{9.73}$$

If none of the tested disks z have LCF cracks, then the previous approach, in which $N_{d,\min}^{(z)}$ will refer to the minimum cycle number of the tested disks, is still valid. However, tests of disks with a small value $N^{(k)}$ conduct to fall of sampling size $N_{d,\min}^{(z)}$; therefore, the value of statistically predicted cycle number $N_{d,\min}$ can start decreasing despite an increase in the number of tested disks. So, subsequent increase of number z becomes useless.

References

Birger IA (1948) Strength margins at variable stresses. Vestnik Mashinostroeniya, No.6. Moscow, pp 5–14 (in Russian)

Birger IA (1970) Damage probability, strength margins and diagnosis. Problems of solid mechanics. Sudostroenie, Leningrad, pp 70–81 (in Russian)

Birger IA, Shorr BF (eds) (1975) Thermal strength of machine parts. Mashinostroenie, Moscow (in Russian)

Birger IA, Shorr BF, Iosilevich GB (1993) Strength design of machine parts, 4th ed. Mashinostroenie, Moscow (in Russian)

Birger IA, Shorr BF, Shneyderovich RM (1969) Designing machine parts for strength. Foreign Technology Division Wright–Peterson AFB OHTO. Ft. Belvoir Defense Technical Information Center (Trans. from Russia, 2nd ed, 1966). Mashinostroenie, Moscow

Bolotin VV (1965) Statistical methods in structural mechanics. Stroyizdat, Moscow (in Russian)

Dul'nev RA (1971) Damage summation and strength condition at thermocycle loading. Problemy Prochnosty 10:101–104 (in Russian)

Karimbaev KD, Starodubtzev VV (2010) To design verification of disk bearing capacity. In: Theses of reports at VI International scientific conference, Orenburgskiy Gos. Universitet, Orenburg, pp 185–191 (in Russian)

Nozhnitsky YuA, Lokshtanov EA, Dolgopolov IN et al (2006) Probabilistic prediction of aviation engine critical parts lifetime. In: Proceedings of GT2006 ASME Turbo Expo 2006; Power for Land, Sea and Air, GT2006-91350, Barselona, Spain

Serensen SV, Kogaev VP, Shneyderovich RM (1963) Loadability and strength calculation of machine structures, 1st ed (3rd ed 1975). Mashinostroenie, Moscow (in Russian)

Servetnik AN (2012) Load-carrying capability simulation of aviation gas turbine engine disks. In: Engineering Journal Handbook, vol 10, pp 44–49 (in Russian)

Shorr BF, Kochukov NS, Porter MA (1970) Accelerated equivalent tests of operational reliability of welded machine parts. J. Problemy Prochnosty 2:63–67

Shorr BF, Lokshtanov EA, Khalatov YuM (1972) On a possible approach to probabilistic estimation of vibration strength of turbomachine details. J Problemy Prochnosty 11:11–14 (in Russian)

Velikanova NP (1990) Effect of engine operating time on turbine disk strength. In: Proceedings of International Conference on ICAE-90, Moscow-Zagorsk, CIAM, pp 90–97

Chapter 10
Elastic and Inelastic Thermal Stability

Abstract Non-isothermal loss of stability of machine and structure elements under uniform or varying external loads and temperatures is considered. Stability at elevated temperatures may depend on a certain critical temperature value along with critical values of stress and strain. Peculiarities of structure buckling under temperature stresses in case of constrained deformation are demonstrated. Approaches for evaluating critical parameters (stress, strain, temperature, and operation duration) are presented for lateral buckling of rod elements with a straight and curvilinear axis under elastoplastic deformation and creep. The loss of a cylindrical equilibrium form of a stretched rod due to increase of creep rate evolution is discussed. Loss of stability, caused by non-uniform heating of the bodies of various shapes, is analyzed. Nonlinear calculation of transition to a new equilibrium state caused by a temperature variation is studied in detail using von Mises modified truss.

10.1 Loss of Stability at Increased Temperatures

Influence of heating on structure stability was studied in many publications (Van-der-Neut 1958; Boley and Weiner 1960; Jones 2006, others). Below, some most important features of non-isothermal stability loss and buckling of structural elements are considered.

10.1.1 Problems of Structure Thermal Stability

Under the common term "structure thermal stability," we shall unite problems of elastic or inelastic stability loss of design elements at increased temperature whose influence appears upon many directions:

B.F. Shorr, *Thermal Integrity in Mechanics and Engineering*,
Foundations of Engineering Mechanics, DOI 10.1007/978-3-662-46968-2_10

- Critical loads, stresses, and strains at constant or gradually increasing external forces become smaller due to the temperature alteration of elastic characteristics and diagrams of plastic deformation;
- There arises an effect of elastic or inelastic stability loss at constant loads under increasing temperature;
- Structure stability loss can appear at constant external loads and temperature as a result of creep development;
- Temperature stresses arising due to external restrictions of displacements can result in loss of the general design stability;
- Internal temperature stresses arising under non-uniform heating can lead to local stability loss of some design members.

Quite often a joint influence of the specified factors is shown.

The extensive scientific and technical literature is devoted to problems of various aspects of stability loss. For the analysis of the stability loss phenomenon of deformable systems at normal temperature, some different approaches are used. Their application at increased temperatures is connected with certain peculiarity.

Some solution programs are available in up-to-date commercial packages of finite-element analysis. Therefore, we shall focus the basic attention not to the computing side, but to the analysis of the engineering aspects of processes of stability loss, of which knowledge is necessary for correct performance of practical computations. For the achievement of the specified purpose, simple models are fully suited.

In the given chapter, compressing stresses and strains are considered positive if the opposite does not stipulate.

10.1.2 Critical Load and Temperature at Compression of Rods with a Straight Axis

10.1.2.1 Euler's Formula

For elastic systems, the loss of stability is usually considered, due to Euler's concept, as a condition of an opportunity of occurrence of a new deflected equilibrium form alongside with the initial one. The system parameters corresponding to this condition are named *critical*. In supercritical condition, the initial form of system equilibrium becomes unstable.

The elementary approach to research of the behavior of a system at gradually increasing load consists of the use of "quasistatic" statements when at each value of load, the states of static equilibrium are considered. The same concerns gradually increasing temperature.

In Fig. 10.1, a straight elastic rod of constant cross section of an area A and moment of inertia J compressed at temperature T by the force F is shown. One of

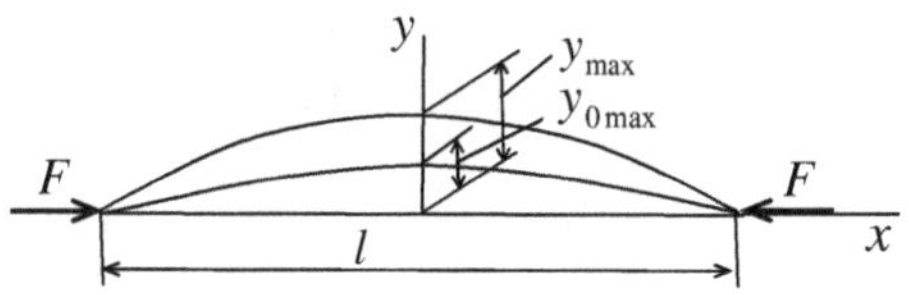

Fig. 10.1 Axial compression of a rod

the ends of the rod can freely move in an axial direction, so thermal stresses at the rod's heating do not arise.

At any moment let the rod obtain a few bends, y_{max}, affected by random lateral loads.

The equation of bending of a compressed rod at small deflections has a known expression

$$EJy'' = -Fy,$$

which has a solution under a boundary condition at the left end $y(0) = 0$ as

$$y(x) = y_{\max} \sin kx,$$

where $k^2 = F/El$, $E = f(T)$, and the maximal deflection y_{max} correspond to the point $x = \pi/2k$.

The second boundary condition (on the right end) is

$$y(l) = y_{\max} \sin kl = 0.$$

It is satisfied in two cases: At any values of force F and temperature T for $y_{max} = 0$ (i.e., for initial rectilinear form of the axis), and also at $y_m \neq 0$ (i.e., at lateral buckling), but only at some discrete (critical) values of the parameter $k = k_{cr} = m\pi/l$, $m = 1, 2, \ldots$. The final condition is satisfied also in two cases:

- At the given temperature T, by *critical force* (Euler's formula)

$$F_{cr} = E(T)Jk_{cr} = \frac{m^2\pi^2 E(T)J}{l^2}; \tag{10.1a}$$

- At the given force F, by Young's modulus at *critical temperature*

$$E(T_{cr}) = \frac{F}{k_{cr}J} = \frac{Fl^2}{m^2\pi^2 J}. \tag{10.1b}$$

The simplest way to find out under which conditions the received forms of balance are steady is to check the balance of the bending moments at the mean section of the rod at $x = 0.5l$. The deflection and the axis curvature at this section are accordingly equal y_m and $\chi_m \approx y''_{x=0.5l} = -(\pi/l)^2 y_m$. For this section, the external bending moment is $M_* = Fy_m$ and the inner bending moment of elastic stress is $M_{**} = -EJ\chi_m$.

At $M_{**} > M_*$, i.e., $F < F_{cr}$ or $E(T) > E(T_{cr})$, after a random bend the rod returns back to its original rectilinear position—retaining stability. At $M_{**} < M_*$, i.e., $F > F_{cr}$ or $E(T) < E(T_{cr})$, the arisen deviation will increase, and rectilinear form of the rod will be unstable. An equality $M_{**} = M_*$ meets the critical conditions $F = F_{cr}$ or $E(T) = E(T_{cr})$ which correspond to the minimum of full potential energy of the system.

A practical value has the first form of stability loss at $m = 1$.

So, the *critical stress* is

$$\sigma_{cr} = \frac{\pi^2 E(T)}{\lambda^2} \tag{10.2a}$$

and accordingly, the Young's modulus corresponding to critical temperature is

$$E(T_{cr}) = \frac{\sigma\lambda^2}{\pi^2}. \tag{10.2b}$$

where the parameter of flexibility of a rod is

$$\lambda = l\sqrt{A/J}. \tag{10.3}$$

Sometimes, it is more convenient to express critical parameters through *critical strain* ε_{cr}

$$\left.\frac{\sigma}{E(T)}\right|_{cr} = \varepsilon_{cr} = \frac{\pi^2}{\lambda^2}, \tag{10.4}$$

which for an elastic condition depends only on the geometrical configuration of a system and boundary conditions, but not on temperature. Then

$$\sigma_{cr} = \varepsilon_{cr} E(T), \quad or \quad E(T_{cr}) = \sigma/\varepsilon_{cr}. \tag{10.5}$$

The given formulas are fair, as long as critical parameters of compression ε_{cr} and σ_{cr} do not exceed the proportionality limits of a material in strain ε_{pr}, or in stress σ_{pr}, i.e., at flexibility of a rod not smaller than

$$\lambda_{\lim} = \frac{\pi}{\sqrt{\varepsilon_{pr}}}. \tag{10.6}$$

For the alloys used in springs and other similar details where very high limits of proportionality are required, it leads to the condition $\lambda_{\lim} \approx 30$. In usual constructive materials, there is $\lambda_{\lim} \approx 50$ and above.

10.1.2.2 Stability of Inelastic Rods

Nonlinear analysis of elastic loss of stability (see e.g., Ponomarev et al. 1952) shows that deviations do not increase once parameters reach critical values. The deviations begin increasing only at subsequent parameters' increase but this process occurs very rapidly. For the example given above, supercritical growth of the maximal deflection $y_{\max}$ in the middle of a rod is determined by the formula

$$\frac{y_{\max}}{l} = \frac{4}{\pi}\sqrt{\sqrt{\frac{F_{cr}}{F}} - \frac{F_{cr}}{F}}.$$

When compressing force is 1 % over critical, the deflection $y_{\max}$ will make 9 % of the length of a rod.

After the loss of elastic stability, bending stresses increase rapidly, which should lead to the destruction of a rod of a material with high fragility. In a more plastic material, the elastic loss of stability occurs only in a case when elastic critical parameters are limited by the flexibility $\lambda_{\lim} = \lambda_y = \pi/\sqrt{\varepsilon_y(T)}$.

The initial site of the elastoplastic deformation diagram can be presented as shown in Figs. 3.2 and 3.10.

In view of decrease of the elasticity modulus with growth of temperature, dependence $\sigma_{cr} = f(\lambda, T)$ on the parameter of flexibility λ for temperatures T_0 and $T_{\max}$ for a material with the expressed yield point σ_y is presented in Fig. 10.2. Here, it is taken into account that the real possible range of decrease of the elasticity modulus at an increase of operational temperature from T_0 to $T_{\max}$ usually does not exceed 50 %. The dependence $f(\sigma_{cr}, \lambda, T) = 0$ can be also expressed like a shaded surface shown in Fig. 10.2.

If the critical elastic stress $\sigma_{cr.e}$ calculated using Eq. (10.2b) exceeds the yield point $\sigma_y(T)$, i.e., $\lambda \leq \lambda_y$, then for non-hardening material (at $E_t = 0$), it is necessary to consider

$$\sigma_{cr}(T) = \sigma_y(T). \tag{10.7}$$

The points λ_y are shown in Fig. 10.2 by the black circle for T_0 and the white one for $T_{\max}$.

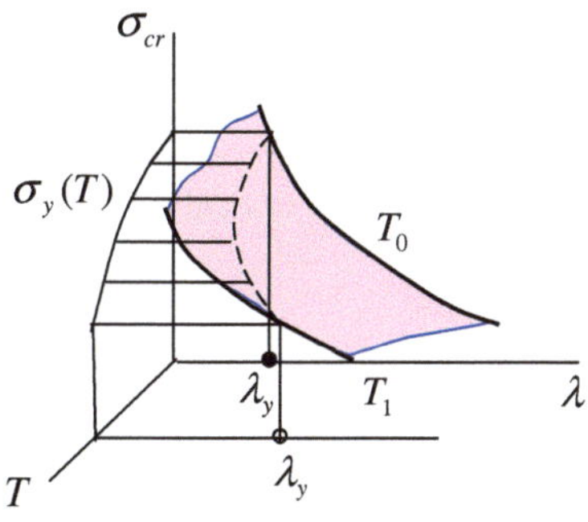

Fig. 10.2 Correlation between critical parameters

In a compressed rod of a hardening material, the rectilinear form of equilibrium can be kept after the occurrence of plastic deformation to the extent of the certain critical conditions. As such, the critical condition of equilibrium between the external and internal moments $M_{**} = M_*$ remains valid, but the bending rigidity, which at elastic condition is $B_e(T) = E(T)J$, should be defined according to the development of plastic deformations.

When the stress in a compressed rod exceeds a limit of proportionality, then at small random bending its concave part receives additional compression with increasing plastic strains, while the opposite convex part is elastically unloaded.

This effect is accounted for by the introduction of the "reduced" modulus of elasticity E_r depending from both Young's modulus $E(T)$ and tangential modulus E_t; the latter depends generally on stress and temperature (see Chap. 3).

For the compressed rod of rectangular cross section, the formula for E_r yields

$$\overline{E}_r(\sigma, T) = \frac{E_r(\sigma, T)}{E(T)} = \frac{4\overline{E}_t(\sigma, T)}{\left(1 + \sqrt{\overline{E}_t(\sigma, T)}\right)^2}, \tag{10.8}$$

where $\overline{E}_t = E_t/E$.

At $\overline{E}_t \ll 1$, the quantity $\overline{E}_r \approx 4\overline{E}_t$. For other kinds of cross section, this correlation differs, but always $\overline{E}_r/\overline{E}_t > 1$.

With use of the reduced modulus of elasticity, a critical force is determined as given below:

$$F_{cr.r} = \frac{\pi^2 E_r J}{l^2}, \tag{10.9a}$$

and a critical stress σ_{cr} as

$$\sigma_{cr.r} = \frac{\pi^2 E_r}{\lambda^2}. \tag{10.9b}$$

The critical stress $\sigma_{cr.r}$ and the reduced modulus of elasticity E_r should concern the same point in the diagram of deformation, i.e.,

$$\sigma_{cr.r}(\varepsilon_{cr.r}, T) = \frac{\pi^2 E_r(\varepsilon_{cr.r}, T)}{\lambda^2}.$$

Therefore, it is more convenient to define an inverse relationship of flexibility parameter versus critical ones

$$\lambda^2 = \pi^2 \frac{E_r(\varepsilon_{cr.r}, T)}{\sigma_{cr.r}(\varepsilon_{cr.r}, T)}, \tag{10.10}$$

and next to find values of these parameters for the necessary magnitude of flexibility corresponding to the given geometry of a rod.

The estimation of critical loading at elastic–plastic deformation of a compressed rod with an originally straight axis using the reduced modulus of elasticity in case of isothermal loading has been offered by Engesser (1895) and Karman (1910) and is applied in many technical appendices, see e.g., Panovko (1988).

10.1.2.3 Stability Margins

The ratio of critical value of stress or strain to its value under operating conditions, $K_\sigma = \sigma_{cr}/\sigma$ or $K_\varepsilon = \varepsilon_{cr}/\varepsilon$, characterizes the stability margin of a design. A limiting design parameter can also be a quantity of strains at lateral buckling, $\varepsilon_{\lim} < \varepsilon_{cr}$, even within the area of elastic deformations. In this case, the stability margin should be defined as $K_{\lim} = \varepsilon_{\lim}/\varepsilon$.

Let us analyze the influence of temperature on the stability margin at fixed loads and stresses.

With a rise in temperature, material deformation diagrams change: Young's and tangential modulus', and consequently the reduced modulus, and also limits of proportionality and elasticity decrease. If at some stress σ_{cr0} and temperature T_0, there was a sufficient stability margin in stress $K_{\sigma 0} = \sigma_{cr0}[E(T_0)]/\sigma$, then at the same stress the margin drops to unit at the rise in temperature up to $T = T_0 + \Delta T$ and the critical stress will decrease to $\sigma_{cr}[E(T)] = \sigma_{cr0}[E(T_0)]/K_{\sigma 0}$.

In elastic area at moderate temperatures, the critical stress varies in temperature not too strongly.

The situation cardinally varies at the stresses exceeding a limit of proportionality, especially at elevated temperatures. In Fig. 10.3a the typical deformation diagrams of a hardening material at two, not too differing temperatures T_0 and T, are presented. Corresponding dependences $\sigma_{cr} = f(\lambda)$ for the same two temperatures are shown in Fig. 10.3b.

Let a rod has an initial temperature T_0 with the elastic stress close to the limit of proportionality, but with sufficient stability margin $K_0 = \sigma_{cr.r}[E_r(T_0), \lambda]/\sigma$, shown by a piece $m_0 - n_0$. As temperature rises a little to $T = T_0 + \Delta T$, the new deformation

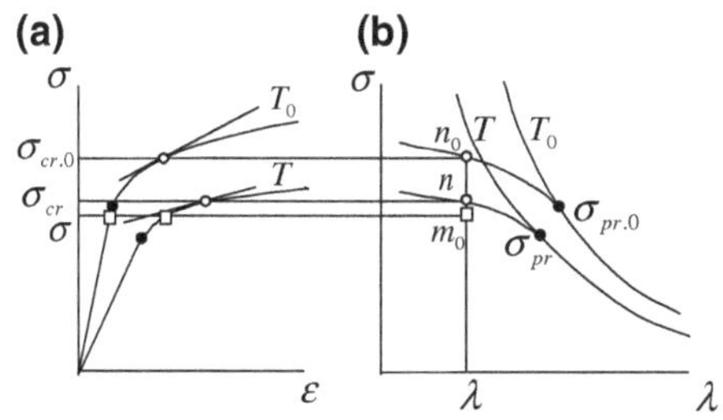

Fig. 10.3 Change of stability margins due to temperature increase

diagram will have a lowered limit of proportionality, and at the same stress plastic strains can appear that are not too large. But the critical stress at the same value of flexibility λ, as it is visible from curves in Fig. 10.3, will considerably decrease because now it will be defined by the smaller reduced modulus $E_r(T)$.

This can lead to the sharp reduction of the stability margin $K = \sigma_{cr.r}[E_r(T), \lambda]/\sigma$ shown by piece $m_0 - n$.

10.1.2.4 Stability at Increasing Loads

Critical parameters assess stability of a deformable body at already achieved values of loads and temperatures, but do not completely reflect processes of their further increase.

The account of influence of increasing load has been proposed by Shanley (1947) and was then investigated in other works applying to problems of isothermal stability, by example in Panovko and Gubanova (1964), Birger et al. (1979), and in some others. The analysis shows that the bent state of initially straight elastic–plastic rod under increasing load and constant temperature $T_{cr} = \tilde{T}$ becomes unstable at the force

$$F_{cr.t}(T) = \frac{\pi^2 E_t(T) J}{l^2}, \tag{10.11}$$

where the critical force $F_{cr.t}$ depends not on reduced modulus E_r, but on tangential E_t, and always $F_{cr.t} < F_{cr.r}$.

In Fig. 10.4a, a black square indicates a stress, which is a bit less than critical. In this case, the stress σ^-, being at bending less than σ^+, continues growing if the mean stress σ increases.

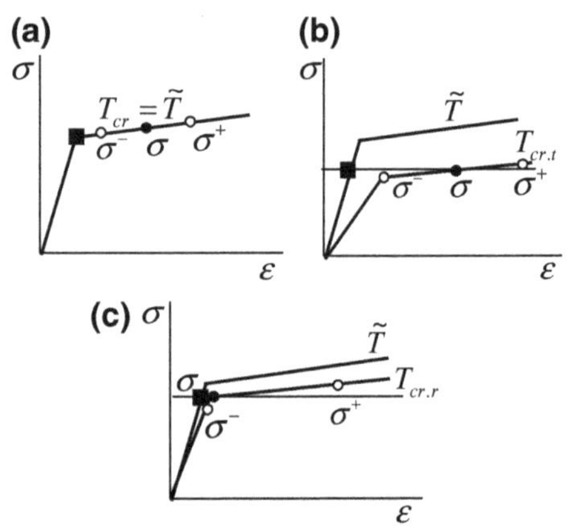

Fig. 10.4 Different situations of non-isothermal stability loss

The analysis of stability at the growth of temperature at constant load F reveals some specific effects. Rate of temperature change, as of loading, is assumed to be rather slow, which allows neglecting dynamic effects.

If a limit of proportionality strongly decreases with a temperature rise from $\tilde{T}$ to T_{cr} (Fig. 10.4b), the pattern remains the same as below, and critical temperature $T_{cr} = T_{cr.t}$ is determined by a quantity

$$E_t(T_{cr.t}) = \frac{l^2}{\pi^2 J} F. \tag{10.12a}$$

However, more reality is the bend beginning at a very small change of deformation diagram (Fig. 10.4c) when unloading is elastic, and critical temperature is determined by reduced modulus

$$E_r(T_{cr.r}) = \frac{l^2}{\pi^2 J} F. \tag{10.12b}$$

10.1.3 Elastoplastic Deformations of Compressed Rods with Initial Deviations from Linearity

10.1.3.1 Lateral Bucking of a Rod

In practice, technological deviations from ideal geometry and some eccentric application of compressing loads are inevitable. As a result, lateral buckling and rapid increase of stresses and plastic strains begin earlier than the loads or temperature will reach critical values. Therefore, even at a loss of stability in elastic area, it is necessary to introduce a rather significant margin of stability.

Under the action of compressing loadings in one (main) direction of no ideal rods, plates, shells, and other similar bodies, displacements can intensively develop in other (lateral) directions.

Contemporary computing means allow modeling of a gradual increase of lateral displacements and their accelerated growth under the approach to critical unstable state, using step-by-step calculation.

For the initiation of lateral deformations in calculation, it is required to introduce small disturbing factors (initial or current deviations in body geometry or directions of forces) though discrepancies of calculations can sometimes serve as such initiators.

As it was specified in previous chapters, in spite of the fact that all physical processes develop in time, it is expedient to separate a study of plasticity and creep influence. The same also concerns the analysis of thermal stability. Therefore, features of "thermal creep" will be further considered separately.

Let a rod shown above in Fig. 10.1 has at initial state bend of the axis

$$y_0 = y_{0\cdot\max} \sin\frac{\pi x}{l} \tag{10.13}$$

with the same boundary conditions as above.

The equation of the rod bending at small elastic deflections $y - y_0$ under the action of a compressing force F yields

$$EJ(y'' - y_0'') = -Fy, \tag{10.14}$$

with solution

$$y(x) = y_{\max} \sin\frac{\pi x}{l} \tag{10.15a}$$

where

$$y_{\max} = \frac{y_{0\cdot\max}}{1 - \left(\frac{kl}{\pi}\right)^2} = \frac{y_{0\cdot\max}}{1 - \frac{F}{F_{cr.e}}}. \tag{10.15b}$$

Here, in addition to the above, $E = f(T)$, $k^2 = F/E(T)l$, and Euler's critical force is $F_{cr.e}(T) = \pi^2 E(T)J/l^2$.

If the force F permanently increases, elastic deflection $y - y_0$ begins appearing at $F > 0$, i.e., at once after applying the compressing force, and theoretically unlimited increases at $F/F_{cr.e} \to 1$ (if the rod remained elastic).

Elastic stresses in outer fibers of the most loaded cross section are given as follows:

$$\sigma^{\pm} = \frac{F}{A} \pm \frac{F y_{\max}}{W} = \sigma\left[1 \pm \frac{k_0}{1 - \sigma/\sigma_{cr.e}}\right], \tag{10.16}$$

where W is the factor of cross-sectional bending resistance, and $k_0 = A y_{0\cdot\max}/W$.

The positive (+) sign concerns a point at the concave surface of a rod where stresses from compression and bending are summarized, and the sign (−) refers to a point at the opposite surface.

For a rod of rectangular cross section with thickness h, there is $k_0 = 6y_{0\max}/h$. At small deflections of relatively rigid rods up to $y_{0\max}/h < 1/6$, the value $k_0 < 1$.

For the deformations $\bar{\varepsilon} = \varepsilon/\varepsilon_{cr.e}$ related to the critical magnitude $\varepsilon_{cr.e} = \sigma_{cr.e}/E(T)$, we obtain

$$\bar{\varepsilon}^{\pm} = \bar{\varepsilon}\left[1 \pm \frac{k_0}{1 - \bar{\varepsilon}}\right]. \tag{10.17}$$

Dependence $\bar{\varepsilon}^{\pm} = f(\bar{\varepsilon})$ at several values of the factor k_0 is displayed in Fig. 10.5. Indices at curves show the values k_0.

Fig. 10.5 Dependence of rod's faces strains $\bar{\varepsilon}^{\pm}$ versus mean strain $\bar{\varepsilon}$ for different initial deflections

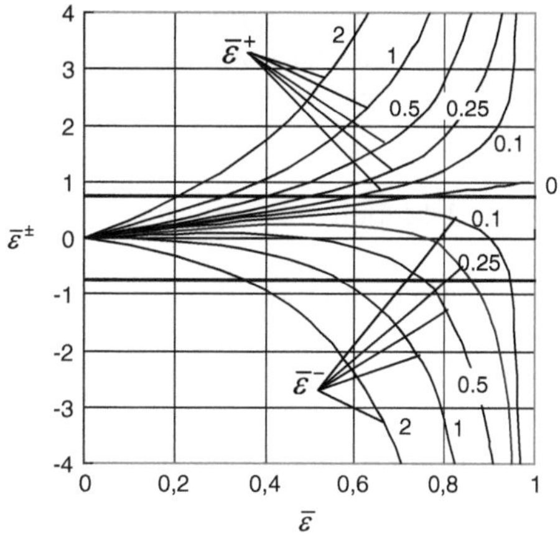

For rather rigid rods at small values $k_0 \ll 1$, the extension at the convex surface ($\varepsilon^{-} < 0$) begins only at $\bar{\varepsilon}^{-} \approx \bar{\varepsilon} \approx 1$, i.e., is close to the moment of the loss of elastic stability. With reduction of bending rigidity, the extension starts earlier at $\bar{\varepsilon} = 1 - k_0$, and at the moment of its beginning (at $\varepsilon^{-} = 0$), the strain at the concave surface reaches a value $\bar{\varepsilon}^{+} = 2\bar{\varepsilon}$. Since the quantity $k_0 = 1$, the extension begins at once at the moment of loading.

In absolute size, there is always $|\bar{\varepsilon}^{-}| < \bar{\varepsilon}^{+}$.

When the compressing strain $\bar{\varepsilon}^{+}$ at the concave surface of a rod achieves a limit of proportionality (or approximates, under kinematic hardening, a yield point), i.e., at $\bar{\varepsilon}^{+} = \bar{\varepsilon}_y$ (see, for example, lines $\bar{\varepsilon}^{\pm} = \bar{\varepsilon}_y = \pm 0.75$ in Fig. 10.5), the rod starts to be deformed plastically though at the convex surface the strain $\bar{\varepsilon}^{-}$ has not yet reached a yield point. During further loading, the plastic zone and rate of the rod's bend grow, and as strain $\bar{\varepsilon}^{+}$ approaches its critical quantities, the deflections might theoretically indefinitely increase. In reality, destruction will begin at once as the stress σ^{+} achieves the strength limit in compression.

The stresses σ^{-} at the convex surface can become at this moment extensible. For many materials, the strength limit in tension happens to be smaller than in compression, which can result in destruction from the convex surface.

At great deformations, a process becomes essentially nonlinear, and calculations at a linear approach should be considered only as qualitative.

If the growth of plastic bending deformation happens without elastic unloading at the convex surface, critical parameters σ_{cr} or ε_{cr} initially accept smaller values as they will be determined by the tangential modulus E_t, but not by the reduced modulus E_r, and the rate of initial development of plastic deformations increases. However at further bending, unloading on a convex surface will begin, and critical parameters of stability will be defined by the reduced modulus E_r.

For the real diagram of deformation $\sigma = f(\varepsilon, T)$ with $\sigma_y > \sigma_{pr}$, when tangential modulus gradually reduces in the process of loading, the estimation of stability by the use of a moment of the first plastic strain appearance, i.e., by the condition

$\sigma_{II} = \sigma_{pr}$, becomes underrated. In this case, the reduced modulus must be calculated by variable tangential modulus $E_t = \varphi(\varepsilon)$ which proves to be true by experiments, for example, Panovko (1988).

10.1.3.2 Stability at Increasing Temperature

Development of lateral buckling at gradual rise in temperature under conditions of action of constant stress can be analyzed in a similar way. With growth of temperature, both moduli $E(T)$ and $E_t(T)$ decrease approximately equally. The yield point $\sigma_y(T)$ also decreases but usually more quickly. Compression strain at $\sigma =$ const increases versus temperature as

$$\varepsilon(T) = \frac{\sigma}{E(T)}. \tag{10.18}$$

Yield point in strains $\varepsilon_y(T) = \sigma_y(T)/E(T)$ and its relative expression

$$\bar{\varepsilon}_y(T) = \frac{\varepsilon_y(T)}{\varepsilon_{cr.e}} = \frac{\sigma_y(T)}{E(T)\varepsilon_{cr.e}}, \tag{10.19}$$

decrease due to temperature rise with smaller rate than a yield point $\sigma_y(T)$.

To simplify the analysis, we assume that the deformation diagram of a material is described by kinematic hardening at any temperatures.

As it has been shown above, the plastic buckling at $\sigma =$ const practically always happens with elastic unloading at the convex surface of the system, i.e., at $T = T_r$.

In Fig. 10.6, an increase of the mean strain $\bar{\varepsilon}(T) = \sigma/[E(T)\varepsilon_{cr.e}]$, corresponding to constant compressing stress σ (curve 1), and strain $\bar{\varepsilon}^+(T, k_0)$ at the concave side of a rod (curves 2 and 3) versus raising temperature $T \geq T_0$ for some initial rod curvature $k_0 \neq 0$ are shown.

Figure 10.6a concerns a case when in all range of considered temperatures, the yield point $\varepsilon_y(T)$ is greater than Euler's critical value ε_e, which does not depend on temperature, i.e., $\bar{\varepsilon}_y(T) > 1$. The increasing strains $\bar{\varepsilon}^+$ displayed by curves 2 and 3 are affected by the identical constant stress $\sigma_0 < \sigma_y(T_0)$. The loss of stability of an originally rectilinear rod, when before this moment would be $\bar{\varepsilon}^+ = \bar{\varepsilon}$, occurs in the elastic area at temperature $T_e < T_y$ at point A, after which the strain $\bar{\varepsilon}^+$ would grow toward point B. In the rods with initial curvature (curves 2 and 3) in which strain at the concave side $\bar{\varepsilon}^+$ from the very beginning exceeds the mean strain $\bar{\varepsilon}$, plastic deformation starts earlier at temperatures T_y^+ in points C or D, after that the critical temperature decreases to the reduced value T_r shown by dashed lines. If $T_y^+ > T_r$, the equilibrium loss occurs at once (curve 2), and if $T_y^+ < T_r^-$ when temperature reaches the value T_r (curve 3).

Figure 10.6b refers to a case when the yield point even at initial temperature T_0 is smaller than Euler's critical value $\bar{\varepsilon}_y(T_0) < 1$. An originally rectilinear rod (curve 1) would lose the stability in point B. A curvature of the rod with small initial bend

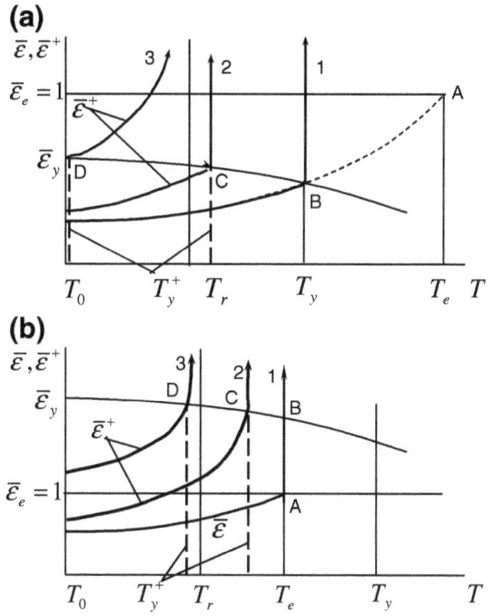

Fig. 10.6 Dependence of mean strain $\overline{\varepsilon}_0$ and maximal compressing strain $\overline{\varepsilon}^+$ versus raising temperature T; **a** $\overline{\varepsilon}_y(T) > 1$; **b** $\overline{\varepsilon}_y(T) < 1$

(curve 2) begins to gradually increase at $T > T_0$ and the rod losses the stability in point C at $T_y^+ = T_r$. A rod with great initial bend, for which there is $T_y^+ = T_0$, begins rapid plastically bulging at once (curve 3).

Due to Eq. (10.16), the temperature T_y^+ can be found from solution of the system of equations

$$\begin{aligned} &\sigma\left[1 + \frac{k_0}{1 - \overline{\varepsilon}(T_y^+)}\right] = \sigma_y(T_y^+), \\ &\overline{\varepsilon}(T_y^+) = \sigma / [E(T_y^+)\varepsilon_{cr.e}]. \end{aligned} \tag{10.20}$$

At simultaneous increase of loadings and temperatures, the process of lateral buckling of the compressed designs can be investigated by numerically step-by-step calculation.

10.2 Loss of Stability Under Creep

10.2.1 Critical Creep Deformations of a Compressed Rod

10.2.1.1 Stability of an Elastically Fixed Pillar

The problem of deformation development of compressed rods under creep at constant temperature and constant loads is rather significantly discussed in the

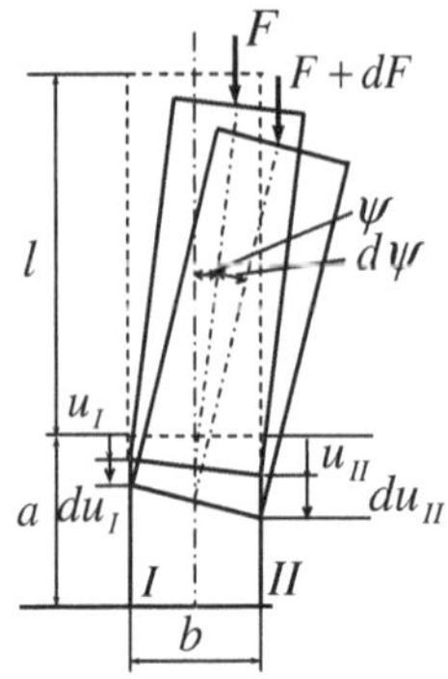

Fig. 10.7 Scheme of a compressed rigid pillar

literature, see Rjanitzin (1946), Ross (1946), Freudenthal (1950), Hoff (1956), Rabotnov (1966), etc.

For the illustration of basic features of any problems referring compression of deformable bodies, it is convenient to apply the simplified models, in particular, a rigid pillar shown in Fig. 10.7. In some works mentioned above in this chapter, a similar pillar has been used for some theoretical analysis.

The rigid pillar leans on two deformable rods I and II having given mechanical characteristics; a cross-sectional area of each rod is $A_I = A_{II} = 0.5$ A. Temperature of rods T_I and T_{II}, as well as their deformation and creep diagrams can be different. Two states of the system are usually recognized: (i) A current arbitrary state with the rods experiencing displacements u_I and u_{II}, and small inclination angle ψ; and (ii) a nearest deviation state differing from the previous one in increments du_I, du_{II}, and dψ.

A rigid pillar model at d$F > 0$ has been used by Shanley (1947) and Birger et al. (1979) to study the effect of increasing load influence on stability loss, and by Panovko and Gubanova (1964) to model stability loss under creep.

Let us recognize a typical pattern of stress and strain changes under creep in compressed pillar with elastic rods of those presented in Fig. 10.7. These changes at constant average stress σ lesser than yield point $\sigma < \sigma_y$ and critical stress ($\sigma < \sigma_{cr.e}$), and also $T_I = T_{II} =$ const, are shown in Fig. 10.8, where

$$\overline{\sigma} = \frac{\sigma}{\sigma_{cr.e}}, \quad \overline{\varepsilon} = \frac{\varepsilon}{\varepsilon_{cr.e}}, \quad \overline{\psi} = \frac{2l}{b}\psi.$$

A creep law for strain increase in time $\varepsilon_c = f(\sigma, t)$ is assumed to be given.

Until any moment t_0, the pillar remains rectilinear with identical stresses $\overline{\sigma}_I = \overline{\sigma}_{II} = \overline{\sigma}$ and full compressing strains $\overline{\varepsilon}_I(t) = \overline{\varepsilon}_{II}(t) = \overline{\sigma} + \overline{\varepsilon}_c(t)$ over both rods and their entire cross sections (curves 1 in Fig. 10.8a, b). At this moment, the pillar is assumed to be affected by a small deviation $\overline{\psi}_0$. Then, the stress $\overline{\sigma}_{II}$ and elastic strain $\overline{\varepsilon}_{II}$ of the rod II (at compression site of the pillar) increase, and the stress $\overline{\sigma}_I$ and strain $\overline{\varepsilon}_I$ at the stretched rod I decrease (respectively, shown by curves 2 and 3 in Fig. 10.8a, b). From this time, the rates of the creep at the opposite sides of the rod

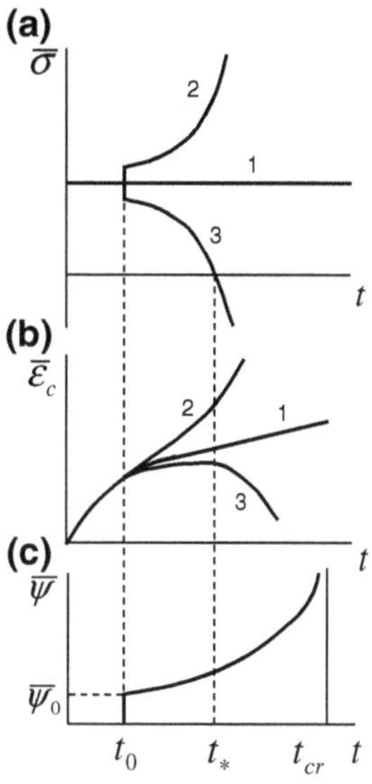

Fig. 10.8 Typical pattern of stresses and strains in the rods of a pressed pillar under creep

vary so that $\bar{\dot{\varepsilon}}_{cII} > \bar{\dot{\varepsilon}}_{cI}$, and the pillar inclination will progressively grow (curve 1 in Fig. 10.8c).

At a moment of time t_*, the stress in rod I alters the sign (i.e., becomes stretching) and creep in it starts to develop in an opposite direction. Further increase of lateral buckling of the rod will lead to its destruction.

As the small random deviation can happen at any moment t_0 or be present from the very beginning of loading, it is natural to consider $t_0 = 0$. In this sense, the buckling calculation of a compressed rod is a special case of non-stationary non-linear creep considered in Chap. 6.

Numerical step-by-step nonlinear methods of creep calculation are basically possible, but their reliability in a strong degree depends on the accepted description of a material behavior at the stresses varying in size and direction.

As under some conditions, the rate of creep theoretically tends to infinity for the final time duration t_{cr}, this time can be considered as a critical parameter of stability.

10.2.1.2 Numerical Solution of a Compressed Pillar Creep

The established creep 1D equation (see Chap. 6) in the simplest form is given as follows:

$$\dot{\varepsilon} = \frac{\dot{\sigma}}{E} + k\sigma^n \operatorname{sign}\sigma, \tag{10.21}$$

where all characteristics of a material E, k, n depend on temperature and is designated $\dot{f} = df/dt$.

The experimental exponent n, as a rule, is more than unit. For quality analysis, it is convenient to let n be a whole odd number, in particular $n = 3$. Such a value is close enough to some experimental data and allows omitting in the Eq. (10.21) the sign of stress, as at $\sigma < 0$ always $\sigma^3 = -|\sigma|^3$.

Physical correlations of the full strain rates for the rods I and II in a dimensionless kind are given as

$$\dot{\bar{\varepsilon}}_I = \dot{\bar{\sigma}}_I + \bar{k}\bar{\sigma}_I^3, \quad \dot{\bar{\varepsilon}}_{II} = \dot{\bar{\sigma}}_{II} + \bar{k}\bar{\sigma}_{II}^3, \tag{10.22}$$

where

$$\bar{t} = t/t_{cr}, \quad \bar{k} = t_{cr}E\sigma^n_{cr.e}k, \quad \dot{\bar{f}} = d\bar{f}/d\bar{t}. \tag{10.23}$$

In this subsection, hyphens above dimensionless parameters are also omitted.

Conditions of deformation compatibility and equilibrium, independent from material properties for a constant average stress $\sigma = \text{const}$ are given as

$$\dot{\varepsilon}_{II} - \dot{\varepsilon}_I = 2\dot{\psi}, \tag{10.24a}$$

$$\dot{\sigma}_I + \dot{\sigma}_{II} = 0, \quad \dot{\sigma}_{II} - \dot{\sigma}_I = 2\dot{M}_\sigma, \quad \sigma\dot{\psi} = \dot{M}_F. \tag{10.24b}$$

where M_σ and M_F are inner and outer bending moments, respectively.

In equilibrium state, there is $\dot{M}_\sigma = \dot{M}_F$, and also

$$\sigma_I = \sigma(1 - \psi), \quad \sigma_{II} = \sigma(1 + \psi). \tag{10.25}$$

Equations (10.21), (10.24a), (10.24b), and (10.25) allow obtaining the differential equation for an inclination angle of a pillar

$$\dot{\psi}(1 - \sigma) = k\sigma^3\psi(3 + \psi^2). \tag{10.26}$$

Having integrated Eq. (10.26) with the initial condition $\psi(0) = \psi_0$, we get the dependence $t(\psi)$, whence we find

$$\psi = \psi_0\sqrt{\frac{3\exp(\nu t)}{3 + \psi_0^2(1 - \exp(\nu t))}}, \tag{10.27}$$

with $\nu = 6k\sigma^3/(1 - \sigma)$.

The inclination angle (in linear approach) unrestrictedly increases at the moment of time

$$t_{cr} = \frac{1}{\nu}\ln\left(1 + \frac{3}{\psi_0^2}\right). \tag{10.28}$$

10.2.1.3 Lateral Buckling of a Rod Under Creep

Equations (10.27) and (10.28) allow determining the basic features of lateral buckling of an elastic compressed rod under creep conditions:

- At an average relative stress $\sigma = 1$, there occurs a usual elastic loss of stability at Euler's critical force at once at the instant of loading ($t_{cr} = 0$).
- At very small stresses $\sigma \ll 1$, critical time increases in inverse proportion to a cube (at $n = 3$) of stress.
- At any small (but distinct from zero) compressing stress and presence of initial deviations, creep leads to unlimited increase of lateral displacements during the finite time.

Obviously, at some moment t_*, the stress in the compressed rod II reaches the value of yield point σ_y, and due to the second Eq. (10.25), the inclination angle becomes $\psi_y = \sigma_y/\sigma - 1$. The magnitude t_* can be calculated from Eq. (10.27) at $\psi = \psi_y$. After this moment, the pillar will incline yet faster.

Thus, in a condition of creep, we must take as criterions of working capacity of a design the times when not only design stress but also deflection achieve maximum admissible magnitudes σ_{ad} or ψ_{ad}.

In the literature, there are a lot of solutions of partial tasks with the analysis of influence of creep on lateral buckling of compressed designs. The basic sides of this process remain the same as for the compressed pillar, but some features additionally come to light.

So, at deflections of thin compressed plates and shells commensurable with their thickness, the effect of increase of rigidity due to the influence of stresses in a median surface is shown. This leads to slower development of creep deformations. In addition to the rods, plates must be calculated in view of initial imperfections (Vol'mir 1967; Vol'mir and Kil'dibekov 1968).

10.2.2 Loss of Stability of an Extended Rod Under Creep

10.2.2.1 Loss of the Cylindrical Form of Equilibrium

Loss of cylindrical form of equilibrium due to specimen necking at elastic or elastoplastic extension of a rod was studied by Rjanitzin (1955), Panovko and Gubanova (1964), and others.

In conditions of creep, the similar phenomenon can occur when stresses do not exceed a yield point that was investigated experimentally by Andrade (1910) and theoretically by Hoff (1953), Rabotnov (1966), etc.

Let, an elastic rectilinear rod be extended by a constant force F; material behavior in conditions of the established creep is described by the equation

$$\dot{\varepsilon} = \frac{\dot{\sigma}}{E} + k\sigma^n, \tag{10.29}$$

where the rate of full strain $\dot{\varepsilon}$ is equal to the sum of elastic $\dot{\varepsilon}_e$ and "creeping" $\dot{\varepsilon}_c$ components.

Full lateral deformation ε_l is

$$\varepsilon_l = -\mu_e \varepsilon_e - 0.5\varepsilon_c, \tag{10.30}$$

where μ_e is the Poisson's factor at elastic deformations, and $\mu_c = 0.5$ answers the condition of material incompressibility at plastic deformations.

The running area of a cross-sectional rod $A(t)$ decreases during time dt by $dA = 2A\dot{\varepsilon}_l dt$. Due to Eqs. (10.29) and (10.30),

$$\dot{A} = -2A(t)(\mu\dot{\sigma}/E + 0.5k\sigma^n). \tag{10.31}$$

On the other hand, at $\sigma(t) = F/A(t)$,

$$\dot{A} = -\frac{F}{\sigma^2(t)}\dot{\sigma}, \tag{10.32}$$

which yields the nonlinear equation

$$\dot{\sigma}(1 - 2\mu\sigma/E) = k\sigma^{n+1}. \tag{10.33}$$

Having divided in Eq. (10.33) variables, we find

$$t = \int_{\sigma_0}^{\sigma} \left(1 - \frac{2\mu}{E}\sigma\right) \frac{d\sigma}{k\sigma^{n+1}}, \tag{10.34}$$

where $\sigma_0 = F/A_0$ is the tensile stress referring to the initial area of the cross-sectional rod.

After integration of Eq. (10.34), we obtain

$$t = \frac{1}{nk\sigma_0^n}\left[1 - q - \left(\frac{\sigma_0}{\sigma}\right)^n (1 - q\sigma)\right], \tag{10.35a}$$

where

$$q = \frac{2\mu n}{(n-1)}\frac{\sigma_0}{E}. \tag{10.35b}$$

As real values are $\sigma/E \ll 1$ and $n > 2$, this quantity is $q \ll 1$.

From Eq. (10.35a), it follows

$$\left(\frac{\sigma}{\sigma_0}\right)^n \left(1 - q - nk\sigma_0^n t\right) + \frac{\sigma}{\sigma_0} q - 1 = 0. \tag{10.36}$$

While stresses σ and σ_0 have one order that corresponds to real cases of loading, the members of Eq. (10.36) with the factor $q \ll 1$ can be neglected, which gives

$$\sigma \approx \frac{\sigma_0}{\sqrt[n]{1 - nk\sigma_0^n t}}. \tag{10.37}$$

Equation (10.37) shows the stress of extended elastic rod in conditions of creep theoretically unrestrictedly increases at finite value of time

$$t_{cr} \approx 1/nk\sigma_0^n. \tag{10.38}$$

With increase of temperature, the coefficient k grows, and a critical time t_{cr} diminishes.

10.2.2.2 Time Prior to Failure

The time t_{cr} essentially depends on the initial stress σ_0. In real conditions, the initial area of a cross-sectional rod A_0 can, at least feebly, vary in the length of a rod. But even such cross-sectional changes cause different stresses and rate of creep along the rod, which finally leads to necking and breaking at the weakest cross section.

Even small changes of temperature along the length of a rod result in a similar effect.

Thus, the time t_{cr} represents a critical parameter of stability loss referred to cylindrical form of rod equilibrium. At the same time, a value t_{cr}, being a function of initial stress and given temperature, estimates in essence a long durability of a rod material (see Chap. 7).

At a big change of length of a rod from l_0 up to l, a true relative strain is defined as

$$\varepsilon_{tr} = \ln\frac{l}{l_0} = \ln(1 + \varepsilon). \tag{10.39}$$

Using Eq. (10.39) with the dependence $\sigma = E\varepsilon_{tr}$, it is possible to estimate time before destruction and increase in length of a rod during creep more precisely. However for real-design deformations, at which there comes destruction at creep, distinction between values ε_{tr} and ε are usually insignificant.

If the initial stress σ_0 surpasses a yield point and a material is deformed with hardening, time before destruction decreases but the general character of creep influence does not change.

10.3 Loss of Stability Under Thermal Stresses Action

General problems of stability loss under action of temperature stresses were discussed by Timoshenko (1936), Prager (1947), Boley and Weiner (1960), etc. A number of partial solutions were studied in numerous publications; however, in the majority of them, the critical loads were calculated at constant physical–mechanical properties of a material without analyzing the influence of further heating. Below, a greater attention is given to the last problem.

10.3.1 Loss of Rod Stability at Uniform Heating

At uniform heating of a body, temperature stresses arise when external conditions prevent its free thermal expanding.

10.3.1.1 Heating of a Rod with Fixed Ends

At heating of an elastic rod with fixed ends on ΔT, its thermal elongation $\Delta l_T = \alpha_T \Delta T\, l$ must be compensated by elastic compression $\Delta l_e = -Fl/EA$ from the reaction force F. It leads to the expression

$$F = E\alpha_T \Delta T A. \tag{10.40}$$

Compressing forces, stresses, and strains are still considered positive. To estimate the stability of the rectilinear equilibrium form of the rod at some heating $\Delta T = T - T_0$, it is necessary to assume, as above, that it will get a small random deviation such as in Eq. (10.13) that must satisfy the boundary conditions at the motionless ends.

The condition of the balance between the internal and external bending moments, expressed through compressing stress $\sigma(T)$ and a deflection at the middle section of the rod $y_{0\max}$, will have a former appearance

$$\left[E(T)(\pi/\lambda)^2 - \sigma(T)\right] y_{0\max} = 0.$$

If $y_{0\max} \neq 0$, then

$$\sigma(T_{cr}) = \frac{E(T_{cr})\pi^2}{\lambda^2}. \quad (10.41)$$

Basic difference of Eq. (10.41) from the similar Eq. (10.2a) is that the critical stress and temperature are now connected by dependence

$$\sigma(T_{cr}) = E(T_{cr})\alpha_{T_0,T_{cr}}(T_{cr} - T_0), \quad (10.42)$$

that follows from Eq. (10.40).

Here, $\alpha_{T_0,T_{cr}}$ is an average value of thermal expansion factor in a range of temperatures $\Delta T = T - T_0$.

By excluding the stress from the last two equations, we find

$$T_{cr} - T_0 = \frac{\pi^2}{\alpha_{T_0,T_{cr}}\lambda^2}, \quad (10.43)$$

after which it is possible to define a critical stress $\sigma_{cr}(T_{cr})$.

Taking $\overline{T} = T/T_0$ and accepting for some range of temperatures linear dependences

$$E(\overline{T}) = E_0(1 - k_E\overline{T})/(1 - k_E)$$

and

$$\alpha_{T_0,T}(\overline{T}) = \alpha_0(1 + k_\alpha\overline{T})/(1 + k_\alpha),$$

at $k_E < 1$ and $k_E\overline{T} < 1$, we obtain

$$\sigma(\overline{T}_{cr}) = \frac{E_0\alpha_0 T_0(1 - k_E\overline{T}_{cr})(1 + k_\alpha\overline{T}_{cr})(\overline{T}_{cr} - 1)}{(1 - k_E)(1 + k_\alpha)}.$$

It is easy to verify that in some cases the function $\sigma(\overline{T}_{cr})$ passes through a maximum at a temperature $\overline{T}_*$(when $1 < \overline{T}_* < \overline{T}_{cr}$). So, at $k_E = k_\alpha = k < 1$

$$\overline{T}_* = \frac{\sqrt{3 + k^2} + k}{3k}.$$

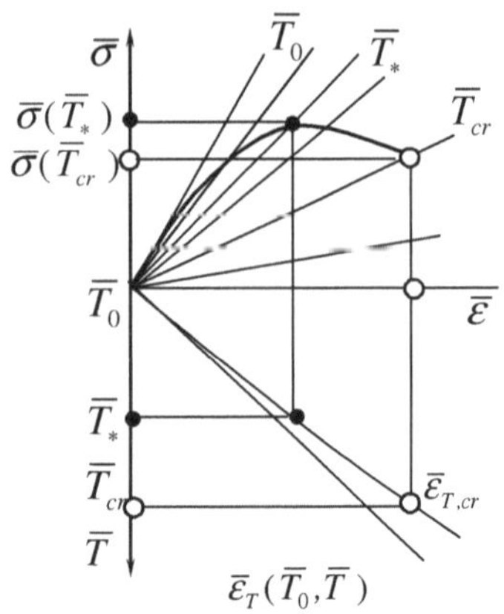

Fig. 10.9 Correlations between critical parameters at stability loss of elastic compressed rod under heating

Qualitative relationships between temperature, strains, and stresses are visible from Fig. 10.9. In the top diagram, rays along which elastic deformation $\overline{\sigma}(\overline{\varepsilon}, \overline{T}) = \overline{E}(\overline{T})\overline{\varepsilon}$ for some area of temperatures occurs are shown. Here, $\overline{\sigma} = \sigma/E_0\alpha_0 T_0$ and $\overline{\varepsilon} = \varepsilon/\alpha_0 T_0$. In the bottom diagram, dependence of relative temperature expansion on temperature growth $\overline{\varepsilon}_T = \overline{\alpha}_T(\overline{T})(\overline{T} - 1)$, where $\overline{\alpha}_T = \alpha_T/\alpha_0$, is resulted.

Having determined for given parameter of flexibility λ, the critical strain

$$\overline{\varepsilon}_{cr} = (\pi/\lambda)^2/\alpha_0 T_0 = \overline{\varepsilon}_{T,cr} = \overline{\alpha}(\overline{T}_{cr})(\overline{T}_{cr} - 1),$$

we find the critical temperature $\overline{T}_{cr}$ and corresponding value $\overline{\sigma}(\overline{T}_{cr})$. The parameters, describing given procedure, are allocated by light points.

At a smaller temperature within the limits of $1 \leq \overline{T} \leq \overline{T}_{cr}$, the stresses can be determined by the condition $\overline{\varepsilon}(\overline{T}) = \overline{\varepsilon}_T(\overline{T})$, whence $\overline{\sigma}(\overline{T}) = \overline{E}(\overline{T})\overline{\varepsilon}_T(\overline{T})$. The presence of maximum $\overline{T}_*$ for one of the values λ is shown by black points and the bold curve.

The stated formulas specify only the critical size of heating $\Delta T_{cr} = T_{cr} - T_0$, at which the rectilinear elastic rod can start to be bent.

The further lateral rod buckling at $\Delta T > \Delta T_{cr}$ can be estimated only by nonlinear calculation considering change of the rod length at the bend, see for example Gatewood (1957).

If deflection of a bended rod with fixed ends is approximately described by Eq. (10.9a, 10.9b), then increase in its length at small deflection, when $(dy/dx)^2 \ll 1$, is

$$\Delta l = 0.5 \int_0^l \left(\frac{dy}{dx}\right)^2 dx = 0.25 y_{\max}^2 \left(\frac{\pi}{l}\right)^2 l. \tag{10.44}$$

Equation (10.44) is applicable at $(dy/dx)^2_{\max} = (y_{\max}/l)^2 \ll 1/\pi^2$, i.e., at $y_{\max}/l \ll 0.3$.

From the equality $\Delta l = \Delta l_T = \alpha_{T_{cr}T}(T - T_{cr})l$, we find

$$\frac{y_{\max}}{l} = \frac{2}{\pi}\sqrt{\alpha_{T_{cr}T}(T - T_{cr})}. \qquad (10.45)$$

Equation (10.45) has kinematic character and does not depend on a kind of a material deformation diagram. In elastic area,

$$\frac{y_{\max}}{l} = \frac{2}{\lambda}\sqrt{\frac{\alpha_{T_{cr},T}\Delta T}{\alpha_{T_0,T_{cr}}\Delta T_{cr}} - 1}. \qquad (10.46)$$

Excess of metal rod parts heating over critical value in 100 °C causes deflection $y_{\max} \approx (0.01 - 0.02)l$. At large length of a design member (for example, a rail) at length $l \approx 10$ m, the bend will reach an inadmissible value of $y_{\max} \approx (10 - 20)$ cm. In shorter machine parts, bending stresses can exceed the ultimate strength of a material and leads to destruction. Therefore, freedom of axial expansion of heated details is usually the necessary requirement of their application.

For a rod which from the very beginning at T_0 has had some axis deviation $y_{0\max}$ described by Eq. (10.13), the form of curved axis at the further moderate increase can be assumed the same.

The equilibrium condition of an elastic rod at a current temperature T and compressing stress $\sigma(T)$ at deformed state with deflection $y_{\max}(T)$ yields

$$E(T)(\pi/\lambda)^2[y_{\max}(T) - y_{0\max}] - \sigma(T)y_{\max}(T) = 0, \qquad (10.47)$$

whence

$$y_{\max}(T) = \frac{y_{0\max}}{1 - E(T_{cr})\sigma(T)/E(T)\sigma(T_{cr})} \qquad (10.48a)$$

or

$$y_{\max}(T) = \frac{y_{0\max}}{1 - \varepsilon(T)/\varepsilon_{cr}} \qquad (10.48b)$$

where values T_{cr} and also $\sigma(T_{cr})$ are defined as well as above by Eqs. (10.40) and (10.42), and $\varepsilon_{cr} = (\pi/\lambda)^2$.

Change of the axis length during increase of a deflection from T_0 up to T is defined due to Eq. (10.44) as

$$\Delta l = 0.25(y_{\max}^2 - y_{0\max}^2)\left(\frac{\pi}{l}\right)^2 l. \tag{10.49}$$

Then

$$\varepsilon(T) = \alpha_{T_0,T}(T - T_0) - 0.25(y_{\max}^2 - y_{0\max}^2)\left(\frac{\pi}{l}\right)^2. \tag{10.50}$$

But from Eq. (10.48b), it follows

$$\varepsilon(T) = \varepsilon_{cr}(1 - y_{0\max}/y_{\max}). \tag{10.51}$$

Having excluded the strain $\varepsilon(T)$ from Eqs. (10.50) and (10.51), we come after some transformations to the equation

$$\left(1 - \frac{\bar{y}_{m0}}{\bar{y}_m}\right)[1 + 0.25\lambda^2\bar{y}_m(\bar{y}_m + \bar{y}_{m0})] = \bar{\alpha}(T)\overline{\Delta T}, \tag{10.52}$$

where

$$\begin{aligned} &\bar{y}_m = y_{\max}/l, \quad \bar{y}_{m0} = y_{0\max}/l, \\ &\bar{\Delta} T = (T - T_0)/(T_{cr} - T_0), \quad \bar{\alpha}(T) = \alpha_{T_0,T}/\alpha_{T_0,T_{cr}}. \end{aligned} \tag{10.53}$$

Under heating, there are $\overline{\Delta T} \geq 0$ and $\bar{y}_m \geq \bar{y}_{m0}$. At $\bar{y}_{m0} = 0$, the real solution of Eq. (10.52) exists at $\overline{\Delta T} \geq 1$ and coincides with the Eq. (10.46).

Having calculated by Eq. (10.52) at preset values λ and $\bar{y}_{m0}$ the dependence $\overline{\Delta T} = f(\bar{y}_m)$, we shall find the inverse function $\bar{y}_m = f(\overline{\Delta T})$ showing increase of the rod bend in process of temperature growth, since $\overline{\Delta T} = 0$. The specified dependence for $\lambda = 30$ and various values of initial deviation is constructed in Fig. 10.10. Even at very small initial curvatures, the rod bulging sharply increases to the moment the critical value $\overline{\Delta T} = 1$ is achieved, but its further increase proceeds approximately the same as for originally rectilinear rod.

In Fig. 10.11 for the same preset values λ and $\bar{y}_{m0}$, the mean stresses $\bar{\sigma} = \sigma(T)/\sigma(T_{cr}) = f(\overline{\Delta T})$ referring to the critical value are shown by continuous lines; it is accepted that $E(T) \approx \text{const}$. The bold line 1 concerns originally rectilinear rod.

Bending stresses in extreme fibers of the compressed rod in view of nonlinear development of deformations and Eq. (10.47) can be found as

$$\sigma^{\pm}(\overline{\Delta T}) = \sigma(\overline{\Delta T}) \pm E(T)J(\pi/\lambda)^2[y_{\max}(T) - y_{0\max}]/W.$$

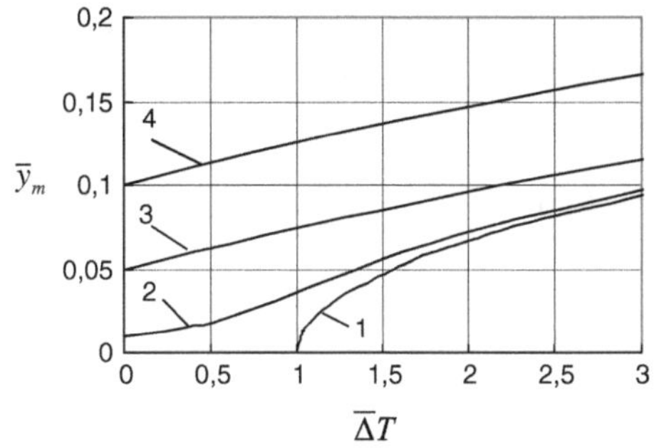

Fig. 10.10 Increase of an elastic rod inclination at heating for different initial bend *1* $\bar{y}_{m0} = 0$, *2* $\bar{y}_{m0} = 0.01$, *3* $\bar{y}_{m0} = 0.05$, *4* $\bar{y}_{m0} = 0.1$

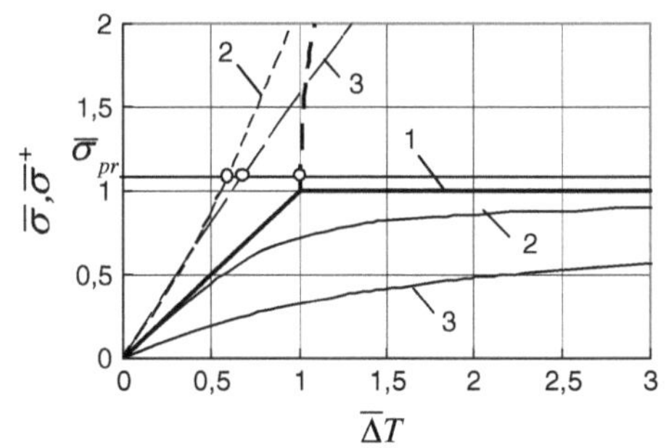

Fig. 10.11 Increase of stresses $\bar{\sigma}$ and $\bar{\sigma}^+$ at heating of an elastic rod for different initial bend: *1* $\bar{y}_{m0} = 0$, *2* $\bar{y}_{m0} = 0.01$, *3* $\bar{y}_{m0} = 0.05$

or in a dimensionless kind

$$\bar{\sigma}^{\pm}(\overline{\Delta T}) = \bar{\sigma}(\overline{\Delta T}) \pm k_1 \bar{y}_m(\overline{\Delta T}), \tag{10.54}$$

where $k_1 = Al/W$. For a rod of rectangular cross section, $k_1 = 6l/h = 1.73\lambda$.

Change of compressing stress $\bar{\sigma}^+$ at the concave side of a rod for $\lambda = 30$ and various values of initial deviation is shown in the same Fig. 10.11 by shaped lines. Very fast increase of bending stresses is visible.

At the convex side of the rod, the tensile stresses develop as quickly as compressing ones, but with small lag. In Fig. 10.11, they are not shown.

After the compressing stresses at the rod's concave side $\bar{\sigma}^+$(and a bit later the tensile stresses at the convex side $\bar{\sigma}^-$) achieve yield points, the bending rigidity of the rod quickly falls, and the rate of buckling in heating develops significantly faster.

The analysis shows that full deformation increases, basically due to thermal expansion, and stresses can start decreasing. Change of the deformation diagram with temperature rise strengthens the specified effect.

The admissible degree of heating $\overline{\Delta}T$ is determined (with margins) by limit bulging or by maximum permissible plastic strain.

10.3.1.2 Heating of a Plate with Fixed Boundaries

In addition to rods, an originally flat plate deprived of the possibility to freely extend in its plane loses stability at heating when temperature compressing stresses reach a critical value for the given system.

So, at uniform heating in ΔT, an elastic circular plate of radius R and thickness h, clamped at its contour, there exists in it temperature compressing stresses

$$\sigma = \frac{E\alpha_T \Delta T}{1-\mu},$$

while the critical stress from the action of contour compressing load is

$$\sigma_{cr} \approx 0.35 \frac{E}{(1-\mu^2)} \left(\frac{h}{R}\right)^2.$$

Having equated $\sigma = \sigma_{cr}$, we obtain

$$\Delta T_{cr} \approx \frac{0.35}{\alpha_T(1+\mu)} \left(\frac{h}{R}\right)^2.$$

Critical temperatures for plates of other configurations, and also for cylindrical shells with motionless end faces, are calculated similarly.

Critical temperatures of elastic plates and shells at stability loss only from heating do not depend on the Young's modulus (Birger et al. 1993).

A number of solutions on the definition of critical temperatures for special cases are expressed in the literature (see for example, Boley and Weiner 1960). Supercritical behavior of plates and shells at nonlinear deformations should be investigated only by numerical methods.

10.3.2 Loss of Stability and Buckling at Non-uniform Heating

10.3.2.1 A Multi-rod System

In some parts of a body containing thin or thin-walled elements which cannot freely extend at non-uniform heating, there arise compressing stresses resulting in some cases from loss of stability and local buckling.

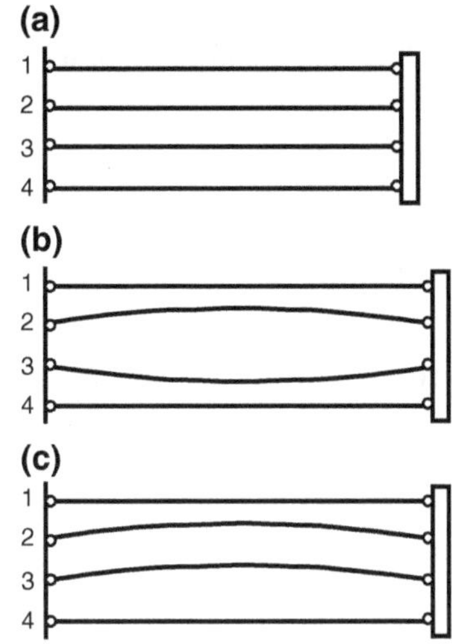

Fig. 10.12 Loss of system equilibrium due to uneven heating

In Fig. 10.12a, system consisting of four identical parallel rods jointly connected at each end is presented. Their reference temperature T_0 is the same. The system has no external restrictions, so temperature stresses in the rods at heating should be self-counterbalanced.

If two outer rods have constant temperature T_0 and two internal ones are heated up to temperature $T = T_0 + \Delta T$, then there occurs in internal rods compressing stresses $\sigma_2 = \sigma_3 = 0.5E\alpha_t\Delta T$ and in outer ones—the same in size tensile stresses. According to Eq. (10.2a), compressing stresses reach critical value at temperature rise

$$\Delta T_{cr} = \frac{2}{\alpha_T}\left(\frac{\pi}{\lambda}\right)^2.$$

At further heating, internal rods will begin bending. Theoretically, their deformed shape can happen in the opposite direction (Fig. 10.12b) and in the identical one (Fig. 10.12c). The actual bending directions depend on possible small inaccuracies of the system.

At joint fastening of the rod ends, external rods remain rectilinear, but in the case of tight coupling of the ends of all rods and bulging of the compressed rods in the common direction, all system, including external elastic rods, will start to be bent.

In addition to the presence of external forces, columns that are not too flexible start buckling at the occurrence of plastic deformations. Under the relation to entire system shown in Fig. 10.12a, the heating is non-uniform in thickness. Such problems are typical for many design members, including thin rods or plates.

Since a source of temperature strains is not non-uniformity of temperature rise ΔT, but non-uniformity of temperature extension

$$\Delta\varepsilon_T = \Delta(\alpha_T T),$$

then the bimetallic rods, which consist of materials with different coefficients of thermal expansion, start to be bent under the uniform temperature field.

10.3.2.2 Other Forms of Stability Loss

At heating of twisted rods of extended cross section, thin plates and shells, rather various forms of stability loss are possible.

In a twisted rod of the extended cross section, at which the temperature of external inclined fibers considerably surpasses the temperature of the central rectilinear part, external fibers are stressed by compression. Calculations show that under certain conditions originally smooth fibers can lose stability and become wavy.

In a thin rectangular plate non-uniformly heated in its plane, there are areas with compressing stresses which can cause local buckling of some sites of a plate. Non-uniformity of a temperature field on width of a plate can lead to loss of stability of some flat form of a bend.

Long cylindrical shells can lose stability in cases when various segments of their circumference are heated differently. Hotter segments are stressed by axial compression, as a result of which on its surface can be formed some dents of rhomboid forms; their occurrence is accompanied by claps (Panovko 1988).

In heated cylindrical shells which ends are connected with cold rigid face plugs, compressing stresses can occur near to end faces. Though these stresses damp in process of removal from the end faces, they can lead to local bugling of a shell.

10.3.3 Thermal Jump of a Mechanical System into Another Equilibrium State

10.3.3.1 Loss of Thermal Stability of von Mises' Girder

Some mechanical systems (flat girders, arches, shells) under action of the loads, causing compression, can lose stability of the initial form of equilibrium by spasmodic transition into other equilibrium state. Isothermal problems were considered to a number of works, since Mises (1923) and Mises and Ratzersdorfer (1925). Beneath is shown how similar effects can arise at heating a system.

In Fig. 10.13, the elementary system of two elastic rods in length l_0 and the area of cross section A, jointly connected among themselves and inclined to the basis under some angle β_0 ("Mises' girder"), is shown. Other ends of the rods can freely turn but the distance between them is constant. The rods are assumed to be rigid

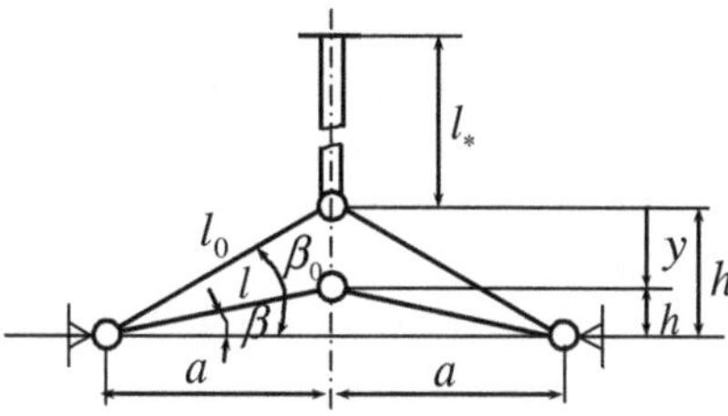

Fig. 10.13 Scheme of the "Mises' girder"

enough to allow taking up only a stretching–compression load without loss of Euler's equilibrium. Unlike approaches in which the behavior of such girder was analyzed under action of the constant vertical load applied to its top, see for example, Panovko and Gubanova (1964), we shall accept that the girder top is connected to the motionless casing through a very rigid stock in length l_* which gradually heats up or cools.

In the beginning, we assume that at any change of the girder configuration, connection between the rods and the stock is provided both at compression, and at extension.

Thermal stock expansion $y_T = \alpha_T \Delta T l_*$ at contact with the girder causes vertical displacement of its top

$$y = h_0 - h = a(\tan\beta_0 - \tan\beta) = h_0(1 - \tan\beta/\tan\beta_0), \tag{10.55}$$

where angles of rod inclination β_0 and β concern, accordingly, the initial and deformed states of the system.

The condition $y = y_T$ yields

$$\tan\beta = \tan\beta_0(1 - \overline{y}_T), \tag{10.56}$$

where $\overline{y}_T = y_T/h_0$.

Elastic strain of each rod under the action of compressing force F is equal to

$$\varepsilon = \frac{\Delta l}{l_0} = \frac{a}{l_0}\left(\frac{1}{\cos\beta_0} - \frac{1}{\cos\beta}\right) = \frac{F}{AE}. \tag{10.57}$$

The equilibrium condition of the system in deformed state

$$F_* = 2F\sin\beta, \tag{10.58}$$

leads to

$$\overline{F}_* = 2\sin\beta\left(\frac{1}{\cos\beta_0} - \frac{1}{\cos\beta}\right), \tag{10.59}$$

where $\overline{F}_* = F_* l_0/EAa$.

Equations (10.56) and (10.59) allow us to find a dependence of the vertical force F_*, affecting the girder at heating of the stock, versus the displacement of its top for a preset value of an angle β_0. The compressing force F_*, besides the initial value $F_* = 0$ at $\beta = \beta_0$ ($y_T = 0$), still reaches zero twice: at $\beta = 0$ ($\bar{y} = h_0/a = tg\beta_0$) when the girder is straightened, and at $\beta = -\beta_0$ ($\bar{y} = 2h_0/a = 2tg\beta_0$) when it is symmetrically outward bent in the opposite direction. For two intermediate extremes of function $F_* = f(y_T)$, there meet values of angles

$$\beta'_{1,2} = \pm\arccos\left(\sqrt{[3]}\cos\beta_0\right), \tag{10.60}$$

from which the maximal force $\overline{F}_{*\max}(\beta'_1)$ can be found.

Dependence $\overline{F}_*/\overline{F}_{*\max} = f(\bar{y}_T)$ for a girder with the initial angle $\beta_0 = 15°$ is shown in Fig. 10.14 by thin line 0–1–2–3–4–5 where characteristic values of the angle β are also specified.

Displacement of girder top at cooling the stock will have a return direction ($\bar{y}_T < 0$), as shown by the dot monotonic line.

The kind of function $F_* = f(y_T)$ displayed in Fig. 10.14 is completely defined by displacement of girder top $\bar{y} = \bar{y}_T$, reflecting "kinematical" loading (see Panovko and Gubanova 1964), and does not depend on the force $\overline{F}_*$ origin.

At a big relative length of the stock when the ratio h_0/l_* has the order of its possible thermal expansion $\varepsilon_T = \alpha_T\Delta T$, heating of the stock under the condition of $y = y_T$ can be a real source of girder loading with alteration of the sign of the rods inclination.

Until the moment of full straightening at point 2, the force $\overline{F}_*$ is compressive. Then, at the site $0 > \beta > -\beta_0$ (curve 2–3–4), it becomes extensive, as the girder tends to take the free configuration at $\beta = -\beta_0$ and the slowly moving stock end at heating opposes this.

At further heating, the force $\overline{F}_*$ will be again compressive (after point 5).

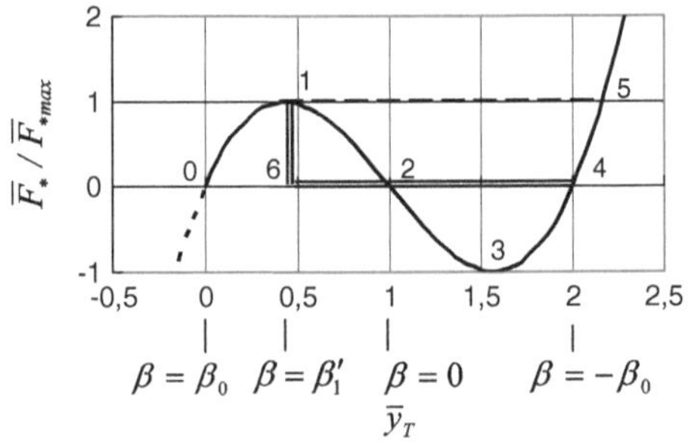

Fig. 10.14 A force affecting a "Mises' girder" due to the stock thermal extension $\bar{y}_T$ under its heating

10.3.3.2 Changes of Equilibrium Forms of a Girder/Stock System

Now we shall assume retention of a contact of heated stock to the girder, $y = y_T$, only at their pressing to each other, i.e., at the force $\overline{F}_* > 0$, resisting to thermal expansion. Otherwise, the girder comes off a rod, their contact disappears ($\overline{F}_* = 0$), and the displacements become different ($\overline{y} \neq \overline{y}_T$).

In the absence of contact to the stock, the girder aspires to occupy one of the three specified positions of balance, meeting the force $\overline{F}_* = 0$. But average (rectilinear) position of compressed rods at $\beta = 0$ is unstable; therefore, in a free state, the girder can borrow one of the two positions—initial $\beta = \beta_0$ or opposite, $\beta = -\beta_0$.

The process of girder deformation with $\beta_0 = 15°$ under the action of thermal expansion of the stock is displayed by a fat curve in Fig. 10.14 and the consecutive configurations are shown in Fig. 10.15.

The force $\overline{F}_*$ achieves the maximum value $\overline{F}_{*\max}$ in point 1 at $\overline{y} \approx 0.43$ and $\beta_1' \approx 8.7°$. In this moment, the girder loses stability and, being affected by the same force, would jump to point 5 as shown in Fig. 10.14 by the dashed line. But at decrease of the force, the contact between the stock and girder remains until the force $\overline{F}_*$ drop to zero (to point 2) and only at this moment the girder jumps to the

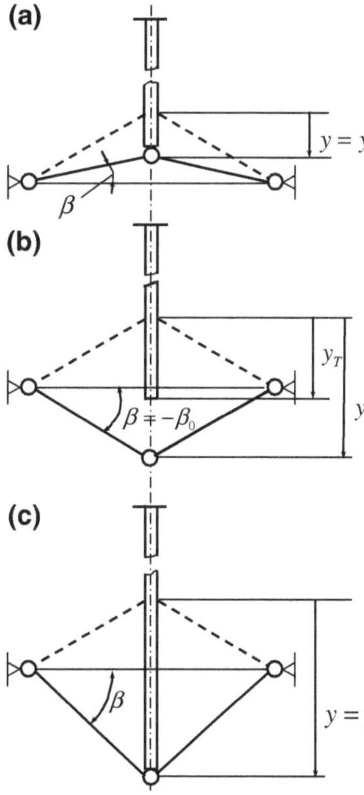

Fig. 10.15 Sequential patterns of "Mises' girder" deformation under heating

opposite equilibrium unstressed state displayed in Fig. 10.15b. When the heated stock will be extended so, that will catch up with top of the girder, the condition $y = y_T$ will be restored, and the further deformation of the system will proceed along the line 4–5 and further as shown in Fig. 10.15c.

At the subsequent cooling the stock, the girder will reach a free state in point 4 at $\overline{F}_* = 0$, and then the shortened stock will come off the girder until the end of the process. So, at $\Delta T = 0$, the girder will not return to the initial condition.

Reduction of the inclination angle β affected by pressure of a heated rod leads to the decrease of a girder rigidity in vertical direction and the increase of compression strains in the rods. Their maximal value at $\beta = 0$ is equal to $\varepsilon_{\max} = 1 - \cos\beta_0$. If $\varepsilon_{\max} < \varepsilon_{pr}$, all strains of rods remain elastic, at $\varepsilon_{\max} > \varepsilon_{pr}$ the rods begin to be deformed plastically, since the angle β_{pr}, at which $\cos\beta_{pr} = \cos\beta_0/(1 - \varepsilon_{pr})$.

At an uninterrupted connection of the stock with a girder, displacement of the girder top will be completely determined by thermal expansion $y = y_T$, but the force F_* will decrease according to the diagram of deformation. Without stock to girder connection, the value of jump angle to the opposite side $|\beta_{jump}|$ will be much smaller than the initial angle β_0.

The picture of the jump in a new equilibrium state, described above, is typical for such mechanical systems, as shallow arches and shells of different configurations.

If a heated up stock has an opportunity of turn, asymmetrical forms of stability loss are possible.

References

Andrade EN da C (1910) On the viscous flow of metals and allied phenomena. Proc Royal Soc Ser A 84, A567

Birger IA, Shorr BF, Iosilevich GB (1979) Design of machine parts for strength, 3rd edn. Mashinostroenie, Moscow (in Russian)

Birger IA, Shorr BF, Iosilevich GB (1993) Design of machine parts for strength, 4th edn. Mashinostroenie, Moscow (in Russian)

Boley BA, Weiner JH (1960) Theory of thermal stresses. Wiley, New York

Engesser F (1895) Schweizerische Bauzeitung №13

Freudenthal AM (1950) The inelastic behavior of engineering materials and structures. Wiley, New York

Gatewood BE (1957) Thermal stresses with applications to airplanes, missiles, turbines and nuclear reactors. McGraw-Hill, New York

Hoff NJ (1953) The necking and rupture of rods subjected to constant tensile loads. J Appl Mech 20:1

Hoff NJ (1956) Creep buckling: the aeronautical quarterly. February VII:1–20

Jones RM (2006) Buckling of bars, plates, and shells. Bull Ridge Publishing, Blacksburg

Karman T (1910) Untersuchungen über Knickfestigkeit, Mitteilunge über For-schungsarbeiten. Verein Deutcher Ingenieure, H.8, Springer, Berlin

Mises R (1923) Ueber die Sfabilitaeatsprobleme der Elastizitaetstejrie. Zeitschr angew Math Mech 206–462

Mises R, Ratzersdorfer J (1925) Die Knicksicherheit von Fachwerkens. Zeitschr angew Math Mech 218–231

Panovko JG (1988) Stability of rods. In: Birger IA, Panovko Ja G (eds) Strength, stability, vibration, V.3. Mashinostroenie, Moscow (in Russian)

Panovko JG, Gubanova II (1964) Stability and vibrations of elastic systems. Nauka, Moscow (in Russian)

Ponomarev SD, Biderman VL, Likharev KK et al (1952) Foundations of contemporary methods of strength design in machinery. Mashgiz, Moscow (in Russian)

Prager W (1947) The general variation principle of the theory of structural stability. Quart Appl Math 4:378–384

Rabinovich VP (1966) Strength of turbine disks. Mashinostroenie, Moscow (in Russian)

Rjanitzin AR (1946) Deformation processes in design of viscous-elastic materials. Proc AN SSSR 53(1) (in Russian)

Rjanitzin AR (1955) Equilibrium stability of elastic systems. GITTL, Moscow (in Russian)

Ross AD (1946) The structure engineering, No. 8

Shanley FK (1947) Inelastic column theory. JAS 14(5)

Timoshenko SP (1936) Theory of elastic stability. McGraw-Hill, New York

Van-der-Neut A (1958) High temperature effects on aircraft structures and buckling caused by thermal stresses. Pergamon Press, New York

Vol'mir AS (1967) Stability of deformable systems. Nauka, Moscow (in Russian)

Vol'mir AS, Kil'dibekov IG (1968) Stability of shells. In: Birger IA, Panovko Ja G (eds) Strength, stability, vibrations. v.3, Chap. 3. Mashinostroenie, Moscow (in Russian)

Chapter 11
Thermal Integrity of Anisotropic Bodies

Abstract Thermal strength for anisotropic bodies is considered in case of single-crystal nickel alloys widely used in manufacturing of high-temperature gas turbine blades. Anisotropy of mechanical properties and failure conditions, typical for single crystals, is described. Experimental data for deformation and damage of single-crystal specimens under extension are compared with results of theoretical predictions. The role of the «Schmid factor» describing the relative level of tangential stresses in different slip systems is discussed. It is shown that its maximum value corresponds to a critical stress state related to the initiation of plastic deformation or material damage only for specimen extension and cannot be considered as a failure criterion of a single-crystal body for other loading orientations. Comparison of calculated maximal tangential stresses, with their critical values, obtained by testing specimens of chosen orientation under appropriate loading gives better results. It is demonstrated that strength analysis of single-crystal machine parts under combined stresses may rely on material properties associated with uniaxial loading at a certain "equivalent" specimen orientation. Several problems of creep, low and high cyclic fatigue, and thermal fatigue of anisotropic bodies are also discussed.

11.1 Bases of Anisotropic Bodies Mechanics

11.1.1 Natural and Artificial Anisotropy

11.1.1.1 Natural Anisotropy

The majority of engineering metal alloys have some natural anisotropy of mechanical properties, i.e., the values of these properties differ depending on the direction within each small element of a body. The term "natural" may be regarded conditional, to some extent. Such anisotropy might be connected not only with the crystal properties initially inherent in solids or with initial heterogeneity of material

B.F. Shorr, *Thermal Integrity in Mechanics and Engineering*,
Foundations of Engineering Mechanics, DOI 10.1007/978-3-662-46968-2_11

structure, but can also arise as a consequence of mechanical–thermal processing of a raw structure part. Under elastoplastic deforming, a body "deformation" anisotropy arises and usually increases until exhaustion of a structure bearing ability is reached (Fridman 1974).

When natural anisotropy is insignificant, or grains with various character of anisotropy are casually allocated all over a body, anisotropy may be neglected and material may be considered as initially isotropic. Thus, the isotropic body can be non-uniform with various mechanical characteristics within its different parts.

When natural anisotropy of the same character extends on the large number of grains, covering an area of a body commensurable to the body l size, it becomes necessary to be taken into account influence anisotropy has on mechanical behavior of a material.

11.1.1.2 Artificial Anisotropy

Sometimes anisotropy can be induced artificially. Technical operations such as sheet metal rolling processes, manufacturing of wire cables by stretching, optimization of structure of a spring material, and many others are widely used. During these operations, the material grains are usually plastically extended in a direction where the largest operating loads are expected. By molding, growth and orientation of all crystals can be directed beforehand.

Such procedure increases the loading ability of a system.

Anisotropic structure of a material can sometimes happen not as an end in itself, but as an additional result of material modification. Considering a problem of ensuring the integrity of gas turbine blades casted from heat-resistant nickel alloys, it has been established that their damage in the long term mainly occurs through the boundaries of the grains disposed transversely to the direction in which tension stresses from centrifugal forces act. To eliminate such grain boundaries, blades are sometimes casted with "columnar" structure, in which crystals are grown only in a radial direction. Such approach is called "the process of directed crystallization."

Further development of technology has led to a blade structure in the form of an individual single crystal, with no grain boundaries. But in a single-crystal blade, the anisotropy inherent in a crystal structure of an alloy covers the entire blade volume, and the anisotropic properties of a material have to be taken into account for blade design and testing.

In real casting, direction of a crystal does not meet the required one perfectly, and there can arise several crystals instead of single one. The material behavior is influenced by the columnar structure of an alloy. There are pores and defects of a crystal lattice; direction of crystallographic axes in a cross section of a detail azimuthal orientation at some technologies turns out to be arbitrary, and so forth (Shalin et al. 1997).

Therefore, experimental data for single crystal structures and even specimens differ significantly, and these data are frequently inconsistent. However, for ensuring durability of real structures manufactured by directed crystallization, it is rather useful to have an insight on anisotropic characteristics of an ideal single crystal, both at normal and elevated temperatures.

In this monograph, we shall consider a problem of thermal integrity of anisotropic bodies mainly for machine parts made from the heat-resistant nickel alloys used in turbine structures.

11.1.2 Mechanics of Single Crystals

Below, we provide only brief information on geometry and mechanics of single crystals that is necessary for understanding the problem by a broad audience of experts on thermal integrity. General theory of crystals can be found in special literature, see for example McClintock and Argon (1966) and Shalin et al. (1997). At the beginning, the material is supposed to be linearly elastic, and the loading is isothermal. More general material characteristics are examined in Sect. 11.3.

11.1.2.1 Symmetry of Crystal Lattices

The theory of materials crystal lattices is considered from various points of view in the works on theoretical physics and crystallography. For problems in mechanics where a crystal is represented as a particle of a solid, geometrical and elastic symmetry have major importance. The presence of symmetry means that in the areas of a body located on the same distance from the axes or planes of symmetry, all characteristics will be identical.

Crystals of nickel-based alloys have three principal axes of symmetry in the three mutually perpendicular directions x, y, z with corresponding three planes of symmetry. The directive cosines of their normal lines form a triple of integers 0,0,1; 0,1,0; 1,0,0. In crystallography, it is accepted to designate directions of axes x, y, z as [100], [010], [001], and planes of cube symmetry, according to the directions of their normal lines, as (001), (010), and (100). Directions of the cube diagonals are also represented in the form of sets of integers which are defined as projections of diagonals onto the principal directions of a crystal. In Fig. 11.1, the diagonals are shown with arrows. Here, the notations [110], [101], and [011] refer to the diagonals of the cube sides and [111]—to the principle diagonal of the cube.

The ends of vectors [001], [011], and [111] form a so-called stereographic triangle of a crystal painted over in Fig. 11.1. The position of its current point A can be obtained by the "scanning" of the ray drawn from the coordinate origin over the area of the specified triangle, so that every crystallographic direction will meet the certain point A. From conditions of cubic symmetry of the crystal, it follows that the distribution of any characteristic over the area of a stereographic triangle completely defines the field of this characteristic over the entire crystal. Scanning angles in planes of cube sides vary from 0 up to 45°, and in a plane of the principal diagonal—from 0 up to $\arctan\sqrt{2} \approx 55°$.

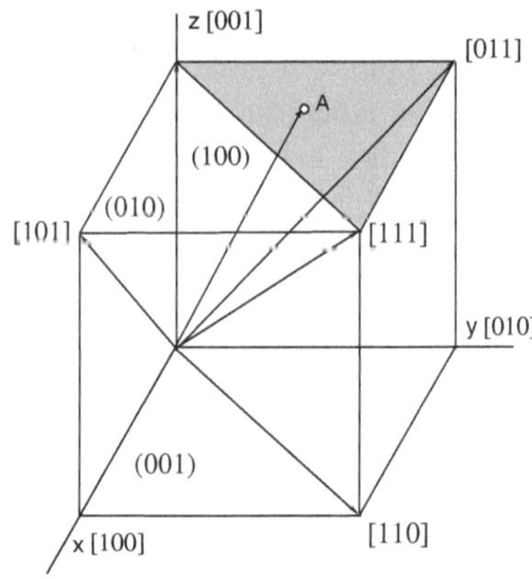

Fig. 11.1 Principal axes and planes in a cubic crystal

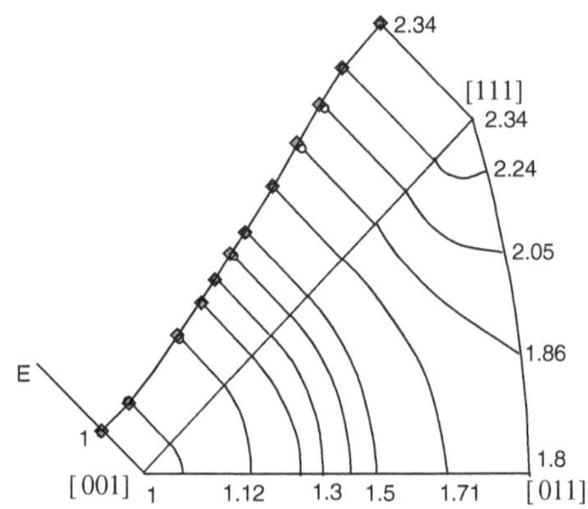

Fig. 11.2 Pattern of Young's modulus distribution over the stereographic triangle

Below, in Fig. 11.2, the stereographic triangle will be used for the analysis of Young's modulus distribution through crystal directions, where for convenience the axis [111] will be directed upwards.

Directions of the crystal principal axes relative to the global coordinate system, in which the machine parts and applied loads are described, specify the common crystal orientation, unlike the crystallographic direction describing properties of the crystal itself. At the analysis, it is sometimes more convenient to fix the position of a crystal axes and to define crystal "orientation" through the directions of the applied load.

11.1.2.2 Anisotropy of Elastic Characteristics

The general theory of elasticity of anisotropic bodies is stated in a number of works, for example, Lekhnitsky (1950), McClintock and Argon (1966), and others.

Components of the stress σ_{ij} and strain ε_{ij}, second-rank tensors in an elastic anisotropic body, are connected by the generalized Hook's law

$$\varepsilon_{ij} = S_{ijkl}\sigma_{kl}, \tag{11.1a}$$

or

$$\sigma_{ij} = C_{ijkl}\varepsilon_{kl}, \tag{11.1b}$$

where S_{ijkl} and C_{ijkl} are material compliance and elastic matrices, respectively.

The matrices S_{ijkl} and C_{ijkl} have generally 21 independent components. Various types of crystal symmetry reduce the number of such components. A crystal with cubic symmetry has three independent components in the directions of the principal axes: Young's modulus $E = E_{[001]} = E_{[010]} = E_{[100]}$, shear modulus $G = G_{[001]} = G_{[010]} = G_{[100]}$, and Poisson's factor $\mu = \mu_{[001]} = \mu_{[010]} = \mu_{[100]}$.

These parameters should be experimentally determined, but unlike isotropic materials, $E \neq 2(1+\mu)G$. For heat-resistant alloys, quantities $\bar{G} = G/E = 0.9-1$ and $\mu \approx 0.38{-}0.39$ are typical, which differ too much from values $\bar{G} = 1/2(1+\mu) \approx 0.36$ at $\mu \approx 0.3{-}0.33$, which is usual for isotropic materials.

Equation (11.1a) presented in the principal crystal axes takes a form

$$\begin{Bmatrix} \varepsilon_1 \\ \varepsilon_2 \\ \varepsilon_3 \\ \gamma_{12} \\ \gamma_{23} \\ \gamma_{31} \end{Bmatrix} = \begin{vmatrix} S_{11} & S_{12} & S_{13} & 0 & 0 & 0 \\ S_{21} & S_{22} & S_{23} & 0 & 0 & 0 \\ S_{31} & S_{32} & S_{33} & 0 & 0 & 0 \\ 0 & 0 & 0 & S_{44} & 0 & 0 \\ 0 & 0 & 0 & 0 & S_{55} & 0 \\ 0 & 0 & 0 & 0 & 0 & S_{66} \end{vmatrix} \begin{Bmatrix} \sigma_1 \\ \sigma_2 \\ \sigma_3 \\ \tau_{12} \\ \tau_{23} \\ \tau_{31} \end{Bmatrix} \tag{11.2}$$

where $S_{11} = S_{22} = S_{33} = 1/E$, $S_{12} = S_{21} = S_{13} = S_{31} = S_{23} = S_{32} = -\mu/E$, and $S_{44} = S_{55} = S_{66} = 1/G$.

In Eq. (11.2), there is no unequivocal connection between parameters of elasticity E, G, and Poisson's factor μ.

Under rotation of coordinates axes with respect to the principal ones, the compliance matrix S_{ijkl} becomes completely filled, which leads to interrelation of all strain and stress components. For a vector $\mathbf{n}_i$ with directive cosines l_1, l_2, l_3 concerning the principal axes, the component $S_{11[n]}$ is defined by the formula (McClintock and Argon 1966)

$$S_{11[n]} = S_{11} - 2L_{(n)}(S_{11} - S_{12} - S_{44}/2) = \frac{1}{E}\left[1 - L\left(2(1+\mu) - \frac{E}{G}\right)\right], \tag{11.3a}$$

where

$$L_{[n]} = l_1^2 l_2^2 + l_2^2 l_3^2 + l_3^2 l_1^2. \tag{11.3b}$$

A strain component in direction of the vector $\mathbf{n}_i$ under extension in this direction is equal to $\varepsilon_{1[n]} = S_{11[n]}\sigma_{1[n]} = \sigma_{1[n]}/E_{[n]}$, hence the elasticity modulus in any direction can be calculated using the formula

$$E_{[n]} = \frac{E}{1 - L_{(n)}[2(1+\mu) - E/G]}. \tag{11.4}$$

As the second term of the denominator of Eq. (11.4) is positive, and the quantity is $L_{[n]} > 0$ for all deviations of the vector $\mathbf{n}_i$ from the principal axes, the elasticity modulus E referred to these axes is minimal. The elasticity modulus $E_{[111]}$ referred to the principal diagonal, for which $L_{[n]} = L_{\max} = 1/3$, accepts the maximal value

$$\bar{E}_{[111]} = \frac{E_{[111]}}{E} = \frac{1}{1 - [2(1+\mu) - E/G]/3}. \tag{11.5}$$

Moduli ratios $\bar{E}_{[n]} = E_{[n]}/E$ and $\bar{G}_{[n]} = \bar{G}_{[n]}/E$ are usually equal to $\bar{E}_{[111]} = (2.0 - 2.5)\,\bar{G}_{[111]} \approx 0.5$.

A typical distribution of Young's modulus over the area of the stereographic triangle is shown in Fig. 11.2 (Shalin et al. 1997). The pattern was obtained for extension of specimens from a heat resisting alloy at temperature 750 °C. Lines of equal relative values of moduli $\bar{E}_{[n]}$ are resulted, and a diagram of these values for the line connecting axes [001] and [111] is displayed by white points.

This diagram can be rather fairly approximated by the formula

$$\bar{E}_{[n]} = 1 + (\bar{E}_{<111>} - 1)\xi^2(3 - 2\xi), \tag{11.6}$$

where $\xi = \theta/\theta_{\max}$ is a relative angular coordinate of any direction. A line referred to the angle $\theta_{\max} = 55°$ calculated by Eq. (11.6) is shown by black points. Equation (11.6) can be used for the estimation of intermediate values of elasticity moduli for various orientations of a crystal.

Anisotropy of elastic characteristics influences the level and distribution of thermal stresses in a body and can make a significant impact on its natural oscillation frequencies (Shorr et al. 2001).

Real orientation of the crystal axes in specimens and machine parts is affected by perfection of a technological process. The orientation of some crystals even at extension of standard specimens can have deviations in several degrees in longitudinal directions, and uncontrollable angle deviations in lateral ones. It concerns especially the specimens cut out of a structure part. The tolerance in axes deviation usually ranges from ±5 up to ±15 grades.

It is visible from Fig. 11.2 that the moduli of elasticity changes at such crystal axes deviations in (20 − 25) %.

11.2 Elastoplastic Deformation and Damage of an Anisotropic Body

11.2.1 Slip Systems in a Crystal

11.2.1.1 Features of Damage Accumulation of an Anisotropic Body

Damage accumulation of an element of the *isotropic* body can be treated as a result of stress works affecting deformations in all directions. Summarizing these works, it is possible to establish a criteria of failure in terms of "equivalent" stress (see Chap. 8), which do not depend on the directions of principal stresses, as their direction for an isotropic body does not matter.

The basic feature of single crystals is anisotropy of all strength characteristics (deformation diagrams, short- and long-term durability, low and high cycle fatigue), which indicates a beginning of a plastic deformation in a crystal and subsequent damage along different directions. As volumetric strain is always isotropic, the anisotropy influences conditions of damage arising from shear stress development.

Study of single crystals has shown that there is a number of "slip (sliding) systems" in a crystal representing an assemblage of planes and directions, where resistance to shear deformation is essentially less, than in other planes and directions. According to the phenomenological approach, the positions of slip systems for a given crystal type are assumed to be established from physical analysis as well as from observations of a damage development in experiments.

However, the accumulation of damage in a structure depends also on the value of stresses acting in slip systems.

11.2.1.2 Slip Systems of Crystals with Cubic Symmetry

Testing of specimens of single crystals with cubic symmetry was carried out, basically, under uniaxial extension. Tests have shown that in most cases, both static and fatigue failures occur against a slip surface coinciding with one of the four octahedral crystallographic planes of the cube, which are generally denoted by $\{111\}$. Orientations of the slip planes are usually defined by X-ray analysis.

In Fig. 11.3, the octahedral slip systems, consisting of the plane (111) with normal [111] and three directions of "easy" sliding $[\bar{1}01]$, $[1\bar{1}0]$, $[01\bar{1}]$ shown by arrows is demonstrated.

The designation $\bar{1}$ specifies that the corresponding projection is negative. In total, there are twelve such octahedral sliding systems in a crystal. But sometimes shears happen along sliding systems in the planes parallel to cube sides in the direction of diagonals of the sides (Shalin et al. 1997).

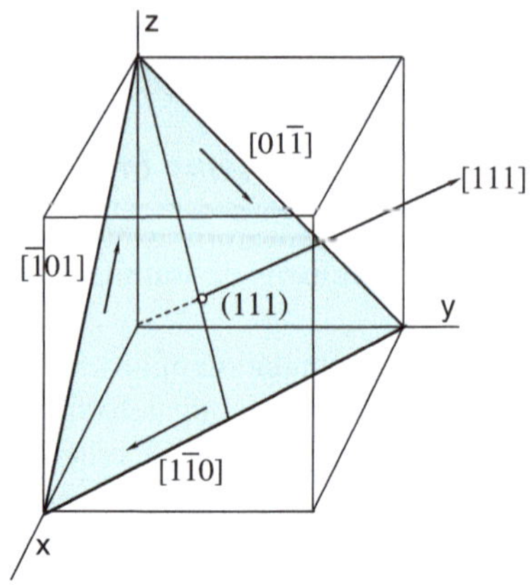

Fig. 11.3 Octahedral slip systems (111)[01$\bar{1}$], (111)[1$\bar{1}$0], and (111)[$\bar{1}$10] of a crystal with cubic symmetry

11.2.2 Longitudinal Loading of a Crystal

11.2.2.1 Tangential Stresses in Slip Systems

In the textbooks on strength of materials, it is shown that under uniaxial extension of a solid with the stress σ, a tangential stress in an inclined plane with the outward normal that makes an angle χ with the lading direction is determined from the equilibrium conditions of a corresponding element part by equation $\tau_\chi = \sigma \sin\chi \cos\chi$. The projection of the stress τ_χ onto other directions located under an angle β in the same inclined plane is equal to

$$\tau_{\chi\phi} = \tau_\chi \cos\beta = \sigma F^*(\chi, \phi), \tag{11.7a}$$

where

$$F^*(\chi, \phi) = \cos\chi \cos\phi \tag{11.7b}$$

and $\cos\phi = \sin\chi \cos\beta$.

Here, ϕ is an angle between directions of the operating stress σ and any tangential stress $\tau_{\chi\phi}$. At angle $\beta = 0$, the factor $F^*(\chi, \phi)$ yields $F^*(\chi, \phi) = 0.5 \sin 2\chi$, and at $\phi = \chi = \pi/4$, it reached the ultimate value $(F^*)_{ult} = 0.5$ that corresponds to maximal tangential stress $\tau_{\max} = 0.5\sigma$ under uniaxial extension.

As it has been specified above, the orientation [111] means that the tensile stress σ is directed along the principal diagonal of a crystal and the orientation [001] corresponds to the stress σ directed along the axis z of a crystal.

Equation (11.7b) allows finding the value of the factor F^* for each of the slip systems. Relation between tangential $\tau_{\chi\phi}$ and normal σ stresses through the factor F^* is geometrical in nature and does not depend on mechanical characteristics of the material. In crystallography, the factor $F(\chi, \phi) = F^*(\chi, \phi)$ is usually referred to as "Schmid factor" for the corresponding slip system (see Schmid and Boas 1935).

At the change of crystal orientation, the value of F for the same slip system will vary. In particular, it is possible to find for given orientation the slip system (or systems) where the Schmid factor and tangential stress acquire the maximal values $F = F_{\max}$ and $\tau_{\chi\phi} = (\tau_{\chi\phi})_{\max}$, respectively.

For "weak" octahedral slip systems, the value $F_{\max}$ varies depending on a direction of stress within the range from $F_{\max} = 0.272$ at orientation [111] up to $F_{\max} = 0.408$ at orientations [001] and [011]. For other slip systems, the factor $F_{\max}$ increases at some orientations and might achieve the ultimate magnitude $F_{\max} = (F^*)_{ult} = 0.5$.

11.2.2.2 Damage Estimation Based on Slip Systems

The beginning of material damage and its development over the slip systems is observed for all types of loading, in which shear deformation plays a key role, including tests for short- and long-term durability, low and high cyclic fatigue, and thermal fatigue. As in a crystal with cubic symmetry, the shears are proportional to the corresponding tangential stresses; the maximum stress $\left(\tau_{\chi\phi}\right)_{\max}$ in one of the slip systems at the beginning of failure should achieve some *critical* value for the given loading type

$$\left(\tau_{\chi\phi}\right)_{\max} = \tau_{cr}. \tag{11.8}$$

The assumption that at the beginning of a plastic deformation, the quantity τ_{cr} is a mechanical characteristic of an anisotropic material, identical to all slip systems, and independent of the orientation of a crystal, usually named Schmid law, was proposed by Schmid and Boas (1935). This condition was assumed to represent a material failure criterion. It is difficult to justify this assumption directly; therefore, special indirect tests had to be conducted. So, according to Eq. (11.8), the "critical" value of an operating stress $\sigma_{cr} = \tau_{cr}/F_{\max}$ for orientation [111] at $\tau_{cr} = \text{const}$ should considerably surpass values σ_{cr} for other orientations.

Experimental justification of this condition using specimens with different loading orientation (Shalin et al. 1997) has shown that such difference in values of σ_{cr} is not observed.

While processing the experimental data, it is needed to take into account the fact that the orientation of real crystals (even for specimens) might sometimes deviate from given quantities up to 10°–15°.

Shorr and Mel'nikova's (2010) analysis shows that with deviation of orientation, the small values of parameter F for nominal orientation [111] change much more quickly than larger values for other nominal orientations. This should lead to smoothing of distinctions in quantity $F_{\max}$, and consequently (at $\tau_{cr} = \text{const}$) also in values σ_{cr}.

On the other hand, circumstances that should lead to the difference in values τ_{cr} for various slip systems affect normal stress components, which differ in both magnitude and sign on the planes of slip. Note also that behavior of the real crystals depends on the presence of the so-called columnar structure and other features connected with concrete technological processes of single crystal melding.

Therefore, though the value of the design factor F_{max} specifies a relative stressing in a slip system, it cannot estimate the critical value of tangential stress τ_{cr} for a material plastic flow or failure along various orientations even under uniaxial loading of a specimen.

11.2.2.3 Failure at Various Orientations of a Single Crystal

For studying experimental strength characteristics of single crystals having significant scatter, it is necessary to average the results based on rather large number of specimens.

In a number of works, the fatigue limits σ_{-1} for heat-resistant nickel-based alloys, were investigated under uniaxial symmetric loading cycle and various temperatures. Tests of two types were conducted. For pure bending of rotating specimens, the stresses were supervised; in other cases, the strains were measured and fatigue limits in terms of strain ε_{-1} were defined, whereupon the stress σ_{-1} was recalculated according to assumption of uniaxial stress state.

In Fig. 11.4, generalized data on single crystal fatigue limits based on Shalin et al. (1997), Kolesnikov et al. (2001), and some others are demonstrated. Diagrams show the fatigue limits for different orientations referred to their values at the orientation [001] and expressed in terms of stresses $\bar{\sigma}_{-1} = \sigma_{-1[n]}/\sigma_{-1[001]}$ (Fig. 11.4a) and in terms of strains $\bar{\varepsilon}_{-1} = \bar{\varepsilon}_{-1[n]}/\bar{\varepsilon}_{-1[001]}$ (Fig. 11.4b). Areas of scattered results obtained in various tests are shaded with black.

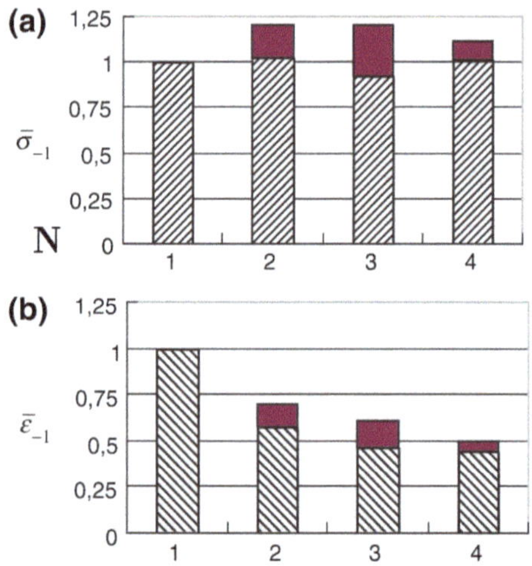

Fig. 11.4 Relative fatigue limits of a single crystal for different loading orientations: *1* [001], *2* [011], *3* [111], *4* [101]

It can be observed from the diagrams that despite the different ways, through which the test results have been obtained, the change in orientation does not too strongly influence the value σ_{-1}.

Therefore, the fatigue limit expressed in maximal tangential stress $\tau_{-1} = 0.5\sigma_{-1}$ weakly depends on orientation. This phenomenon may be explained by the fact that at any nominal orientation in view of real deviation, a fatigue crack "finds" a slip system, in which the Schmid factor is close to ultimate value $(F^*)_{ult} = 0.5$. At the same time, the fatigue limits ε_{-1} and σ_{-1} in all cases appear to be connected by the formula $\varepsilon_{-1} = \sigma_{-1}/E_{[n]}$, where the moduli of elasticity at different orientations were defined experimentally for the tested specimens or were calculated according to the formulas of Sect. 11.1.2.2.

11.2.2.4 Maximal Tangential Stress as a Criterion of Single-Crystal Damage

Experiments show that failure might occur not only along the basic slip systems, but also along others.

The maximal calculated value of Schmid factor is $F_{\max} = 0.471$ and at small deviations in orientation, it can reach the ultimate value 0.5. Therefore, the use the extremely possible for a single crystal quantity of critical stress—*maximal tangential stress—is* assumed being the simpler and at the same time proved enough criterion of an anisotropic solid failure. In case of extension, it is equal to

$$\tau_{\max} = 0.5\sigma. \tag{11.9a}$$

The values $\tau_{cr} = \tau_{\max}$ should be defined for each loading type (static, low, and high cyclic fatigue, thermal cyclic damage) by carrying out corresponding tests.

11.2.3 Damage Models for Combine Stress–Strain State

11.2.3.1 Anisotropy at a Combined Stress–Strain State

Formulating criteria of failure for machine parts of single-crystal structure, it is necessary to take into consideration the following.

The machine parts, for example, turbine blades cast of heat-resistant alloys by methods, which allow theoretically obtaining a single-crystal structure, can be actually considered as single crystals only conditionally (some reasons have been specified in the beginning of this chapter).

In a combined stress state, each pair of the principal stresses defines tangential stresses in the planes normal to a plane of these principal stresses. In particular, in the plane, with outward normal making the angles χ and $\chi + \pi/2$ with directions of stresses σ_1 and σ_2, accordingly, the tangential stress is equal to

$\tau_\chi = (\sigma_1 - \sigma_2)\sin\chi\cos\chi$. The projections τ_χ in the same plane onto other directions with the angle β, as well as under uniaxial extension, are represented in the form $\tau_{\chi\phi} = (\sigma_1 - \sigma_2)F^*(\chi, \phi)$. The expression for the function $F^*(\chi, \phi)$ remains the same as defined above, where ϕ is an angle between directions of stresses σ_1 and $\tau_{\chi\phi}$. If two of the crystal axes coincide with the directions of algebraically maximal σ_1 and minimal σ_3 stress, the greatest tangential stress in the slip system at the orientation χ, ϕ is determined as $(\tau_{\chi\phi})_{\max} = |\sigma_1 - \sigma_3| F_{\max}$, where $F_{\max} = F^*(\chi, \phi)$.

Generally, directions of the principal stresses in the body elements are variable and do not coincide with crystal axes. In principle, it is possible to find the maximal value $(\tau_{\chi\phi})_{\max}$ for each element, consecutively obtaining tangential stresses in all systems of slight slip as the algebraic sum affected by all acting loads. But this task is extremely laborious and useless in the light of the reasons stated above concerning the drawback of quantity $(\tau_{\chi\phi})_{\max}$ as the bases for estimation of critical value of tangential stress τ_{cr}.

11.2.3.2 Design Using Averaged Crystal Characteristics

The above given experimental data confirm a possibility and expediency of using the maximum tangential stress $\tau_{\max}$ as a criterion of damage in a single crystal (Shorr and Mel'nikova 2010). In a combined stress state,

$$\tau_{\max} = 0.5|\sigma_1 - \sigma_3|. \tag{11.9b}$$

As for isotropic materials, equivalent stress σ_{eqv} can be used as the averaged criterion of a damage in a single crystal. Along with the equivalent stress σ_{eqv}, the quantity $\sigma_{int} = 2\tau_{\max} = |\sigma_1 - \sigma_3|$ is sometimes calculated with the software packages.

Some mechanical characteristics depend on the orientation slightly (see Fig. 11.4), and the others might depend more strongly. Reliable definition of stress quantity τ_{cr} or σ_{eqv} corresponding to single crystal failure can be obtained by conducting appropriate tests of machine parts, not only cast on the established technology for maintenance of investigated orientation, but also loaded the same as in operation. If a structure's stress state is close to uniaxial, it is possible to use directly the sample test results for corresponding orientation.

If the control of the applied loading is realized, as usual, by measurement of longitudinal strain ε, the value of the operating stress σ should be determined by the elasticity modulus $E_{[n]}$ for the corresponding crystal orientation, i.e., $\sigma = E_{[n]}\varepsilon$.

Under considerable influence of orientation on mechanical characteristics, the problem of 3D structure calculation based on 1D specimen extension tests becomes more difficult (even without the analysis of slip systems and retrieving maximum values $(\tau_{\chi\phi})_{\max}$). Possible approaches to analyze the influence of anisotropy on damage development in a combined stress state will be considered below.

11.2.3.3 Design at Various Crystallographic Orientation

Mechanical characteristics of a material (yield and break points, plasticity modulus, and so forth) are usually defined using specimen extension tests in a direction of the crystallographic axes [001], [011], and [111]. Mechanical characteristic for any direction $[n]$, conditionally designated as $\Psi_{[n]}$, can be approximately constructed using their three experimental values. A mean quantity of $\Psi_{m[n]}$ is accepted by the value $\Psi_{[n]}$ averaged over the area of a stereographic triangle.

According to Eq. (11.3b), the values of parameter $L_{[001]} = 0$, $L_{[011]} = 0.25$, and $L_{[111]} = 0.333$ correspond to the directions of axes [001], [011], and [111], relatively. The values $L_{[n]}$ for all other directions of the stereographic triangle change are in the range $0 - 0.333$. It is seen from Eq. (11.4) that the elasticity modulus $E_{(n)}$ at known values of E and G is an unequivocal function of $L_{[n]}$ (though magnitudes $E_{(n)}$ can be the same at different values $L_{[n]}$). Let us assume based on Magerramova and Vasil'ev (2012), other mechanical characteristics $\Psi_{[n]}$ to be also unequivocally dependent on the specified parameter $L_{[n]}$.

Vasil'ev (2013) proposed that a combined stress state affects the single-crystal element the same as a stretching "force" vector $\mathbf{F}$ equal to the vectorial sum of the principal stresses $\mathbf{F} = \boldsymbol{\sigma}_1 + \boldsymbol{\sigma}_2 + \boldsymbol{\sigma}_3$. Direction of this vector is defined by parameter $L_{[F]}$ depending on directive cosines l_{1F}, l_{2F}, l_{3F}. Direction of an equivalent vector $\mathbf{F}_0 = \boldsymbol{\sigma}_0$ for uniaxial tensile test defined by parameter $L_{[n]}$ must be the same. Having taken $L_{[n]} = L_{[F]}$, it is possible to define "equivalent" orientation $\mathbf{n}$ and characteristics of a material $\Psi_{[n]}$ answering to it.

First approximation of a stress state of structure elements under given outer loads is usually computed using characteristics of the single-crystal orientation, in which the detail has been cast. This allows defining the equivalent orientation in each element of a body and (if it considerably differs from initial) to carry out the improved calculations of the stress state. The equivalent orientation of each element at a proportional loading remains constant during all deformation process.

This approach was experimentally checked up in the quoted work, but it demands further verification.

11.3 Thermomechanical Characteristics of an Anisotropic Body

11.3.1 Thermoelasticity of a Single Crystal

11.3.1.1 Equations of Thermoelasticity

In crystals with cubic symmetry, thermal expansion is isotropic and identical in all directions, i.e., a factor α_T depends only on temperature. Therefore, the equations of thermoelasticity in the principal axes of crystal symmetry look like the Eq. (2.7) of

Chap. 2 as for an isotropic body. However, as it has been mentioned above, elasticity modules *E*, *G*, and Poisson's factor μ should be determined experimentally.

Equations of thermoplasticity connecting components of the stress σ_{ij} and strain ε_{ij} tensors with thermal expansion $\varepsilon_{T,ij} = \delta_{ij}\alpha_T\Delta T$ are

$$\varepsilon_{ij} = S_{ijkl}\sigma_{kl} + \varepsilon_{T,ij}, \tag{11.10a}$$

or

$$\sigma_{ij} = C_{ijkl}(\varepsilon_{kl} - \varepsilon_{T,kl}) \tag{11.10b}$$

11.3.1.2 Influence of Anisotropy on Thermoelastic Behavior of a Body

When deformations of a crystal with cubic symmetry are limited in any direction, the arising temperature stresses are managed by characteristics of the anisotropic body, which by now depend both on temperature and orientation of a crystal.

In Fig. 11.5, temperature dependence of moduli $E_{[n]}$ and $G_{[n]}$ for various crystallographic orientations of a heat resisting alloy, generated according to Shalin et al. (1997), is shown. The moduli decrease with temperature rise, but their relations remains practically constant and anisotropy increases a little.

Comparison of the moduli for various crystallographic directions calculated due to Eq. (11.4) and experimental values shows that an error does not exceed some percent. But for calculating temperature stresses, we have to take into account possible deviations in crystal orientation.

Some examples of thermal stresses calculations in application to simple bodies of single-crystal materials are shown in Krivko et al. (1989), Shalin et al. (1997), and others.

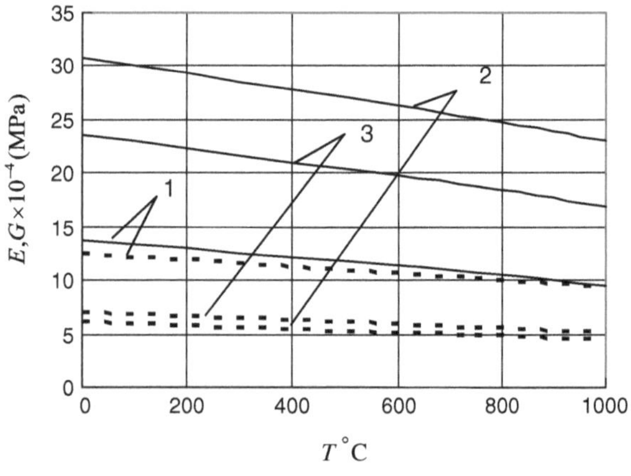

Fig. 11.5 Temperature dependence of elastic moduli for various crystal orientations: *1* [001], *2* [111], *3* [011]; *solid line E, dotted line G*

11.3.2 Static Strength and Durability

11.3.2.1 Short-Time Strength and Plasticity

In Fig. 11.6, a temperature dependence of ultimate stresses σ_B (shown by solid lines) and yield points $\sigma_{0,2}$ (shown by dashed line) for crystal orientations [001] and [111] of a heat-resistant alloy is presented. Stress $\bar{\sigma}$ is referred to the value of $\sigma_{B\{001]}$ at normal temperature.

At normal temperature, strength properties of the alloy with orientation [111] are usually higher than with orientation [001]. As temperature grows, the distinction of strength characteristics becomes smooth.

Yield points of some single-crystal alloys at orientation [011] happen to be a little lower than at orientation [001] (see Kablov and Golubovsky 1998).

11.3.2.2 Creep and Long-Term Strength

Experimental characteristics of creep and long-term strength (durability) of a number of heat-resistant alloys with nickel basis are presented in Kablov and Golubovsky (1998).

A long-term strength of a single-crystal structural element at elevated temperature can be estimated using Larson–Miller's parameter (see Chap. 6). For this goal, a function

$$\tau_{\max} = f(P_{LM}) \tag{11.11}$$

must be established for the given orientation due to tests, and a maximum tangential stress $\tau_{\max}$ in an must element be calculated for an operating regime.

By comparison of long-term durability for the specimens of different orientation, it is necessary to take into account that the scatter of durability referred to deviation of orientation [111] is usually more significant than for other orientations.

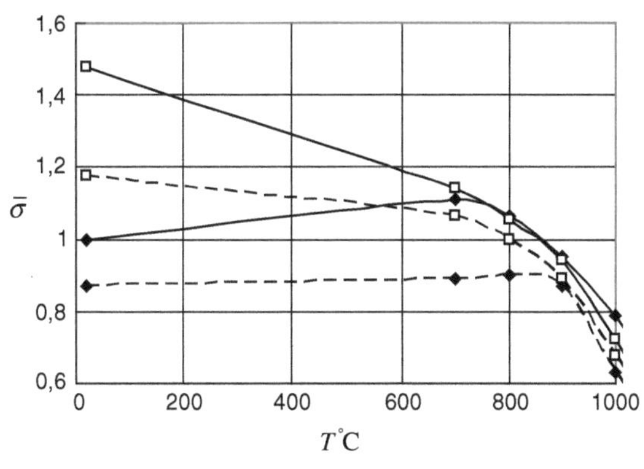

Fig. 11.6 Temperature dependence of ultimate stress σ_B and yield point $\sigma_{0,2}$ for different crystal orientations of a heat-resistant alloy; [001] *black rhombuses*, [111] *light squares*

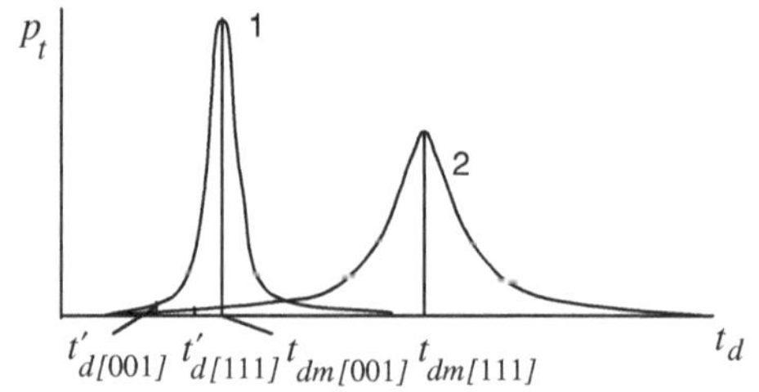

Fig. 11.7 Probability density of durability for different crystal orientations: *1* [001], *2* [111]

In Fig. 11.7, a principal pattern of probability density of distributions p_t versus time t_d prior to failure for specimens with orientations [111] and [001] is demonstrated. The ratio of their mean durabilities is $t_{dm[111]}/t_{dm[001]} \approx 1.7$. However, the durability for small probability of failure becomes nearly identical for both orientations as $t'_{d[111]} \approx t'_{d[001]}$.

During long operation under high temperature, durability for different orientations can vary with temperature not monotonously.

According to Kear and Piearcey (1967) and Shalin et al. (1997), the minimal durability of some alloys at temperatures 750−800 °C was found out at the orientation close to [011], while the maximal one—at [111]. However, in some cases, the maximum of durability passes to orientation [001] depending on heat treatment regimes.

11.3.3 Durability Under Alternating Loads

11.3.3.1 Low Cyclic (LCF) and Thermal Fatigue

Isothermal LCF

Tests of single-crystal materials under repeated static loadings show that they possess an increased resistance to LCF as they are free from the weakened grain boundaries and have higher strength characteristics.

Rather limited tests of a crystal orientation influence on isothermal LCF lead to not quite certain results for various temperatures.

Test results obtained by Sayyah and Schonberg (2002) are presented in Fig. 11.8, where black points show dependences between strains ranges $\Delta\bar{\varepsilon} = \Delta\varepsilon/\Delta\varepsilon_m$ and number of cycles N_d until the occurrence of LCF cracks for tension/compression of specimens cast from PWA alloy with orientations [001] and [111]. The points, concerning orientations [011] and [213], are allocated between plotted points. The ranges $\Delta\varepsilon$ are referred to the value $\Delta\varepsilon_m$—the mean for all samples.

There is a large distinction in cyclic durability for different orientations represented by the dependence $N_d = f(\Delta\varepsilon)$.

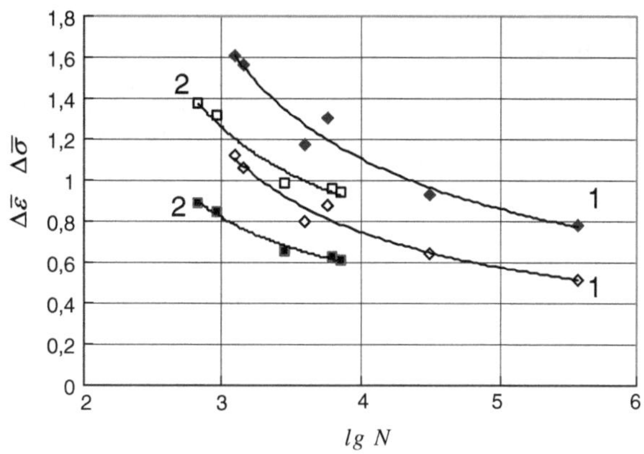

Fig. 11.8 Curves of LCF of single-crystal specimens for different loading orientations: *1* [001], *2* [111]

In the specified work, attempts have been made "to pull together" the points, referred to cyclic durability for various orientations. Six LCF criteria offered in the literature were used, but no positive result was obtained.

At the same time, the values obtained by recalculation of stress ranges $\Delta\sigma_{[n]} = E_{[n]}\Delta\varepsilon_{[n]}$ (shown by light dots), using moduli of elasticity taken for each corresponding orientation, have resulted in a less distortion of points forming the LCF curves for different orientations.

Similar results have been shown in the tests of MAR-M200 alloy by Chieragatti and Remy (1991).

This fact does not contradict the results on fatigue tests presented in Fig. 11.4.

At elastoplastic deformations, the role of the elasticity moduli might become less important, but the common situation will basically remain the same.

Thermal Fatigue

Experimental data on thermal fatigue of single crystals are even more limited. In Dul'nev et al. (1988), test results for a heat-resistant alloy with four different orientations under temperature alteration within the range $\Delta T = 100 \leftrightarrow 950$ °C are presented. Tests were carried out at the installation described in Sect. 4.3.1.2 with regulated rigidity of the specimen fixation, which allows changing and measuring the ranges of elastoplastic strains and stresses independently.

The obtained averaged experimental dependences between number of cycles prior to failure N_d and ranges of stresses $\Delta\sigma$ or elastoplastic strains $\Delta\varepsilon$ are shown in Fig. 11.9a, b, accordingly. Figure 11.9 demonstrates the results scatter over the several tested specimens.

Estimating thermal fatigue in terms of the strain ranges $\Delta\varepsilon$, the specimens with orientations [111] and [001], of which elasticity moduli differ more than twice, have shown close results and higher cyclic durability than the specimens with orientations [011] and [112]. Probably, these results may be connected with the fact that

Fig. 11.9 Dependence of stresses (**a**) and elastic–plastic strains (**b**) ranges on damage cycles N_d for single crystals with different orientation: *1* [001], *2* [111], *3* [011], *4* [112]

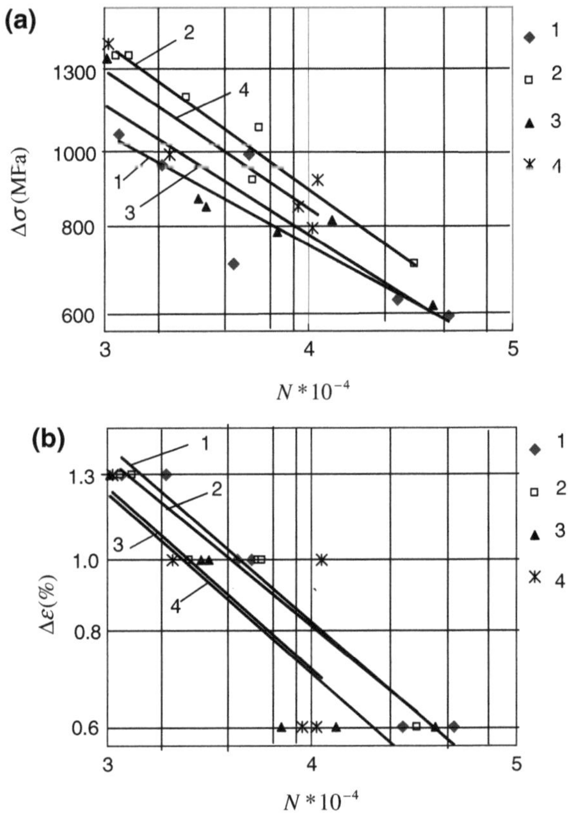

the common strain range under thermocyclic tests is defined, basically, by thermal expansion $\Delta\varepsilon_T = \alpha_T \Delta T$, while the role of the elastic strains $\Delta\sigma/E$ (hence, the elasticity modulus) is significant.

For estimation in terms of stress ranges $\Delta\sigma$, orientation [001] has appeared the least durable.

A wide scatter of durability values for the specimens with nominally identical orientation may be connected both with usual disorder of material durability and with deviations in orientation that has reached 13°.

11.3.3.2 High Cyclic Fatigue (HCF)

Fatigue test data of specimens with various orientations, described above in Sect. 11.2.3, and also similar results obtained by some other researches, have shown that the most objective criterion for anisotropic body failure from HCF in view of orientations scattering is the maximal tangential stress $\tau_{\max}$, or a quantity $\sigma_{int} = 2\tau_{\max} = |\sigma_1 - \sigma_3|$ used in some software packages.

As well as at LCF, the highest values usually have the limits of endurance at orientation [111]. However, at high temperatures, the anisotropy of single crystals decreases.

Single crystals possess high sensitivity to stress concentration. The absence of grain boundaries, increasing the durability, at the same time, facilitates propagation of fatigue cracks along the slip systems. While temperature increases, the single crystals sensitivity to stress concentration relaxes.

References

Chieragatti R, Remy I (1991) Influence of orientation on the low cycle fatigue of MAR-M200 single crystals at 650 °C. Mater Sci Eng 141:1–9

Dul'nev RA, Svetlov IL, Bychkov NG et al (1988) Oriental dependence of thermal fatigue for single crystals of nickel alloy (in Russian). J Problemy Prochnosty 11:3–9

Fridman JaB (1974) Mechanical properties of metals (in Russian). 3rd edn, v.I Deformation and damage. Mashinostroenie, Moscow

Kablov EN, Golubovsky ER (1998) High-temperature strength of nickel alloys (in Russian). Mashinostroenie, Moscow

Kear BH, Piearcey BJ (1967) Tensile and creep properties of single crystals of the nickel-base superalloy MAR-M200. Trans Met Soc AIME 239:1209–1215

Kolesnikov VI, Pridopojny RP, Sheremet'ev AV (2001) Influence of crystallographic orientation on static strength and fatigue resistance of single crystal blades (in Russian). In: Proceedings of the Scientifical Papers "Aviokosmicheskaja Tekhnika I Tekhnologiya", No. 26, Kharkov: 132–135

Krivko AI, Epishin AI, Svetlov IL et al (1989) Calculation of thermal stresses and heat resistance of anisotropic materials (in Russian). J Problemy Prochnosti Materialov, No. 2: 3–9, No. 4: 43–48

Lekhnitsky SG (1950) Theory of elasticity of anisotropic body (in Russian). Gostekhizdat, Moscow

Magerramova LA, Vasil'ev BE (2012) Influence of single crystal orientation on stress-strain state and strength of gas turbine blades (in Russia). Trudy Moskovskogo Aviatzionnogo Instituta, T.12, No. 5: 89–97

McClintock FA, Argon AS (eds) (1966) Mechanical behavior of materials. Addison-Wesley, Boston

Sayyah MT, Schonberg WP (2002) New failure criterion for space shuttle main engine turbine blades. J Spacecraft Rockets 39(1):140–145

Schmid E, Boas W (1935) Plasticity of crystals with special reference to metals. F.A.Hughes & Co. Ltd

Shalin RE, Svetlov IL, Kachanov EB et al (1997) Single crystals of nickel heat resistant alloys (in Russian). Mashinostroenie, Moscow

Shorr BF, Mel'nikova GV (2010) Some problems on dynamics of turbine single crystal blades (in Russian) J Problemy Mashinostroenija I Nadejnosti Mashin, No. 4: 33–40. Trans J Mach Manuf Reliab. Allerton Press Inc, New York

Shorr BF, Mel'nikova GV, Magerramova LA et al (2001) Influence of single crystal orientation and its technologic scatter on frequency spectrum of turbine blades (in Russian). In: Proceedings of the scientifical papers "Aviokosmicheskaja Tekhnika I Tekhnologiya", No. 26, Kharkov: 140–144

Vasil'ev BE (2013) Elaboration of plasticity model of single crystal materials for strength design of perspective engines turbine blades (in Russian). J Tekhnologija Legkikh Splavov, No. 3, Moscow

Chapter 12
Special Problems of Thermal Integrity

Abstract The chapter contains information on the special problems of thermal integrity, which can be of interest for the readers aimed to gain deepened understanding about some peculiarities of materials' and constructions' behavior at elevated temperatures. For problems of non-isothermal dynamics (Sect. 12.1), it is shown that influence of the material thermal extension on detail natural oscillation frequencies with temperature increase can be different. Alteration of contact conditions between machine parts (for example, change negative clearances on tightness or on the contrary) and occurrence of the temperature stresses, especially in thin-walled structures, can change their rigidity and the frequency spectrum. A possibility of directional change of a body shape by serial variable heating of its separate body sections having various elastoplastic temperature-dependent mechanical properties is demonstrated in Sect. 12.2. This effect, used in a number of technological processes, is illustrated on simple examples. Though in the majority of practical thermal strength problems, the temperature state of a body is supposed known and not dependent on its stress–strain state, and for an explanation of some effects, it is necessary to address to the "coupled" theory of the thermoelasticity considering thermodynamic link of mechanical and thermal processes. The bases of this theory in rather simple treatment are stated in Sect. 12.3. The coupling effect is illustrated by calculations of energy dispersion during elastic vibrations and influence of convective heat exchange on material deformation at tensile trials. In Sect. 12.4, the fundamentals of the wave theory of thermal conductivity considering (unlike the classical theory) a finite speed of thermal stream and temperature propagation over a solid are briefly stated, and the influence of this factor on unsteady temperatures and stresses propagation at a heat shock on a semi-infinite body is shown.

B.F. Shorr, *Thermal Integrity in Mechanics and Engineering*,
Foundations of Engineering Mechanics, DOI 10.1007/978-3-662-46968-2_12

12.1 Non-isothermal Dynamics

We shall use the term "non-isothermal dynamics" in a broad enough sense to cover a number of problems of dynamics concerning the influence of a body temperature and temperature change on mechanical behavior of a body, including conditions of non-uniform temperature distribution over the body and its thermal expansion. In this section, we assume that calculations of a temperature field and mechanical behavior of a body can be performed independently. Questions of the "coupled" theory of thermoelasticity including its dynamic aspects are considered in Sect. 12.3. It is supposed that the reader is familiar with the basics of solid dynamics.

The problems of non-isothermal dynamics may be classified into four major groups.

The first group concerns influences of body's element temperatures on elastic characteristics independently of thermal expansion resulted in "temperature heterogeneity" of the body. In case of such heterogeneity, the body's dynamic response might be considered as a behavior of a system consisting of various materials.

The next two groups take into account the influence of thermal expansion which, on the one hand, can lead to the change of mutual position of details (due to disclosing of some joints and so forth), and on the other hand, to the occurrence of thermal stresses. In some cases, it might change the dynamics of a mechanical system. At the moderate temperature, it is possible to neglect the variation of mechanical characteristics of a material.

Finally, the fourth group covers problems in which the influence of the previously specified factors appears simultaneously.

All these problems will be discussed to some extent below.

12.1.1 Influence of a Uniform Temperature Field

12.1.1.1 Temperature-Dependent Frequencies

Dynamic characteristics and processes in a solid are connected with the speed of deformation wave propagation (i.e., speed of a sound) which in a one-dimensional case is defined by the well-known formula

$$c(T) = \sqrt{\frac{E(T)}{\rho(T)}}, \tag{12.1}$$

where $E(T)$ is the modulus of longitudinal elasticity (Young's modulus) and $\rho(T)$ is density of a material.

For example, the lowest natural frequency of longitudinal oscillation of a rod with constant cross section, length L, and fixed ends is equal to

$$f(T) = \frac{c(T)}{2L}. \tag{12.2}$$

Dynamic processes in a combined stress state also depend on the shear modulus $G(T)$ connected with Young's modulus through Poisson's factor $\mu(T)$ by the formula

$$G(T) = \frac{E(T)}{2[1 + \mu(T)]}.$$

All the specified physical characteristics of a material depend on a body temperature. As temperature grows, moduli of elasticity, as a rule, decrease, and according to Eq. (12.1), the deformation wave speed $c(T)$ and natural frequencies also decrease. An insignificant reduction of body density and some change of Poisson's factor at heating can usually be neglected.

Therefore, in the absence of thermal expansion (at $\alpha_T = 0$) and also if a body has an opportunity to freely expand at uniform heating, natural modes of oscillation (distribution of relative displacements, strains, and stresses) remain constant.

If thermal expansion factor is $\alpha_T \neq 0$, then the dynamic behavior of the uniformly heated body can vary depending on both its current temperature T, and also on the temperature change $\Delta T = T - T_0$ relative to some reference value T_0.

Thermal expansion of metal rod under the change of temperature within the limits of 1000 °C at usual values of linear expansion factor might cause relative strain up to $\varepsilon \approx 10^{-2}$. In case of free expansion of a body, such deformation influences frequency and oscillation modes insignificantly. If the expansion of a body is constrained, such deformation might result in thermal stresses and change dynamic response of a system.

Under alternating stresses, displacements and other parameters are summarized with corresponding constant (static) values. In a linear statement of a problem, the system of governing equations for many problems splits into independent static and dynamic parts, and then, it is assumed that preliminary static state does not influence dynamic behavior (this is true if temperature insignificantly changes a body configuration).

There are problems in which a change of even uniform temperature field causes essential influence on dynamics of a mechanical system.

12.1.1.2 Structures with Changed Gaps

During assembling of a structure, there might provide gaps (backlashes), $\Delta_0 > 0$, between details. Thermal expansion can close the gaps and lead to the occurrence of tightness with contact stresses between details. The reverse situation, when under

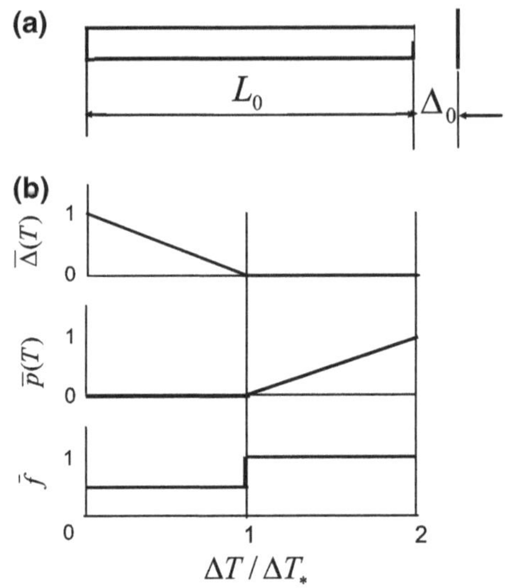

Fig. 12.1 A rod with initial gap (**a**) and alteration of its parameters at heating (**b**)

temperature change the assembling tightness $\Delta_0 < 0$ disappears with the formation of a gap, is also possible.

Dynamic processes in structures with initial small gaps between members become nonlinear due to alteration of members' connections. The general methods of analytical and numerical calculations of nonlinear oscillations in structures with varying boundary conditions are covered in a number of works, for example Babitsky (1978) and Shorr (2004).

The influence of temperature change on the dynamics of such systems can be illustrated by the example of oscillations of the elastic rod. There is an initial gap Δ_0 between one of the rod's ends and the wall at temperature T_0 (Fig. 12.1a).

As temperature rises from T_0 to $T = T_0 + \Delta T$ and the length L_0 of the rod increases in $\Delta L = \alpha_T \Delta T L_0$, an initial gap between the elements will decrease up to $\Delta(T) = \Delta_0 - \Delta L$. At the "threshold" temperature $T_* = T_0 + \Delta T_*$, where $\Delta T_* = \Delta_0 / \alpha_T L_0$, the gap disappears and the tightness $\Delta(T) < 0$ occurs between the elements resulting in a contact pressure $p(T)$ equal at a rigid wall to $p(T) = E\alpha_T(\Delta T - \Delta T_*)$.

The first natural oscillation frequency f_{II} of the rod with clamped ends is defined by Eqs. (12.1) and (12.2), while at the presence of a gap, it will be $f_I = 0.5 f_{II}$.

The change in relative values of the gap $\bar{\Delta}(T) = \Delta(T)/\Delta_0$, the contact pressure $\bar{p}(T) = p(T)/p_0$, where p_0 corresponds to the heating $\Delta T = 2\Delta T_*$, and the first natural rod frequency $\bar{f} = f/f_{II}$ versus temperature change are shown in Fig. 12.1b. In this case, the change of frequency at heating is not connected with thermal stresses, but it depends on the change of boundary conditions.

At forced vibrations of a rod when it will be affected in the longitudinal direction by variable force with a frequency f, dynamic reaction appears to be essentially various depending on excitation intensity.

Under weak excitation within the temperature range $T_0 < T < T_*$, the resonant oscillations arise at congruence of frequencies $f \approx f_I$, as long as the oscillation amplitude u of the rod's free end does not exceed a current gap $\Delta(T)$. At temperatures $T > T_*$, the resonance occurs at $f \approx f_{II}$, as long as the amplitude of compressing stress σ at the right rod's end does not exceed current contact pressure $p(T)$. At temperature $T = T_*$, the conditions of a resonance might sharply change.

Under stronger excitation at $T < T_*$, the oscillation amplitude u of the free end can achieve the size Δ_0 at some temperature $T_I < T_*$. Therefore, fluctuations become especially nonlinear with change of boundary conditions within each period and the occurrence of compression stress in a rod. Similarly, under strong excitation at $T > T_*$, the gap can start to appear at some temperature $T_{II} > T_*$. In these cases, the resonant frequency will be gradually reconstructed from f_I up to f_{II} and back depending on temperature value.

12.1.1.3 Stretched Heated Strings

In the previous example, change of oscillation frequency at temperature expansion has been connected with change of boundary conditions under transition from a gap to tightness and vice versa, but rigidity of the structure is not varied. In some cases, for example, at lateral deformation of stretched strings or beams loaded in the axial direction, natural frequencies may vary as a result of temperature change of the structure rigidity under constant boundary conditions.

Equation of lateral oscillations of a string stretched by a longitudinal force is

$$\frac{\partial^2 y}{\partial t^2} - c_0^2 \frac{\partial^2 y}{\partial x^2} = 0. \tag{12.3}$$

where $c_0 = \sqrt{F_{x0}/\rho A}$ is a speed of lateral disturbances in a string with length L, stretched by force F_{x0}.

Natural oscillation frequencies of the string are

$$\omega_m = \frac{\pi m}{L_0} \sqrt{\frac{F_{x0}}{\rho A}} \quad \text{at } m = 1, 2, \ldots \tag{12.4}$$

If one of the string ends remains free in the longitudinal direction, its length at temperature rise in ΔT increases in $\Delta L = \alpha_T \Delta T L_0$. At real values of temperature expansion $\varepsilon_T = \alpha_T \Delta T \ll 1$, the length of the string and natural oscillation frequencies change very poorly.

In the case when both ends of the stretched string are fixed in the longitudinal direction, the actual length of the string remains constant at heating (certainly until the string remains in a straight line owing to permanent extension). The picture will then essentially change. As full deformation is

$$\varepsilon_x(T) = \sigma_x(T)/E + \varepsilon_T = \varepsilon_{x0} = F_{x0}/EA = \text{const},$$

that at temperature rise up to $T = T_0 + \Delta T$ stretch of the string will decrease up to size

$$F_x(T) = \sigma_x(T)A = F_{x0}(1 - \varepsilon_T/\varepsilon_{x0})$$

and, hence, the natural frequencies will decrease in the ratio

$$\frac{\omega(T)}{\omega(T_0)} = \sqrt{\frac{F_x(T)}{F_{x0}}} = \sqrt{1 - \frac{\varepsilon_T}{\varepsilon_{x0}}}. \tag{12.5}$$

At heating in $\Delta T_* = \varepsilon_{x0}/\alpha_T(T)$ when $\varepsilon_T/\varepsilon_{x0} = 1$, the stretch of the string vanishes and frequencies fall down to zero.

At cooling of a string, its extension and natural frequencies increase.

This sample shows that the tuning of musical string tools depending on the natural string oscillation frequencies is temperature-sensitive.

12.1.1.4 Heating of Two-Fixed Beam

A longitudinal force F_x at longitudinal-transversal loading of a beam can be both stretching and compressing.

The differential equation of transverse vibrations of a beam with constant cross-sectional area A and moment of inertia J affected by static longitudinal force F_x looks like (see Timoshenko 1955)

$$EJ\frac{\partial^4 y}{\partial x^4} - F_x\frac{\partial^2 y}{\partial x^2} = -\rho A\frac{\partial^2 y}{\partial t^2}. \tag{12.6}$$

For a two-fixed beam with boundary conditions at the ends,

$$y(0,t) = 0, \quad \frac{\partial^2 y}{\partial x^2}(0,t) = 0, \quad y(L,t) = 0, \quad \frac{\partial^2 y}{\partial x^2}(L,t) = 0$$

the solution of Eq. (12.6) is

$$y(x,t) = Y_m \sin\frac{\pi m x}{L}\cos\omega t \quad \text{at } m = 1, 2, \ldots. \tag{12.7}$$

Equations (12.6) and (12.7) yield

$$EJ\left(\frac{\pi m}{L}\right)^4 + F_x\left(\frac{\pi m}{L}\right)^2 - \rho\omega^2 A = 0. \tag{12.8}$$

According to Eq. (12.8), natural oscillation frequencies ω of the beam under the action of longitudinal force can be presented in the form

$$\omega = \omega_0 \sqrt{1 + \frac{F_x}{m^2 F_{cr}}}, \tag{12.9}$$

where the frequencies at $F_x = 0$ are equal to

$$\omega_0 = \left(\frac{\pi m}{L}\right)^2 \sqrt{\frac{EJ}{\rho A}}, \tag{12.10}$$

and F_{cr} is Euler's critical force (see Eq. (9.1a)) in Chap. 9.

Axial compression (at $F_x < 0$) reduces the frequencies.

Free ends of the beam under transversal bending approach on a quantity ΔL that for small ratio $(Y_m/L)^2 \ll 1$ are

$$\Delta L \approx \frac{1}{2} \int_0^L \left(\frac{dy}{dx}\right)^2 dx.$$

If the ends of a beam have been fixed at some temperature T_0 which remains constant under oscillations, then the arising longitudinal stretching force $F_x = EA\Delta L/L$ introduces insignificant amendment to frequencies for not too flexible beams.

However, increase of temperature ΔT in a beam with the fixed length provokes the compressing axial force $F(T) = -EA\alpha_T \Delta T$. The ratio

$$\frac{F_x}{F_{cr}} = -\frac{\alpha_T \Delta T A L^2}{\pi^2 J} = -\frac{\alpha_T \Delta T \lambda^2}{\pi^2}$$

in Eq. (12.9) for a beam with cross-sectional height h is equal to

$$\frac{F_x}{F_{cr}} = -\frac{12\alpha_T \Delta T}{\pi^2} \left(\frac{L}{h}\right)^2.$$

This ratio in flexible beams becomes commensurable with unit even at the moderate heating.

According to Eq. (12.9), this leads to significant reduction of natural frequencies. At $\Delta T_* = F_{cr}/E\alpha_T$, the lowest frequency of beam-bending oscillations falls down to zero, and the beam loses stability (see Chap. 9).

Under cooling of a beam, the axial force becomes extensive and natural frequencies increase.

12.1.1.5 Stretched Heated Membranes

Change of temperature renders similar influence on transversal oscillations of the stretched membranes and plates if their contour $r = R$ remains immobile in the radial direction.

The lowest natural frequency of transversal axial-symmetric oscillations of a membrane with radius R, stretched by a stress σ_0 at temperature T_0, is determined by the formula (Timoshenko 1955)

$$\omega_0 \approx \frac{2.4}{R}\sqrt{\frac{\sigma_0}{\rho}}.$$

Under heating in ΔT, the tensile stresses in membrane are diminished to the magnitude

$$\sigma = \sigma_0 - \frac{E\alpha_T \Delta T}{1-\mu}, \tag{12.11}$$

and the first natural frequency decreases to size

$$\omega = \omega_0\sqrt{1 - \frac{\Delta T}{\Delta T_{cr}}}, \tag{12.12}$$

where

$$\Delta \mathrm{T}_{cd} = \frac{(1-\mu)\sigma_0}{E\alpha_T}$$

is the critical rise in temperature at which frequency drops to zero.

The frequency ω_0 does not depend on the elasticity modulus E, and the value ΔT_{cr} increases with reduction of E. Therefore, reduction in the elasticity modulus with temperature rise decreases the natural frequencies of membranes less strongly than in the case of strings.

For a solid plate stretched at temperature T_0 by a stress $\sigma_r = \sigma_0$ and fixed in transversal direction at the outer contour $r = R$, frequency of umbrella-type oscillations is equal to (see Timoshenko 1955)

$$\omega_0 = \omega_{00}\sqrt{1 + \frac{\sigma_0 h R^2}{16D}}, \tag{12.13}$$

where

$$\omega_{00} \approx \frac{10.3}{R^2}\sqrt{\frac{D}{\rho h}}, \quad D = \frac{Eh^3}{12(1-\mu^2)}.$$

Under heating, the stress is defined as before by Eq. (12.11) and frequency with $\sigma_0 \neq 0$—by the formula

$$\omega_0 = \omega_{00}\sqrt{1 + \frac{\sigma_0 h R^2}{16D}\left(1 - \frac{E\alpha_T \Delta T}{(1-\mu)\sigma_0}\right)}. \tag{12.14}$$

If temperature stresses depend on product $E\alpha_T$, then, in view of temperature reduction in the modulus E and some increase of temperature expansion in factor α_T, the influence of the absolute body temperature on its dynamics becomes perceptible only for the temperatures when the elasticity modulus falls sharply.

12.1.2 Influence of a Non-uniform Temperature Field

12.1.2.1 Thermal stresses in a uniformly heated body might arise only during it interaction with other bodies. Under non-uniform heating, thermal static stresses arise during a force interaction of separate body parts. In the frame of dynamic concepts, a system becomes "preliminary stressed." If the displacements of the body boundaries are limited, then non-uniform heating of internal parts usually leads to the occurrence of reactions from adjacent bodies. In these situations, frequency characteristics of a body change in heating, as has been described above in Sect. 12.1.1.

Thermal stresses in a non-uniformly heated body can also arise in some cases when its external boundaries are free, for example, at various temperatures of the interconnected parallel rods or strings, and also as a result of temperature change along the radius of the circular solid plate fixed only at the center. The influence of such thermal stresses on a body dynamics becomes apparent under the conditions, similarly to specify in Sect. 12.1.1.5.

If ΔT_k is a deviation of temperature of separate body part k in volume V_k from average temperature under local non-uniformity of heating, then

$$\sum_k \Delta T_k V_k = 0. \tag{12.15}$$

Let us consider some typical problems.

12.1.2.2 Let two identical parallel rods connected at the ends by rigid weightless bridges have an opportunity to freely, but only jointly, extend in the longitudinal direction, as shown in Fig. 12.2a. Temperature along the length of the rods is

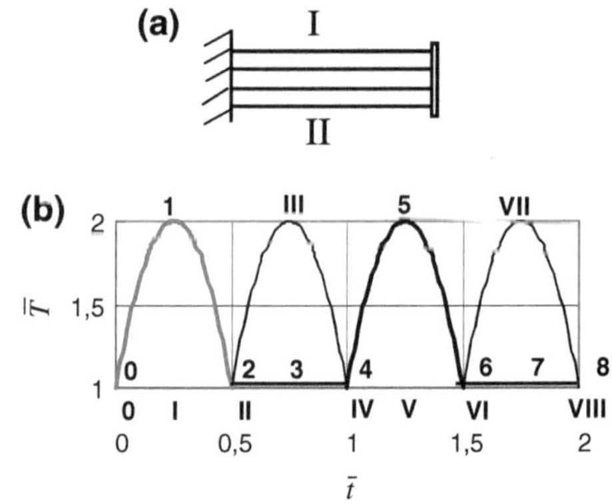

Fig. 12.2 Two coupled rods under cyclic moving temperature: **a** a model, **b** rods' temperature under alternate heating

constant, while temperature increments ΔT_I and ΔT_{II} under heating are various, and $\Delta T_I + \Delta T_{II} = 0$.

From conditions of total strains equality $\varepsilon_I = \varepsilon_{II}$ and balance of boundary bridges $\sigma_I + \sigma_{II} = 0$, longitudinal thermal stresses in rods are defined as

$$\sigma_I = -\sigma_{II} = -\frac{E_I E_{II}}{E_I + E_{II}}(\alpha_I + \alpha_{II})\Delta T_I, \tag{12.16a}$$

where the moduli of elasticity and thermal expansion factors are dependent on the current temperature.

The common longitudinal strain of the rods is equal to

$$\varepsilon_I = \varepsilon_{II} = \frac{E_I \alpha_I - E_{II}\alpha_{II}}{E_I + E_{II}}\Delta T_I. \tag{12.16b}$$

First, we examine *coupled longitudinal oscillations* of the rods. In this case, in linear approximation, the full system of rods' deformations breaks up to the static and dynamic equation systems.

At constant material characteristics, it would be $\sigma_I = -\sigma_{II} = -E\alpha\Delta T_I$ and $\varepsilon_I = \varepsilon_{II} = 0$; hence, in this case, each rod would be in the static conditions similarly accepted in Sect. 12.1.1. Both beams could oscillate either in phase, as console, with frequency $f = 0.25\ c/L$, or in anti-phase, as two fixed at the ends with frequency $f = 0.5\ c/L$.

At temperature-dependent material characteristics, longitudinal displacements of rods at natural oscillation with frequency ω must satisfy the system of the differential equations

$$\left.\begin{array}{l} E_1 \frac{d^2u_1}{dx^2} + \rho_1\omega^2 u_1 = 0 \\ E_2 \frac{d^2u_2}{dx^2} + \rho_2\omega^2 u_2 = 0 \end{array}\right\} \tag{12.17a}$$

with boundary conditions

$$u_1 = u_2 = 0 \quad \text{at } x = 0, \quad u_1 = u_2 \text{ at } x = L, \tag{12.17b}$$

$$E_1 \frac{du_1}{dx} + E_2 \frac{du_1}{dx} = 0 \quad \text{at } x = L.$$

The system (12.17a) with the first equality (12.17b) is satisfied by substituting

$$u_1 = B_1 \sin \lambda_1 x, \quad u_2 = B_2 \sin \lambda_2 x,$$

where $\lambda_1 = \omega/\omega_1 = \omega\sqrt{\rho_1/E_1}$; $\lambda_2 = \omega/\omega_2 = \omega\sqrt{\rho_2/E_2}$.

The second equality (12.17b) demands that $B_1 \sin \lambda_1 L = B_2 \sin \lambda_2 L$, and the third yields the characteristic equation

$$\cos v\lambda_2 L + v \frac{\sin v\lambda_2 L}{\sin \lambda_2 L} \cos \lambda_2 L = 0, \tag{12.18}$$

where $v = \lambda_1/\lambda_2 = \sqrt{E_2/E_1}$.

The system behaves as anisotropic in which the natural frequencies of vibrations do not depend on temperature expansion.

12.1.2.3 Great static displacements or plastic deformations can lead to significant changes of body size and shape. In this case, dynamic calculations should be carried out, as for any strongly nonlinear system, for the deformed state according to the altered geometry and material properties.

In some cases, non-uniform heating of a body, despite small change of the body size and shape, can provoke noticeable change of rigidity and frequencies.

In Appendix A, deformation of a twisted rod with strongly elongated rectangular cross section of sides' ratio $(b/h)^2 \gg 1$ will be considered.

At non-uniform temperature distribution along the cross-sectional width b, when the temperature at the section's central part exceeds the temperature at the section's rear ends, rigidity of the rod in tension decreases and in torsion increases.

This is reflected in change of the oscillation spectrum of the rod. Calculations show that in the rod with sides' ratio $b/h = 10 - 20$ and $\gamma = \theta_{in} b^2/h = 5 - 10$, where θ_{in} is a relative pre-twist angle, frequencies of torsion oscillation modes can change in (3 − 6) % if temperature non-uniformity is $\Delta T = 300 - 500$ °C.

12.2 Change of a Body Shape Under Thermocyclic Loading

12.2.1 Directional Change of a Body Shape Under Thermocyclic Loading

Large number of machine parts and power plants loading cycles can be provided only if the behavior of structure materials is stable enough and structure's size and form are preserved under operating conditions (including creep).

It is known that sizes and shapes of the bodies of anisotropic materials and materials with phase transformations may significantly change under a small number of repeated heating and cooling.

However, under certain conditions, such a phenomenon, named sometimes *thermocyclic change of a body shape*, is also possible for bodies of isotropic materials with stable high plastic properties. This effect was investigated theoretically and experimentally by a number of researchers—see for example Gokhfeld (1961), Gokhfeld and Laptevsky (1968), and others.

Alongside undesirable consequences, shaping effect of thermocyclic loading can be successfully used in some technological processes, for example, for tubes squeezing, receptions of corrugated shells, and so forth.

For rather plastic materials, especially with the well-defined yielding platform, the effect of hardening within the limits of the first several cycles might be neglected, and the use of approximate model of ideally plastic material with the yield point σ_y depending only on temperature is available.

The shaping effect of thermocyclic loading is described below in more details for some problems.

12.2.2 Directional Change of Length of a Rod System Under Moving Temperature Field

12.2.2.1 In Fig. 12.2a, a system consisting of two identical parallel rods I and II has been presented. Left ends of the rods are fixed, while right ones can move together. Now, let us assume that the temperature of rods T_I and T_{II} periodically change from T' up to T'' and back due to any law $T(t)$ with shift in time for half of period τ, as shown in Fig. 12.2b where dependence of dimensionless quantities $\bar{T} = T/T' = f(\bar{t}),\quad \bar{t} = t/\tau$ is taken for $T''/T' = 2$.

During each first half of a period, the temperature of the first rod $T_I(t)$ varies, while the temperature of the second rod $T_{II} = T'$ remains constant. During the second half of the period—on the contrary, $T_{II}(t)$ varies at $T_I = T'$. Temperature alternation of the rod T_I is shown with a bold line, and characteristic points of a curve are marked with the Roman numerals; temperature alternation of the rod T_{II} is shown by a more thin line, and characteristic points of a curve are numbered using the Arabian figures. Time, as usually in the theory of non-stationary static problems, acts as a parameter of a process.

Below we consider a problem, where for each half-cycle $0 \le \bar{t}_1 \le 0.5$, the parabolic law of temperature alternation

$$\bar{T}(\bar{t}_1) = 1 + 8(\bar{T}'' - 1)\bar{t}_1(1 - 2\bar{t}_1) \tag{12.19}$$

is taken, where $\bar{t}_1 = \bar{t} - \bar{t}_{i0}$ and $\bar{t}_{i0}$ are the initial moment of the half-cycle i.

For some temperature swing, we shall use the linear laws of temperature alternation with the elasticity modulus $E(T)$, a yield point $\sigma_y(T)$, and the factor of linear expansion $\alpha_T(T)$. In a dimensionless form

$$\bar{E}(\bar{T}) = \frac{E(T)}{E(T')} = 1 + \frac{k_E - 1}{k_T - 1}(\bar{T} - 1), \tag{12.20}$$

where $k_T = T''/T', \quad k_E = E(T'')/E(T')$.

A similar expression is available for the dimensionless function of $\bar{\sigma}_y(\bar{T})$, which is determined by the factor $k_\sigma = \sigma_y(T'')/\sigma_y(T')$.

We also introduce dimensionless stresses $\bar{\sigma} = \sigma/\sigma_y(T')$ and strains $\bar{\varepsilon} = \varepsilon/\varepsilon_y(T')$, where $\varepsilon_y(T') = \sigma_y(T')/E(T')$ is the yield point expressed in terms of strain. Thermal expansion at heating from temperature T' up to T, equal to $\varepsilon_T(T) = \alpha_{T,m}(T - T')$, is represented in a dimensionless form as $\bar{\varepsilon}_T(\bar{T}) = k_\varepsilon(\bar{T} - 1)$, where $k_\varepsilon = \alpha_{T,m}T'/\varepsilon_y(T')$.

It is required to find laws of rod stresses and strains alternation in time.

In the example, we use parameters $k_T = 2, \ k_E = 0.5, \ k_\sigma = 0.3$, and $k_\varepsilon = 3$.

Further, it is more convenient to represent the elasticity law as

$$\bar{\varepsilon} = \bar{a}\bar{\sigma}, \tag{12.21}$$

where $\bar{a} = 1/\bar{E}$ is the compliance factor of a material.

12.2.2.2 During the common deformation of two rods, the equalities

$$\bar{\sigma}_I = -\bar{\sigma}_{II}, \quad \bar{\varepsilon}_I = \bar{\varepsilon}_{II}. \tag{12.22}$$

hold true for any moment of the process.

At the beginning of each "time" step from initial conditions $\bar{\sigma}_I(0) = -\bar{\sigma}_{II}(0) = 0, \quad \bar{\varepsilon}_I = \bar{\varepsilon}_{II} = 0$ at $\bar{T}_0 = 1$ or from the previous calculation steps at $\bar{T}_0$, all the strain components for both rods (elastic $\bar{\varepsilon}_{e0} = \bar{a}_0\bar{\sigma}_0$, plastic $\bar{\varepsilon}_{p0}$, and thermal $\bar{\varepsilon}_{T0}$) and temperature-dependent material characteristics $\bar{\alpha}_0$ and $\bar{\sigma}_{y0}$ are known.

The first half of any period, $\bar{T}_{II} = 1, \ \bar{a}_{II} = 1, \ \bar{\sigma}_{IIy} = 1$. From Eq. (12.22) follows the basic equation of elastoplastic deformation of the system

$$(\bar{a}_I + 1)\bar{\sigma}_I + \bar{\varepsilon}_{Ip} + \bar{\varepsilon}_{IT} = \bar{\varepsilon}_{IIp} + \bar{\varepsilon}_{IIT}. \tag{12.23}$$

Plastic strain is zero $\bar{\varepsilon}_{IIp} = 0$ only in the first half-cycle.

Calculation of each step begins with a verification of the condition of elastic deformation in the first rod

$$|\bar{\sigma}_{I0}| < \bar{\sigma}_{Iy0}. \tag{12.24}$$

If the condition (12.24) is satisfied, we take $\Delta\varepsilon_{Ip} = 0$. Then, the change of the stresses is determined according to Eq. (12.23) by the difference of stress/strain states at the end and beginning of the halfperiod

$$\Delta\bar{\sigma}_I = -\Delta\bar{\sigma}_{II} = -\frac{\bar{\sigma}_{I0}\Delta\bar{a}_I + \Delta\bar{\varepsilon}_{IT}}{\bar{a}_I + 1}. \tag{12.25}$$

Here, $\bar{a}_I = \bar{a}_{I0} + \Delta\bar{a}_I$. Note that despite the fact that the ratio $\Delta\bar{a}/\bar{a}_0$ is small, neglecting this term contributes to error accumulation and infringement of the equality for total deformation of rods.

At the end of a step

$$\bar{\sigma}_I = \bar{\sigma}_{I0} + \Delta\bar{\sigma}_I. \tag{12.26}$$

If the condition

$$|\bar{\sigma}_I| \leq \bar{\sigma}_{Iy}, \tag{12.27}$$

is also satisfied, then we pass to the next step.

Otherwise, it is necessary to take into account plastic yielding that begins to develop at this particular step.

If condition (12.24) is not satisfied, i.e. $|\bar{\sigma}_{I0}| = \bar{\sigma}_{Iy0}$, then during further loading, the change of plastic deformation of the rod I at a constant stress $\bar{\sigma}_I = \bar{\sigma}_{I0}$ is calculated as

$$\Delta\bar{\varepsilon}_{Ip} = -(\bar{\sigma}_{I0}\Delta\bar{a}_I + \Delta\bar{\varepsilon}_{IT}). \tag{12.28}$$

The increase in plastic strain according to Eq. (12.28) is possible under condition

$$\bar{\sigma}_{I0}\Delta\bar{\varepsilon}_{Ip} > 0. \tag{12.29}$$

Otherwise in the given step, elastic unloading begins.

Taking into account a transition from elastic deformation to plastic and back within the small step demands its partition into two sub-steps. This is in principle possible, but it is easier to refer the specified transitions to the beginning or the end of the current step. Such approach corresponds to the principle of "floating" spatial

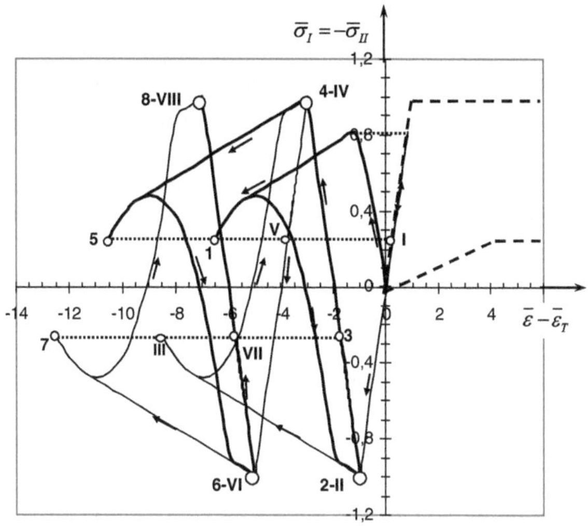

Fig. 12.3 Deformation chart of the two-rod system at a moving temperature field

or time boundary conditions and material characteristics (see Shorr and Melnikova 1988, Shorr (2004).

The second half of any period, $\bar{T}_I = 1,\ \bar{a}_I = 1,\ \bar{\sigma}_{Iy} = 1$. Calculation is carried out in absolutely the same way as for the first half of the period replacing the rod indexes from I to II.

12.2.2.3 Calculation results are presented in Fig. 12.3 in the form of the deformation diagram which shows in what manner the stresses change depending on a "mechanical" part of the strains, i.e., $-\bar{\sigma}_I = f(\hat{\varepsilon}_I)$ and $\bar{\sigma}_{II} = f(\hat{\varepsilon}_{II})$ where $\hat{\varepsilon} = \bar{a}\bar{\sigma} + \bar{\varepsilon}_p = \bar{\varepsilon} - \bar{\varepsilon}_T$. For convenience of comparison, the stress $\bar{\sigma}_I$ in the first rod is plotted with a minus sign. A number of points and thickness of lines in Figs. 12.2 and 12.3 meet each other. The accepted deformation diagrams at temperatures T' and T'' are shown in the top-right quadrant of Fig. 12.3.

In the first half-cycle when temperature of rod I rises and the yield point $\bar{\sigma}_{Iy}$ decreases, the stresses $\bar{\sigma}_I$ and $\bar{\sigma}_{II}$ increase at the beginning elastically. After the stress in the rod I reaches the yield point, the stresses fall down to value $\bar{\sigma}_{\text{ymin}} = 0.3$ corresponding to temperature $\bar{T}''$ (points 1 and I). At subsequent reduction of temperature $\bar{T}_I$ accompanied by the decrease of the compliance factor $\bar{a}_I$, both rods are elastically unloaded to $\bar{\sigma}_I = \bar{\sigma}_{II} = 0$. Residual compression strain in the first rod becomes equal to $\bar{\varepsilon}_I \approx -2$.

Then, the loading takes the opposite direction. At the end of the first half-cycle at $\bar{T}_I = \bar{T}_{II}$, the stresses reach their greatest possible values $\bar{\sigma}_I = \bar{\sigma}_{\text{ymax}}$ and $\bar{\sigma}_{II} = -\bar{\sigma}_{\text{ymax}}$, and the strains of the rods become compressing with identical values $\bar{\varepsilon}_I = \bar{\varepsilon}_I = -1$ (points 2 and II).

In the second half-cycle, the temperature of rod II increases and the area of elastic strain duplicates. Due to growth of temperature $\bar{T}_{II}$, yield point $\bar{\sigma}_{IIy}$ reduces while the plastic strain $\bar{\varepsilon}_{II}$ increases until it achieves a maximal value in point III.

At the end of the first cycle, the stresses become equal to $\bar{\sigma}_I = -\bar{\sigma}_{y\max} = -1$ and $\bar{\sigma}_{II} = \bar{\sigma}_{y\max} = 1$, and full compression strains reach the value $\bar{\varepsilon}_I = \bar{\varepsilon}_I = -3$ (points 4 and IV). During the subsequent thermocycles, the process repeats with alternation of stress signs. The full compression strain will grow under each cycle (cf. points 8.VII with 4.IV).

If at any moment the rods are separated, then their stresses become zero while residual shortenings of the rods increase proportionally to the number of thermocycles.

12.2.2.4 Described irreversible change of the body sizes under temperature cyclic alteration can occur in cases when plastic yielding in various points of a body develops in the same direction at different time instants.

In Birger et al. (1975), periodic heating/cooling of a system of the same type as in Fig. 12.2a, but for connected rods of different cross-sectional area, was considered. A graph-analytic solution of the problem shows that plastic compression strain can arise in a thick, less stressed rod during its heating because of decrease of yield point with rise in temperature. As the thick rod is then cooled, plastic compression strain arises in the thin, more stressed rod. As a result, both rods are gradually shortened.

12.3 Coupled Theory of Thermoelasticity

12.3.1 Defining System of Equations

12.3.1.1 Fundamental equations of thermoelasicity are based on assumptions that the temperature field can be found independently from the stress–strain state of a body. Then, the stresses and strains may be calculated using a previously determined temperature state of the body so as it was recognized in prior chapters.

Although this assumption does not conform to the second law of thermodynamics, it results sufficiently acceptable solutions for most engineering problems. At the same time, it appears to be useful familiarizing with the more accurate theory of thermoelasticity that is known as "coupled" theory (Duhamel 1837; Boley and Weiner 1960; and others). Moreover, there are some problems for which only the coupled theory gives an acceptable solution.

Further, we assume constant values of material characteristics E, α_T, μ and use index notations (see Chap. 1).

12.3.1.2 According to the Fourier' law of heat conductivity and a condition of thermal balance of an elementary body volume, one turns out the known equation

$$\dot{Q} = \lambda T_{,ii} + q^*, \tag{12.30}$$

where Q is a heat quantity in the volume, λ is a factor of heat conductivity, and q^* is a heat evolved by internal sources.

If the increase in inner energy dU of a body is completely defined by the introduced heat, i.e., $dU = dQ = c_v dT$, we come to the equation of classical theory of heat conductivity

$$\lambda T_{,ii} - c_v \dot{T} + q^* = 0, \tag{12.31}$$

where c_v is a volumetric thermal capacity of a material at constant strains.

According to the first law of thermodynamics, inner energy increases due to the sum of introduced heat dQ and applied work dA

$$dU = dA + dQ, \tag{12.32}$$

where $dA = \sigma_{ij} d\varepsilon_{ij}$, and along to the second law of thermodynamics, the increment of heat is connected with entropy differential dS by correlation

$$dQ = TdS. \tag{12.33}$$

To take these laws into account, we consider two states of a material. In an initial state, the absolute temperature in a given point of the body is T_0, and the stresses and strains are absent, $\sigma_{ij} = 0,\ \varepsilon_{ij} = 0$. Under deformation and heating, the temperature becomes $T = T_0 + \Delta T$, thermal expansion is $\varepsilon_T = \alpha_T \Delta T$, and stress and strain tensors $\sigma_{ij}, \varepsilon_{ij}$ are connected among themselves and with temperature expansion by Eq. (2.17a) of Chap. 2. For convenience, we will result it once again:

$$\sigma_{ij} = \frac{E}{1+\mu}\left(\varepsilon_{ij} + \frac{3\mu}{1-2\mu}\delta_{ij}\varepsilon - \frac{1+\mu}{1-2\mu}\delta_{ij}\varepsilon_T\right), \tag{12.34}$$

Stress tensor σ_{ij} is represented as a function of independent state parameters ε_{ij} and T. Inner energy U and entropy S are also the functions of these state parameters. Related to the volume unit, the functions U and S are equal to densities of corresponding functions. Their total differentials are

$$dU = \frac{\partial U}{\partial T} dT + \frac{\partial U}{\partial \varepsilon_{ij}} d\varepsilon_{ij}; \tag{12.35}$$

$$dS = \frac{\partial S}{\partial T} dT + \frac{\partial S}{\partial \varepsilon_{ij}} d\varepsilon_{ij}, \tag{12.36}$$

that results

$$\frac{\partial^2 U}{\partial T \partial \varepsilon_{ij}} = \frac{\partial^2 U}{\partial \varepsilon_{ij} \partial T}; \quad \frac{\partial^2 S}{\partial T \partial \varepsilon_{ij}} = \frac{\partial^2 S}{\partial \varepsilon_{ij} \partial T}, \tag{12.37}$$

as in the linear setting total differentials do not depend on a sequence of strain and temperature components change.

Then,

$$dS = \frac{dQ}{T} = \frac{dU - \sigma_{ij} d\varepsilon_{ij}}{T} = \frac{1}{T}\frac{\partial U}{\partial T} dT + \frac{1}{T}\left(\frac{\partial U}{\partial \varepsilon_{ij}} - \sigma_{ij}\right) d\varepsilon_{ij}, \tag{12.38}$$

and

$$\frac{\partial S}{\partial T} = \frac{1}{T}\frac{\partial U}{\partial T}; \quad \frac{\partial S}{\partial \varepsilon_{ij}} = \frac{1}{T}\left(\frac{\partial U}{\partial \varepsilon_{ij}} - \sigma_{ij}\right). \tag{12.39}$$

Using Eq. (12.37) and differentiating Eq. (12.34) in T at $\varepsilon_{ij} = \text{const}$, we find after some reduction

$$\frac{\partial U}{\partial \varepsilon_{ij}} = \sigma_{ij} - T\frac{\partial \sigma_{ij}}{\partial T} = \sigma_{ij} + \frac{E\alpha_T T}{1 - 2\mu}\delta_{ij}. \tag{12.40}$$

The partial derivative of inner energy in temperature at $\varepsilon_{ij} = \text{const}$ yields

$$\frac{\partial U}{\partial T} = c_v. \tag{12.41}$$

Hence,

$$dU = c_v dT + \sigma_{ij} d\varepsilon_{ij} + \frac{E\alpha_T T}{1 - 2\mu} d\varepsilon_{kk}. \tag{12.42}$$

Now, substituting Eq. (12.42) in Eq. (12.32), we obtain

$$dQ = c_v dT + \frac{E\alpha_T T}{1 - 2\mu} d\varepsilon_{kk}, \tag{12.43}$$

and Eq. (12.30) leads to the basis equation of the coupled theory of heat conductivity

$$\lambda T_{,ii} - c_v \dot{T} - \frac{E\alpha_T T}{1 - 2\mu}\dot{\varepsilon}_{kk} + q^* = 0. \tag{12.44}$$

Equation (12.44) combined with Eq. (12.34) forms a system of the equations of coupled theory of thermoelasticity.

12.3.1.3 In order to transfer from initial to the current state of the body, we represent $dA = \sigma_{ij} d\varepsilon_{ij}$ and dU in Eq. (12.42) as functions either of tensor ε_{ij} or σ_{ij}. Integration of this equation results in

$$\Delta A = \int\limits_{T_0}^{T} dA = \frac{E}{2(1+\mu)}\left(\varepsilon_{ij}\varepsilon_{ij} + \frac{\mu}{1-2\mu}\varepsilon_{kk}^2\right) - \frac{E\alpha_T\Delta T}{1-2\mu}\varepsilon_{kk}; \tag{12.45a}$$

$$\Delta U = \int\limits_{T_0}^{T} dU = c_v\Delta T + \frac{E}{2(1+\mu)}\left(\varepsilon_{ij}\varepsilon_{ij} + \frac{\mu}{1-2\mu}\varepsilon_{kk}^2\right) + \frac{E\alpha_T T_0}{1-2\mu}\varepsilon_{kk} \tag{12.45b}$$

or

$$\Delta A = \frac{1+\mu}{2E}\left(\sigma_{ij}\sigma_{ij} - \frac{\mu}{1+\mu}\sigma_{kk}^2\right) + \alpha_T\Delta T\sigma_{kk} \tag{12.46a}$$

$$\Delta U = c_p\Delta T + \frac{1+\mu}{2E}\left(\sigma_{ij}\sigma_{ij} - \frac{\mu}{1+\mu}\sigma_{kk}^2\right) + \alpha_T T\sigma_{kk}, \tag{12.46b}$$

where

$$c_p = c_v + \frac{3E\alpha_T^2 T_0}{1-2\mu}.$$

is a volumetric thermal capacity of a material at constant stresses.

The first term in Eq. (12.45b) accounts for the increase of inner energy from heating of a rigid body, the second term in Eq. (12.45b) or Eq. (12.46b) is equal to potential energy of elastic deformation at constant temperature, and the last term reflects the change of inner energy due to a necessary input of additional heat ΔQ_V induced by relative change of volume $\Delta V = \varepsilon_{kk}$. If the volume increases, additional heat must be brought to the body, and if the volume decreases, it shall be taken away.

In case of adiabatic process when there is no heat exchange ($dQ = 0$), Eq. (12.43) might be integrated to

$$T = T_0\exp\left(-\frac{E\alpha_T\varepsilon_{kk}}{(1-2\mu)c_v}\right), \tag{12.47a}$$

from which for $\Delta T \ll T_0$, it follows

$$\Delta T \approx -\frac{E\alpha_T T_0\varepsilon_{kk}}{(1-2\mu)c_v}. \tag{12.47b}$$

Equation (12.47a) shows that an adiabatic increase of the volume always leads to decrease of temperature (and vice versa). In case of alternating volume changes with the symmetric cycle, the temperature increase at compression is always greater than temperature decrease at extension.

Equation (12.46b) at $\sigma_{ij} = 0$ corresponds to heating of a deformable body when its volume increases only due to thermal expansion. The difference $c_p - c_v$ in heat capacity reflects a role of additional heat that is brought to the body in this case.

Basis Eq. (12.44) may be also written in an equivalent form

$$\lambda T_{,ll} - c_p \dot{T} - \alpha_T T \dot{\sigma}_{kk} + q^* = 0. \tag{12.48}$$

Some examples of solutions using the coupled theory of thermoelasticity are presented below.

12.3.2 Examples of Elementary Solutions

12.3.2.1 Adiabatic and Isothermal Solutions for Extension of a Rod

The equation system of the coupled theory of thermoelasticity for 1D stress and thermal states yields

$$\varepsilon_1 = \frac{\sigma_1}{E} + \alpha_T \Delta T, \tag{12.49}$$

and

$$\lambda \frac{\partial^2 T}{\partial x^2} - c_p \frac{\partial T}{\partial t} - \alpha_T T \frac{\partial \sigma_1}{\partial t} + q^* = 0 \tag{12.50}$$

Here, q^* includes heat input through lateral surfaces of a rod which is assumed to be constant along its length; hence, $\partial T/\partial x = 0,\ \partial q^*/\partial x = 0$.

Adiabatic Extension $(\Delta Q = 0)$

It follows from Eq. (12.50) for $q^* = 0$

$$\frac{dT}{T} = -\frac{\alpha_T}{c_p} d\sigma_1,$$

that results in temperature increase

$$\ln \frac{T}{T_0} = -\frac{\alpha_T}{c_p}\sigma_1 \quad \text{or} \quad \Delta T = T_0[\exp(-\alpha_T \sigma_1/c_p) - 1]. \tag{12.51a}$$

Approximately,

Table 12.1 Comparison of adiabatic and isothermal solutions

	Parameter		Uncoupled theory	Coupled theory	
				Adiabatic solution	Isothermal solution
A	Strain	%	0.5	0.496	0.5
	Volume increase	%	0.2	0.19	0.2
	Temperature change	°C	0	−3.07	0
	Applied work	MJ/m^3	2.5	2.46	2.5
	Input heat	MJ/m^3	0	0	12
	Increase of inner energy	MJ/m^3	2.5	2.46	14.5
B	Strain	%	−0.2	−0.189	−0.2
	Volume thermal increase	%	0	0.03	0
	Volume pressure increase	%	−0.6	−0.6	−0.6
	Temperature change	°C	0	8	0
	Applied work	MJ/m^3	3	2.67	3
	Input heat	MJ/m^3	0	0	−36
	Increase of inner energy	MJ/m^3	3	2.67	−33

$$\Delta T \approx -T_0 \frac{\alpha_T}{c_p} \sigma_1 \tag{12.51b}$$

and

$$\varepsilon_1 = \frac{\sigma_1}{E}\left(1 - \frac{E\alpha_T^2 T_0}{c_p}\right). \tag{12.52}$$

Isothermal Extension ($\Delta T = 0$)

It follows from Eq. (12.50) that a heat

$$Q = \int_0^t q^* dt = \alpha_T T_0 \sigma_1 \tag{12.53}$$

must be introduced into the body to provide for condition $\partial T / \partial t = 0$.

The comparative characteristics of both extension conditions calculated for a steel rod with $\sigma_1 = 1000$ MPa and $T_0 = 1000$ K are presented below in section A of Table 12.1.

12.3.2.2 Adiabatic and Isothermal Solutions for Uniform Static Compression of a Body

In this case,

$$\sigma_1 - \sigma_1 = \sigma_1 = -p, \quad \sigma_{kk} = -3p, \tag{12.54a}$$

$$\varepsilon_1 = \varepsilon_1 = \varepsilon_1 = -(1 - 2\mu)\frac{p}{E} + \alpha_T \Delta T, \tag{12.54b}$$

and

$$\varepsilon_{kk} = -\frac{3(1 - 2\mu)}{E}p + 3\alpha_T \Delta T. \tag{12.55}$$

Adiabatic Compression

Analogously to the previous solution, we obtain from Eq. (12.48) the temperature increase

$$\Delta T \approx \frac{3\alpha_T T_0}{c_p}p, \tag{12.56}$$

and the change of an elementary volume due to pressure and heating

$$\Delta V = -\left(\frac{3(1 - 2\mu)}{E} - \frac{9\alpha_T^2 T_0}{c_p}\right)p. \tag{12.57}$$

Isothermal Compression

In this case, a heat to be input is

$$Q = -3\alpha_T T_0 p \tag{12.58}$$

and applied work is

$$\Delta W = \frac{3(1 - 2\mu)}{2E}p^2. \tag{12.59}$$

The comparative characteristics of both compression conditions calculated for a steel rod with $\sigma_1 = 1000$ MPa and $T_0 = 1000$ K are presented in section B of Table 12.1.

From this table, one can see that the elastic deformation very weakly influences the change of temperature due to thermodynamic coupling (and vice versa), although energetic balance at adiabatic and isothermal loadings differs notably.

12.3.3 Thermodynamic Coupling of Cyclic Stresses and Temperatures

Let stresses $\sigma(t)$ in a rod vary in time with amplitude σ_0 and period t_0 as sawtooth function shown in Fig. 12.4a, where $\bar{t} = t/t_0$. The rod is surrounded by a medium with constant temperature T_0; a heat transfer factor from the rod to medium referred to unit volume is h. So, the input of heat flow into the rod at its varying temperature T is

$$q^* = h(T_0 - T). \tag{12.60}$$

As above, we assume $\partial T/\partial x = 0$, $\partial q^*/\partial x = 0$.

Inserting Eq. (12.60) in Eq. (12.50) and using non-dimensional parameters

$$\bar{T} = \frac{T}{T_0}, \quad \bar{t} = \frac{t}{t_0}, \quad \bar{\sigma} = \frac{\alpha_T \sigma}{c_p}, \quad \bar{h} = \frac{t_0 h}{c_p} \tag{12.61}$$

we come to the differential equation

$$\frac{\partial \bar{T}}{\partial \bar{t}} + \left(\frac{\partial \bar{\sigma}}{\partial \bar{t}} + \bar{h}\right)\bar{T} = \bar{h} \tag{12.62}$$

with initial conditions $\bar{T}(0) = 1$ and $\bar{\sigma}(0) = 0$.

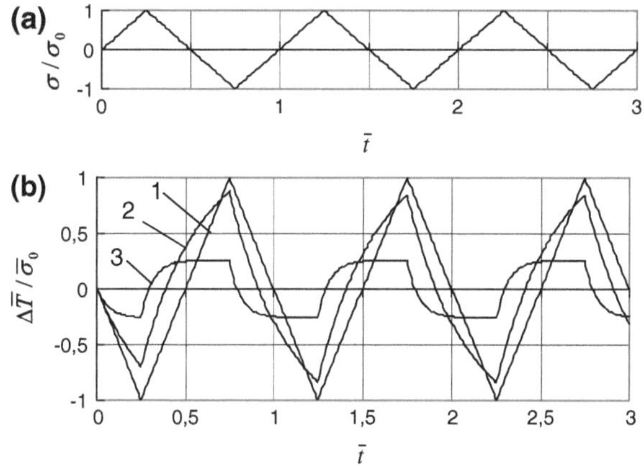

Fig. 12.4 Alteration of temperature (**b**) induced by stress cyclic change (**a**); *1* $\bar{h} = 0$, *2* $\bar{h} = 10^3$, *3* $\bar{h} = 5 \times 10^3$

For each rectilinear part of function $\sigma(t)$ from $\bar{t}'$ to $\bar{t}''$, a derivate

$$\dot{\bar{\sigma}} = \frac{d\bar{\sigma}}{d\bar{t}} = \frac{\bar{\sigma}'' - \bar{\sigma}'}{\bar{t}'' - \bar{t}'} = \pm 4\bar{\sigma}_0$$

is constant and the solution of Eq. (12.62) takes a form

$$\bar{T} = \bar{T}' + \left(\bar{T}' - \frac{\bar{h}}{\dot{\bar{\sigma}} + \bar{h}}\right)(\exp[-(\dot{\bar{\sigma}} + \bar{h})(\bar{t} - \bar{t}')] - 1), \tag{12.63a}$$

where notations $\bar{f}' = \bar{f}(\bar{t}')$ and $\bar{f}'' = \bar{f}(\bar{t}'')$ for $\bar{\sigma}$ and $\bar{T}$ are used, and is expected $\dot{\bar{\sigma}} + \bar{h} \neq 0$. For a specific case where $\dot{\bar{\sigma}} + \bar{h} = 0$, Eq. (12.63a) yields

$$\bar{T} = \bar{T}' + \bar{h}(\bar{t} - \bar{t}'). \tag{12.63b}$$

Having written down Eqs. (12.63a, b) consecutively for ascending and descending branches of the function $\sigma(t)$, we find the full change of temperature during a loading cycle.

A process of temperature alternation in time depends on correlation between speed of stress change and heat transfer as it is demonstrated in Fig. 12.4b where it is taken $\Delta\bar{T} = \bar{T} - 1$ and $\hat{h} = \bar{h}/\dot{\bar{\sigma}} = 0.25\, ht_0/\alpha_T\sigma_0$.

At $\hat{h} < 100$, heat transfer is insufficient and temperature and stress cycles practically coincide up to signs (line 1 in Fig. 12.4b). At $\hat{h} \to \infty$, temperature oscillations decay with $\Delta\bar{T} \to 1$ due to high heat output. At intermediate cases (curves 2 and 3), temperature changes are periodical; maximum difference $\Delta\bar{T}_{\max}$ gradually decreases as heat output grows.

Real maximally possible temperature change $\Delta T_{\max} = \alpha_T\sigma_0 T_0/c_p$ at typical for metals values $\alpha_T/c_p = (3 - 10) \times 10^{-6}\mathrm{m}^3/\mathrm{MJ}$, temperature $(T_0 = 300 - 1000)$ K and stresses amplitudes $\sigma_0 = (100 - 1000)$MPa does not exceeds several degrees.

12.3.4 Energy Dissipation Due to Thermodynamic Coupling

12.3.4.1 In previous problems, a level of alternating stress amplitude has been maintained constant in spite of thermodynamic coupling and the heat output from the system; at that some external force needs to be applied.

When mechanical energy of an elastic system is limited, the thermodynamic coupling leads to energy dissipation. This problem was discussed by Boley and Weiner (1960) who obtained an analytical solution for plane oscillation of an elastic half-space after an initial sudden displacement of its surface.

To illustrate a physical process in which the initial mechanical energy of an elastic system can be dissipated due to thermodynamic coupling, we consider the next example.

A mass M connected to one of the ends of an elastic thin rod of length L and constant cross-sectional area A begins oscillating after a suddenly applied longitudinal impulse. Initial temperature of the rod T_{in} is equal to the temperature of ambient medium. Denoting displacement of the rod's end by $u = \varepsilon L$ and force between the rod and the mass by $F = \sigma A$, where the full strain ε and stress σ are connected by 1D thermoelastic equation $\varepsilon = \sigma/E + \alpha_T \Delta T$ with $\Delta T = (T - T_{in})$, we come to equation

$$\ddot{\varepsilon} + \omega_0^2 \varepsilon = \alpha_T \omega_0^2 \Delta T \tag{12.64a}$$

with initial conditions

$$\varepsilon(0) = 0, \quad \dot{\varepsilon}(0) = \dot{\varepsilon}_0, \quad \Delta T(0) = 0. \tag{12.64b}$$

Here, $\omega_0 = \sqrt{EA/ML}$ is a natural frequency of the system at constant temperature T_{in}. In this case for the conditions (12.64b), we have

$$\varepsilon(t) = \frac{\dot{\varepsilon}_0}{\omega_0} \sin \omega_0 t$$

with maximal strain $\varepsilon_{max} = \dot{\varepsilon}_0/\omega_0$ and oscillation period $\tau = 2\pi/\omega_0$.

Using non-dimensional parameters

$$\bar{\varepsilon} = \frac{\varepsilon}{\varepsilon_{max}}, \quad \dot{\bar{\varepsilon}} = \frac{\dot{\varepsilon}}{\dot{\varepsilon}_0}, \quad \bar{T} = \frac{T}{T_{in}}, \quad \bar{t} = \frac{t}{\tau},$$

Equation (12.64a) yields

$$\ddot{\bar{\varepsilon}} + 4\pi^2 \bar{\varepsilon} = 4\pi^2 K_\alpha \Delta \bar{T}, \tag{12.65}$$

where $\Delta \bar{T} = \bar{T} - 1$ (as $\bar{T}_{in} = 1$) and

$$K_\alpha = \frac{\alpha_T T_{in}}{\varepsilon_{max}}. \tag{12.66}$$

It follows from the above equations

$$\bar{\sigma} = \frac{\sigma}{E\varepsilon_{max}} = \bar{\varepsilon} - K_\alpha \Delta \bar{T} \quad \text{and} \quad \dot{\bar{\varepsilon}}_0 = 2\pi.$$

Assuming temperature and other parameters to be constant along the rod, the equation of coupled theory of thermoelasticity (12.44) for 1D case referred to the entire rod yields

$$c_v\dot{T} + \frac{E\alpha_T T}{1-2\mu}\dot{\varepsilon}_{kk} - q^* = 0,$$

where relative change of volume expressed through the full longitudinal strain is

$$\varepsilon_{kk} = \varepsilon(1-2\mu) + 2(1+\mu)\alpha_T\Delta T$$

and the heat input through lateral rod surfaces is $q^* = -h\Delta T$.

Then,

$$c_v\dot{T} + \frac{2(1+\mu)E\alpha_T^2}{1-2\mu}T\Delta T + E\alpha_T T\dot{\varepsilon} + h\Delta T = 0,$$

or in the equivalent non-dimensional form,

$$\Delta\dot{\bar{T}} + K_c(1+\Delta\bar{T})\Delta\bar{T} + K_\sigma(1+\Delta\bar{T})\dot{\bar{\varepsilon}} + K_h\Delta\bar{T} = 0, \tag{12.67}$$

where additionally to Eq. (12.66) is noted

$$K_c = \frac{2(1+\mu)E\alpha_T^2 T_{in}}{(1-2\mu)c_v}, \quad K_\sigma = \frac{E\alpha_T\varepsilon_{\max}}{c_v}, \quad K_h = \frac{h\tau}{c_v}. \tag{12.68}$$

Equations (12.65) and (12.67) constitute nonlinear differential system in strain velocity $\dot{\bar{\varepsilon}}$ and temperature change $\Delta\bar{T}$ in the rod which takes into account their thermodynamic coupling.

12.3.4.2 The equation systems (12.65) and (12.67) can be integrated in time using numerical step-by-step procedure. Although nonlinear terms in Eq. (12.67) depending on small temperature change $\Delta\bar{T} \ll 1$ seem to be inessential, they determine the effect of energy thermodynamic dissipation. Therefore, they must be saved through any numerical calculations.

A possible procedure is to take strain acceleration $\ddot{\bar{\varepsilon}} = \text{const}$ in Eq. (12.65) within each short-time step $\Delta\bar{t}$. Determining it using mean values of $\Delta\bar{T}_m$ for the step, we find

$$\ddot{\bar{\varepsilon}} = \frac{4\pi^2(K_\alpha\Delta\bar{T}_m - \bar{\varepsilon}_* - 0.5\dot{\bar{\varepsilon}}_*\Delta\bar{t})}{1 + 2\pi^2\Delta\bar{T}^2/3}, \tag{12.69}$$

where lower asterisk notes the beginning of running step.

A mean step velocity is

$$\dot{\bar{\varepsilon}}_m = \dot{\bar{\varepsilon}}_* + 0.5 * \ddot{\bar{\varepsilon}}\Delta\bar{t}. \tag{12.70}$$

Integrating Eq. (12.67) within the step using the nonlinear term $(1+\Delta\bar{T})\Delta\bar{T} \approx (1+\Delta\bar{T}_m)\Delta\bar{T}$, we obtain

$$\Delta \bar{T}(\bar{t}_1) = \left(\Delta \bar{T}_* + \frac{K_\sigma \dot{\bar{\varepsilon}}_m}{K_\sigma \dot{\bar{\varepsilon}}_m + K_h}\right) \exp\left(-\frac{(K_\sigma \dot{\bar{\varepsilon}}_m + K_h)\bar{t}_1}{1 + K_c(1 + \Delta \bar{T}_m)}\right) - \frac{K_\sigma \dot{\bar{\varepsilon}}_m}{K_\sigma \dot{\bar{\varepsilon}}_m + K_h}, \tag{12.71}$$

where $0 \le \bar{t}_1 = \bar{t} - \bar{t}_* \le \Delta \bar{t}$.

A mean value $\Delta \bar{T}_m$ corresponds to $\bar{t}_1 = 0.5$.

The mean step values of $\dot{\bar{\varepsilon}}_m$ and $\Delta \bar{T}_m$ can be calculated from Eqs. (12.69)–(12.71) by sequential approximations.

For the purpose of computation control, it is useful to calculate full heat Q which is outputted from the system during the process. It can be calculated as a sum of elementary heat outputs $q^* = h\Delta T \Delta t$ over all the steps $n = n_1 n_\tau$ where n_1 is a number of steps within each oscillation period and n_τ is a number of periods. Then,

$$Q = LA \sum_n h \Delta T_m \Delta t$$

where $\Delta t = \tau / n_1$.

Its dimensionless form $\bar{Q} = Q/K_{in}$ referred to the initial kinetic energy $K_{in} = 0.5\,\mathrm{M}(\dot{u}_{in})^2$ is

$$\bar{Q} = K_q \sum_n \Delta \bar{T}_m \tag{12.72}$$

where

$$K_q = \frac{2ET_{in}\tau}{\sigma_{\max}^2 n_1} h. \tag{12.73}$$

Obviously, $\bar{Q} \to 1$ when the initial oscillations go to the end.

12.3.4.3 As examples of possible elastic energy dissipation due to thermodynamic coupling, calculation results for the dimensionless factors $K_\alpha = \sim 2.5$, $K_c \approx 5 \times 10^{-2}$, and $K_\sigma \approx 3 \times 10^{-3}$ are presented. Dimensionless heat transfer factor is varied from $K_h = 0$ till $K_h = 0.25$.

Figure 12.5a shows alteration of stress $\bar{\sigma}$ and relative temperature change $\Delta \bar{T} / \Delta \bar{T}_0$ versus time $\bar{t}$ for the first cycle $n_\tau = 1$ and the cycle with $n_\tau = 1000$ (curves are referred to maximum value of temperature change $\Delta \bar{T}_0$ in the first cycle; $K_h = 0.125$).

Calculations show that maximum of positive temperature curve in each period is few greater than absolute value of extreme point of negative temperature curve; so, the mean temperature changes in all periods are positive, $\Delta \bar{T}_m > 0$. Although these changes are very small, they cause permanent transition of heat from the rod to ambient medium and decrease of rod's oscillation stresses and temperature that is illustrated by Fig. 12.5b.

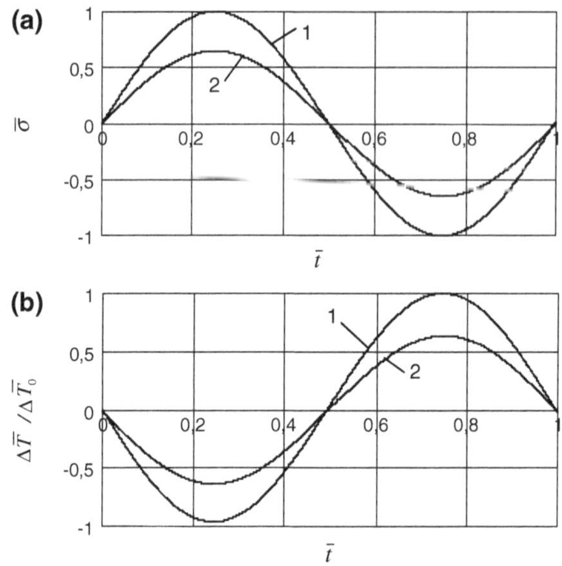

Fig. 12.5 Alteration of stress $\bar{\sigma}$ (**a**) and relative temperature change $\Delta\bar{T}/\Delta\bar{T}_0$ (**b**) versus time $\bar{t}$ for different cycle indices $n_\tau : 1 - n_\tau \ldots n_\tau = 1000$

In Fig. 12.6a, change of stress amplitudes $\bar{\sigma}_{\max}$ versus number of cycles n_τ is shown. At moderate values of heat transfer factor till $K_h = \sim 10^{-3}$, energy dissipation is imperceptible (within calculated cycle number $n_1 = 4000$). At more

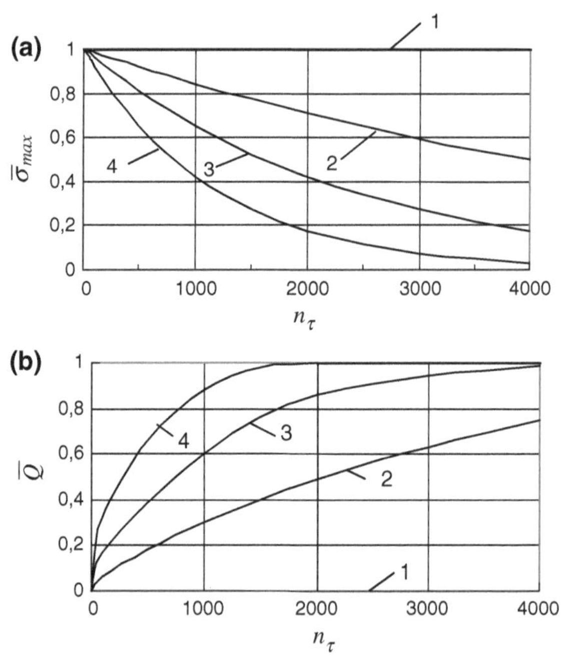

Fig. 12.6 Dependence of stress amplitude $\bar{\sigma}_{\max}$ decrease (**a**) and total relative heat output $\bar{Q}$ increase (**b**) for cycle's number n_τ for heat outputs: *1* $K_h = 0 - 10^{-3}$; *2* $K_h = 0.05$; *3* $K_h = 0.125$; *4* $K_h = 0.25$

intensive heat output, stress amplitudes significantly decrease as a number of cycles is increased.

Figure 12.6b demonstrates a rise of heat transfer $\bar{Q}$ to ambient medium equal to decreasing of mechanical energy of the system. The heat $\bar{Q}$ arrives to unit when oscillations go to end.

The effect of elastic energy dissipation depends on a multitude of different parameters. Above-mentioned factors approximately correspond to more or less typical parameters of metals in mechanics and engineering applications.

This effect is usually rather weak and to separate it experimentally from other dissipative sources is difficult. However, it is useful to keep it in mind.

12.4 Wave Theory of Thermoelasticity

12.4.1 Wave Theory of Heat Conductivity

12.4.1.1 Many dynamic processes in machines, structures, and various deformable continua possess a highly unsteady (non-stationary) character. Some examples of such processes are explosive, shock, and vibro-impact loading, oscillation of structures with strongly nonlinear damping elements, and contact dynamic problems.

Non-stationary dynamic processes in solids are described by the wave equations of hyperbolic type which reflect the experimentally confirmed fact that stresses and strains propagate over a body in the form of waves with the finite speed. Finite speed of wave propagation is inherent also in other branches of mechanics of continuous medium—aerodynamics, hydrodynamics, and electromagnetics. Waves in electric circuits also propagate with finite speed. At the same time, in the classical statement of heat conductivity described by Fourier's equation of parabolic type, speeds of temperature and thermal stream propagation are infinitely large (see, for example, Boley and Weiner 1960). In spite of the fact that infinitely large speeds in the nature are impossible in principle, results of calculations of the heat conductivity based on Fourier's equation are justified by numerous experiments and are widely used in practice, including problems of thermal strength.

However, in some cases, for example, when the wave front of stresses is caused by a non-stationary temperature field spreading over a body, the account of finite speed of thermal process becomes important.

In this connection, there is a certain interest in representing a description of heat conductivity as wave process and transition from it to the classical theory.

12.4.1.2 There are some works in which the problem of heat conductivity is formulated in such a way that leads to finite speed of temperature and thermal stream propagation. Without going into the discussion of physical preconditions and validity of such approaches, we shall accept given below statements as the working hypothesis:

- The wave equation of heat conductivity should be mathematically similar to the equations of other physical processes arising in a continuous medium (in solid mechanics, acoustics, hydraulics considering compressibility of a liquid, etc.).
- Thermal streams in a body are naturally described by Euler's coordinate system connected with body points, instead of Lagrange's system concerning with coordinates of a body's particles as in solid mechanics. However, the linear approximation distinction between these two descriptions does not play any role.
- Speed of heat distribution in a solid cannot surpass speed of distribution of elastic wave propagation that will be agreed with conclusions of the coupled theory of thermoelasticity.

We consider one-dimensional dynamic equations for solids in the form

$$\frac{\partial L}{\partial t} = \frac{\partial \sigma}{\partial x}; \quad \frac{\partial v}{\partial x} = \frac{1}{E}\frac{\partial \sigma}{\partial t} + \eta\sigma; \quad L = \rho v, \tag{12.74}$$

where along with previously used parameters σ, E, and ρ, there have also been introduced a particle speed $v = \dot{u}$, momentum L, and also a factor of viscosity η.

By analogy with Eq. (12.74), the one-dimensional dynamic equations of heat conductivity should look like

$$\frac{\partial U}{\partial t} = -\frac{\partial q}{\partial x}; \quad \frac{\partial T}{\partial x} = -\frac{1}{B}\frac{\partial q}{\partial t} - \frac{1}{\lambda}q; \quad U = \rho c_p T, \tag{12.75}$$

where U is internal energy in volume unit, q is a thermal stream through unit of the area, c_p is a specific thermal capacity, λ is a factor of heat conductivity and T is temperature, and B is an additional characteristic of a material.

In the linear case, all material characteristics are considered to be constants.

Each of the mechanical parameters $v, \sigma, L, \rho, E, \eta$ corresponds to the thermal parameter $T, -q, U, \rho c_p, B, 1/\lambda$.

From Eq. (12.74), the one-dimensional wave equation of mechanics with viscous resistance turns out as

$$\frac{\partial^2 v}{\partial t^2} - \frac{E}{\rho}\frac{\partial^2 v}{\partial x^2} + E\eta\frac{\partial v}{\partial t} = 0, \tag{12.76}$$

The similar one-dimensional wave equation of heat conductivity follows from Eq. (12.75)

$$\frac{\partial^2 T}{\partial t^2} - \frac{B}{\rho c_p}\frac{\partial^2 T}{\partial x^2} + \frac{B}{\lambda}\frac{\partial T}{\partial t} = 0. \tag{12.77}$$

Having set $B \to \infty$, we shall come to the Fourier's equation of heat conductivity

$$\frac{\partial^2 T}{\partial x^2} - \frac{1}{\kappa}\frac{\partial T}{\partial t} = 0, \tag{12.78}$$

where $\kappa = \lambda/\rho c_p$ is the factor of thermal conductivity (diffusivity).

According to Eq. (12.77), the finite speed of thermal wave propagation must be defined as $c_T = (B/\rho c_p)^{0.5}$. As the dimension of parameter B is [J/ms^2 K] and dimension of volumetric thermal capacity ρc_p is [J/m^3 K], their ratio has dimension of a squared speed [(m/s)^2]. Note that the term used in some works "time of a thermal relaxation τ[s]" is equal to the ratio of λ/B.

The physical sense of the additional member in the right-hand side of the second Eq. (12.75) indicates that besides the usual (stationary) temperature gradient proportional to the thermal stream $-q/\lambda$, there should also be an additional (dynamic) temperature gradient proportional to the change of heat transfer speed $-\dot{q}/B$. The first gradient provides overcoming of thermal resistance; the second one is connected with overcoming of thermal inertia.

Classical Fourier's Eq. (12.78) is thermal analogue of the equation of mechanics (12.76) for a case when viscous forces essentially exceed elastic ones and the basic process of strain propagation proceeds with a speed, essentially smaller than speed of the elastic wave propagation $c = (E/\rho)^{0.5}$. Similar model might be used, in particular, for calculation of plastic impact under large deformations. Obviously, this circumstance insures efficiency of application of Eq. (12.78) for solution of the majority of problems on non-stationary heat conductivity at speeds of temperature propagation essentially smaller than $c_T = (B/\rho c_p)^{0.5}$.

Now, there are still no reliable experimental or theoretical data for the definition of the factor B value or directly the speed c_T. But considering the condition accepted above, namely that speed of thermal wave propagation cannot exceed the speed of elastic wave $c_T \le c$, it is possible to study influence of finite speed c_T, assuming $c_T = c$.

Below, the example of the wave theory in heat conductivity is used for calculation of a non-stationary temperature field and stresses under a heat shock.

12.4.2 Heat Shock

12.4.2.1 We shall consider a classical 1D problem on sharp rise in temperature at one of the ends $x = 0$ of a long rod from initial value $T(0) = 0$ up to temperature T_* which then is maintained constant (a problem of *heat shock*). Lateral surfaces of the rod are assumed to be heat insulated. The similar problem in "non-wave" statement with reference to sharp rise in temperature of a free surface of a semi-infinite body was examined by many researchers, for example, Parkus (1959). For bodies of the finite dimensions (for example thick plates), such consideration was limited to an initial stage of process.

Let us introduce a distance L_* from the end of a rod up to some point $x_* = L_*$ and dimensionless parameters

$$\bar{T} = T/T_*, \quad \bar{q} = q/\rho c_p c_T T_*, \quad \bar{x} = x/L_*, \quad \bar{t} = tc_T/L_*, \quad \bar{\lambda} = \lambda/\rho c_p c_T L_*. \tag{12.79}$$

The size L_* can be taken arbitrarily, but for calculation convenience, we shall connect it with thermal relaxation time $\tau = \lambda/B = \lambda/\rho c_p c_T^2$ for a material, so that $L_* = c_T\tau$. Then, $\bar{x} = x/c_T\tau$ and $\bar{t} = t/\tau$.

Equation (12.75) becomes

$$\frac{\partial\bar{T}}{\partial\bar{t}} = -\frac{\partial\bar{q}}{\partial\bar{x}} = 0; \quad \frac{\partial\bar{T}}{\partial\bar{x}} = -\frac{\partial\bar{q}}{\partial\bar{t}} - \bar{q}. \tag{12.80}$$

Initial conditions are $\bar{T}(\bar{x},0) = 0$, $\bar{q}(\bar{x},0) = 0$, and we take the boundary condition at the end $x = 0$ in the form $\bar{T}(0,\bar{t}) = 1$. At $\bar{x} \to \infty$, the initial conditions remain constant.

The analytical solution of the analogous *non-wave* problem given in our designations yields (see Parkus 1959; and also Boley and Weiner 1960),

$$\bar{T}(\bar{x},\bar{t}) = 1 - 2\Phi(z), \quad z = \bar{x}/\sqrt{2\bar{t}}, \tag{12.81}$$

where

$$\Phi(z) = \int_0^z \exp(-0.5\,\xi^2)d\xi/2\pi$$

is the integral of probability.

Solution of the 1D *wave* problems of solids referred to various problems can be effectively carried out by a numerical method of the direct mathematical modeling stated in Shorr (Shorr 1984, 1995) which represents a special case of the general method of wave finite elements (Shorr 1998, 1999, 2004). In Shorr and Mel'nikova (1988), the specified method was adopted for the problem of heat conductivity.

Below we shall briefly state the specified approach with the reference to the examined problem.

12.4.2.2 The referred length L_* is divided into n finite elements of equal quantity $\Delta x = L_*/n$, or $\Delta\bar{x} = 1/n$. The time necessary for a wave to pass the length L_* with speed c_T is $t_* = L_*/c_T$ and to propagate over an element is $\Delta t = \Delta x/c_T$, or $\Delta\bar{t} = \Delta\bar{x}$. At numerical calculation, it is necessary to assign the maximal calculated time $\bar{t}_{\max} = (i_{\max} - 1)\Delta\bar{t}$ where $i_{\max}$ is a number of the maximal time steps which may be taken as both smaller and greater than n. Each element is assumed to be consisting of the inner continual part, which has some thermal capacity c_p and infinite large thermal conductivity $\lambda \to \infty$, and two outer discrete parts with real thermal conductivity λ. The inner part reflects the wave nature of the process, and the outer parts describe the usual thermal conductivity. To simulate heat conduction of the whole element, the conduction of each outer part is referred to conventional length 0.5 $\Delta\bar{x}$.

At some time instant t^{-}, the temperature $\bar{T}_{j,0}$ and heat flow $\bar{q}_{j0}$ within inner part of the element j are assumed to be constant along the full element length that is possible due to condition $\lambda \to \infty$. Then, discontinuities of temperature and heat flow arise between the next elements. Such discontinuities at the element boundaries must immediately (i.e., at the infinitely near instant t^{+}) decay to values $\bar{T}_j^{\pm}$ and $\bar{q}_j^{\pm}$ where indices $\pm$ refer to element borders $j \pm 1$; the disturbances $\bar{T}_j^{\pm} - \bar{T}_{j,0}$ and $\bar{q}_j^{\pm} - \bar{q}_{j,0}$ can be "strong" and remain constant during time interval $\Delta\bar{t}$.

The outer domains, being thermally unbalanced, begin expanding and propagating into the inner element part as waves with speed c_T and reach the opposite element borders in the time $\Delta t = \Delta x / c_T$, or $\Delta\bar{t} = \Delta\bar{x}$. Applying to a finite element $\Delta\bar{x}$ and time interval $\Delta\bar{t}$, Eq. (12.80) with $\lambda \to \infty$ and $c_T = (B/\rho c_p)^{0.5}$ is transformed to the single equation

$$\bar{T}_j^{\pm} - \bar{T}_{j0} = \mp(\bar{q}_j^{\pm} - \bar{q}_{j0}). \tag{12.82}$$

Temperatures at the outer element borders $\bar{T}_{j*}^{\pm}$ differ from $\bar{T}_j^{\pm}$ and provoke heat flows

$$\bar{q}_j^{\pm} = \mp 2(\bar{T}_{j*}^{\pm} - \bar{T}_j^{\pm})/\Delta\bar{x}. \tag{12.83}$$

It follows from Eqs. (12.82) and (12.83) that

$$q_j^{\pm} = \frac{q_{j0} \mp (\bar{T}_{j*}^{\pm} - \bar{T}_{j0})}{1 + 0.5\,\Delta\bar{x}} \tag{12.84a}$$

and vice versa

$$\bar{T}_{j*}^{\pm} = \bar{T}_{j0} \pm q_{j0} \mp q_j^{\pm}(1 + 0.5\,\Delta\bar{x}). \tag{12.84b}$$

Conditions of temperature and heat flow continuity between outer borders of next elements within time interval $\Delta\bar{t}$ require the equalities

$$\bar{T}_{j*}^{\pm} = \bar{T}_{j\pm1,*}^{\mp}; \quad \bar{q}_j^{\pm} = \bar{q}_{j\pm1}^{\mp}. \tag{12.85}$$

Substituting Eq. (12.84a, b) for adjacent boundaries of elements j and $j \pm 1$ into the first Eq. (12.85), we obtain the boundary heat flows

$$\bar{q}_j^{\pm} = \frac{\bar{q}_{j0} + \bar{q}_{j\mp 1,0} \pm (\bar{T}_{j0} - \bar{T}_{j\mp 1,0})}{2(1 + 0.5\,\Delta\bar{x})}, \tag{12.86}$$

and then, from Eq. (12.82), the boundary temperatures

$$\bar{T}_j^{+} = \bar{T}_{j0} \mp (\bar{q}_j^{\pm} - \bar{q}_{j0}). \tag{12.87}$$

New values of heat flow and temperature for the inner part of the element are

$$\bar{q}_j = \bar{q}_j^{-} + \bar{q}_j^{+} - \bar{q}_{j0}; \quad \bar{T}_j = \bar{T}_j^{-} + \bar{T}_j^{+} - \bar{T}_{j0}. \tag{12.88}$$

Boundary conditions at $\bar{x} = 0$ yield

$$\bar{T}_{1*}^{-} = 1, \quad \bar{q}_{1*} = \frac{q_{10} + 1 - \bar{T}_{10}}{1 + 0.5\,\Delta\bar{x}}. \tag{12.89}$$

Now, the problem of thermal wave propagation comes to the step-by-step calculation of heat flow $\bar{q}_j^{\pm}$ and temperature $\bar{T}_j^{\pm}$ (at borders of all elements) and then $\bar{q}_j$ and $\bar{T}_j$ (in all elements), where time index varies within $1 \le i \le i_{\max}$ and distance index within $1 \le j \le i$.

12.4.2.3 In Fig. 12.7, curves of temperature T distribution along the rod length $\bar{x}$ at various time instants $\bar{t}$, calculated numerically up to $i_{\max} = 5$, are displayed. Continuous lines show the calculation results due to wave model, dashed lines—due to non-wave Eq. (12.81). The bold lines allocate the curves concerning the time instant $\bar{t} = 1$ when the head wave reaches the point $\bar{x} = 1$.

It is visible that within the initial phase of the process when time t is comparable to the time of relaxation τ (i.e., $\bar{t}$ is of the order of unity), a final speed of temperature propagation renders a great influence. Within this area, temperature drop along the length of the rod occurs more slowly than for infinite wave speed. At the time $\bar{t} \gg 1$, wave effects become negligibly small.

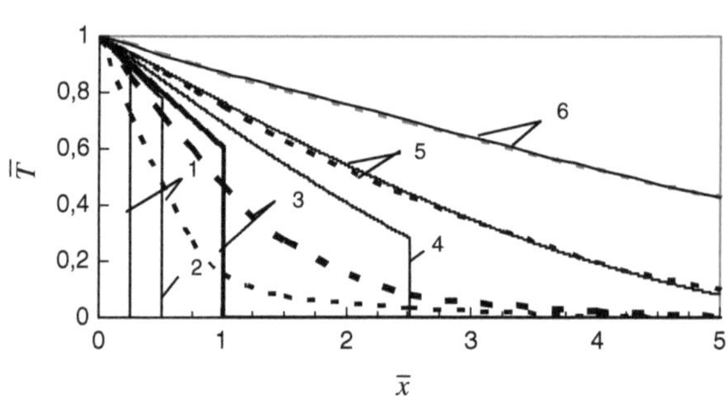

Fig. 12.7 Temperature distribution along the rod after thermal shock at time instants: *1* $\bar{t} = 0.25$, *2* $\bar{t} = 0.5$, *3* $\bar{t} = 1$, *4* $\bar{t} = 2.5$, *5* $\bar{t} = 5$, *6* $\bar{t} = 20$

Curves 5 and 6 confirm that a numerical solution by the method of direct mathematical modeling assures high accuracy. The method easily extends to the analysis of temperature wave propagation in a semi-infinite body with arbitrary initial and boundary conditions and a surrounding temperature varying in any law.

12.4.2.4 The problem of heat shock in a semi-infinite body was examined by Danilovskaya (1950) and Mura (1956). Non-steady stresses were calculated due to constant or varying temperature field, which had been found beforehand in the "non-wave" statement.

Using numerical step-by-step calculation by the method of direct mathematical modeling, it is possible to define changes in time of temperature and stresses simultaneously, including the terms of thermal expansion in the equations of mechanics. In this statement, the opposite influence of stresses on temperature field is not considered, but in most cases, this does not play any role (see previous Sect. 12.3).

Equations of mechanics (12.74) in a discrete form, accounting influence of thermal expansion in the absence of any external loads, lead to expression (Shorr and Mel'nikova 1988):

$$\bar{v}_{j*}^{\pm} = \bar{v}_{j0} \pm (\bar{\sigma}_{j*}^{\pm} - \bar{\sigma}_{j0}) \pm 0.5(\bar{T}_{j*}^{\pm} - \bar{T}_{j0}), \tag{12.90}$$

where $\bar{v} = v/c\alpha_T T_*$ and $\bar{\sigma} = \sigma/E\alpha_T T_*$.

Here, each mechanical element is also assumed to be consisting of inner continual elastic-inertial part reflecting the wave nature of the process and two outer discrete parts affected by thermal expansion.

In the positive area of Fig. 12.8, curves of temperature $\bar{T} \geq 0$ distribution along the rod length in the various moments of time $\bar{t}$ are presented. The curves of compressing stresses $\bar{\sigma} \leq 0$ distribution referred to the same times are shown in negative area of Fig. 12.8. For the specified problem statement, thermal and mechanical wave speeds are assumed to be identical. The curves concerning the moments of time $\bar{t} = 5$ and $\bar{t} = 10$ are shown with the bold curves.

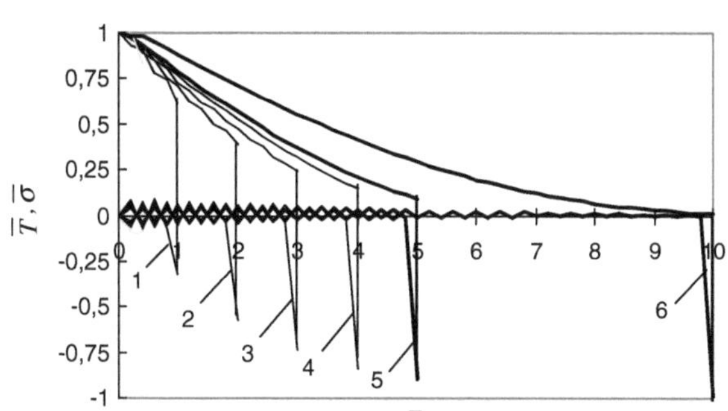

Fig. 12.8 Temperature $T \geq 0$ and stress $\sigma \leq 0$ distributions along the rod length after thermal shock at time instants: *1* $\bar{t} = 1$, *2* $\bar{t} = 2$, *3* $\bar{t} = 3$, *4* $\bar{t} = 4$, *5* $\bar{t} = 5$, $\bar{t} = 10$

In the first time interval $\bar{t}<1$ after the heat shock, the rod area with elevated temperature where strains arise due to the thermal expansions is very small and thermal stresses are also small. As the heated area becomes greater, the thermal stresses increase and reach the maximal compressing value $|\bar{\sigma}|_{max}=1$ or $|\sigma|_{max}=E\alpha_T T_*$ to the time $\bar{t}\approx 10$ at the point $\bar{x}\approx 10$. These results differ from the conclusion of Danilovskaya (1950), Mura (1956), and others obtained on the base of non-wave theory of thermal conductivity in which the temperature propagates with infinitely large speed and must "wait" arrival of the elastic wave of thermal stress which is improbable.

At a heat shock on semi-infinite space in case when lateral deformations are absent, the maximal compression stress is

$$|\sigma_y|_{max}=|\sigma_z|_{max}=\frac{E\alpha_T T_*}{1-\mu}.$$

Propagation of longitudinal 1D thermal and mechanics waves after a heat shock along continuous elastic medium with various restrictions in lateral directions can be calculated using the same algorithm by replacing Young's modulus by the reduced modulus of elasticity (Shorr 2004).

References

Babitsky VI (1978) Theory of vibro-impact systems. Approximated methods (in Russian). Nauka, Moscow

Birger IA, Shorr BF (eds) (1975) Thermal strength of machine parts (in Russian). Mashinostroenie, Moscow

Boley BA, Weiner JH (1960) Theory of thermal stresses. Wiley, New York

Danilovskaya VI (1950) Thermal stresses in elastic half-space induced by sudden heating of its boundary (in Russian). J PMM 14(3)

Duhamel JMC (1837) Deuxième memoire sur les phénomènes thermo-mécaniques. Journal de l'Ecole Polytechnique 15(25):1–57

Gokhfeld GA (1961) On possibility of plastic strains increasing as result of temperature cyclic effects (in Russian). In: Proceedings of "Raschety na Prochnost", vol 7. Mashgis, Moscow

Gokhfeld DA, Laptevsky AG (1968) Forming under thermal alternation and some technological applications (in Russian). In: Problemy Prochnosty Tekhnicheskikh Konstruktsiy, No. 45, Chelyabinsk Polytechnic Institute

Mura T (1956) Dynamical thermal stresses due to thermal shocks. Research reports, Faculty of Engineering, Meiji University 8

Parkus H (1959) Instationäre Wärmespannungen. Springer, Wien

Shorr BF (1984) Mathematical modelling of wave processes in elastic-viscoplastic solids (in Russian). Izv. AN SSSR, MTT 4: 144–151 (trans. Allerton Press, 1984: 141)

Shorr BF (1995) Analysis of wave propagation in elastic-plastic rods of a variable cross section using direct mathematical modeling. J Arch Appl Mech 65:537–547

Shorr BF (1998) Mathematical modelling of wave propagation in a rod of time-dependent material. J Mech of Time-Dependent Mater 1(4):397

Shorr BF (1999) A wave approach to finite element analysis of solids. J Mech Res Commun 26 (2):191–196

Shorr BF (2004) The wave finite element method. Spriger, Berlin
Shorr BF, Mel'nikova GV (1988) Calculation of mechanical systems using method of direct mathematical modelling (in Russian). Mashinostroenie, Moscow
Timoshenko SP (1955) Vibration problems in engineering, 3rd edn. In collaboration with D.H. Young DVan Nostrand Company, Toronto New York London

Appendix
Stresses and Strains in Non-uniformly Heated Rods, Plates, and Trusses

The Appendix gives examples of analytical solutions of some thermal strength problems for basic elements of mechanical structures (rods of a different kind and plates and trusses), which can be useful for demonstrating some sides of non-uniformly heated structure behavior and testing of numerical programs. There is a very vast literature on analytical solutions of thermal strength problems (see references to Monograph Chapters). Only the most necessary additional information is stated below.

A.1 Non-uniformly Heated Rods

A.1.1 Prismatic Rods

A.1.1.1 Tension and Bending

A prismatic rod of constant cross-section A is stretched along the axis z by longitudinal force F_z and bent in two planes XZ and YZ by moments M_ξ and M_η (Fig. A.1a). The principal axes of rod cross sections have identical directions. A temperature field $T(\xi, \eta)$, arbitrary over the section and constant along length of the rod, is set. Fields of the elasticity modulus $E(\xi, \eta)$ and temperature expansion $\varepsilon_T(\xi, \eta)$ corresponding to the temperature field are also known.

According to the flat cross-section hypothesis, a total longitudinal strain of a rod along its axis z is presented by the equation

$$\varepsilon(\xi, \eta) = \varepsilon_0 + \chi_\xi \eta - \chi_\eta \xi, \tag{A.1}$$

B.F. Shorr, *Thermal Integrity in Mechanics and Engineering*,
Foundations of Engineering Mechanics, DOI 10.1007/978-3-662-46968-2

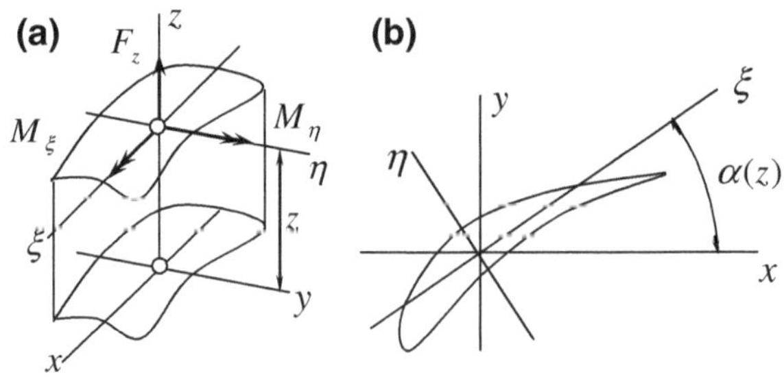

Fig. A.1 Tension and bending of a rod; **a** loads, **b** section of a pre-twisted rod

where ε_0 is a strain at the coordinates ξ, η origin, and χ_ξ and χ_η are components of the axis elastic curvature in the planes ηz and ξz.

Equation (A.1), as well as other kinematic relations

$$\varepsilon_0 = u_z', \chi_\xi = -u_\eta'', \chi_\eta = u_\xi'', \tag{A.2}$$

do not depend on material properties.

Here, u_ξ, u_η and u_z are displacements of the point of coordinates ξ, η origin for the current section along the corresponding axes, and $f' = df/dz$ is a derivative of any function f in coordinate z.

Total strain is equal to the sum of the elastic, temperature, and additional (inelastic) parts. In case of uniaxial stress state it is

$$\varepsilon(\xi, \eta) = \frac{\sigma}{E} + \varepsilon_T + \varepsilon^\circ. \tag{A.3}$$

Having input stress from Eq. (A.3) into three equations of the rod equilibrium

$$\int\limits_A \sigma dA = F_z, \quad \int\limits_A \sigma \xi dA = -M_\eta, \quad \int\limits_A \sigma \eta dA = M_\xi,$$

and having defined the coordinates origin and direction of axes ξ, η from conditions

$$\int\limits_A E\eta dA = 0, \int\limits_A E\xi dA = 0, \int\limits_A E\xi\eta dA = 0,$$

we obtain an expression for longitudinal normal stress at any point of a non-uniformly heated cross-section under tension and bending

$$\sigma(\xi, \eta) = \bar{E}\left(\frac{F + \tilde{F}}{A} + \frac{M_\xi + \tilde{M}_\xi}{J_\xi^*}\eta - \frac{M_\eta + \tilde{M}_\eta}{J_\eta^*}\xi - E_m\tilde{\varepsilon}\right), \tag{A.4a}$$

where elastic-geometrical characteristics are defined by integration of the function $E(\xi, \eta)$ over the entire area of the cross section:

$$E_m = \frac{1}{A}\int_A E dA, \quad J_\xi^* = \int_A \bar{E}\eta^2 dA, \quad J_\eta^* = \int_A \bar{E}\xi^2 dA, \quad \bar{E} = E/E_m,$$
$$\tilde{F} = \int_A E\tilde{\varepsilon} dA, \quad \tilde{M}_\xi = \int_A E\tilde{\varepsilon}\eta dA, \quad \tilde{M}_\eta = -\int_A E\tilde{\varepsilon}\xi dA. \tag{A.4b}$$

For thermal expansion it is accepted that $\tilde{\varepsilon} = \varepsilon_T$, and for additional deformations of any origin is $\tilde{\varepsilon} = \varepsilon^\circ$.

At constant temperature there is $\bar{E} = 1$, the origin of the coordinates ξ, η coincides with the center of cross-section mass, and parameters J_ξ^* и J_η^*—with the principal moments of inertia.

For statically determinable mechanical systems the force F_z and the bending moments M_ξ and M_η induced by external loads can be defined for any cross section due to equilibrium conditions

$$F_z' + q_z = 0, \quad M_\xi'' + q_\eta = 0, \quad M_\eta'' - q_\xi = 0\,, \tag{A.5}$$

where q_ξ, q_η, q_z are outer loads distributed on a rod length in corresponding directions.

Generally, Eqs. (A.1)–(A.3) yield

$$u_z' = \frac{F_z + \tilde{F}_z}{E_m A}, \quad u_\xi'' = \frac{M_\eta + \tilde{M}_\eta}{E_m J_\xi^*}, \quad u_\eta'' = -\frac{M_\xi + \tilde{M}_\xi}{E_m J_\eta^*}. \tag{A.6}$$

Statically, indeterminable problems can be solved separately for each of the directions by integrating Eqs. (A.5) and (A.6) under the set boundary conditions.

A.1.1.2 Torsion

The relative cross-section rotation angle θ of a longitudinally non-uniformly heated rod induced by a torque M_z is defined as

$$\theta = \frac{d\alpha}{dz} = \frac{M_z + \tilde{M}_z}{G_m J_z^*}, \tag{A.7a}$$

where J_z^* is a reduced rigidity in torsion, and $\tilde{M}_z$ is a torque moment from additional deformations.

For a circular cylinder with an axisymmetric temperature field there is

$$J_z^* = \int\limits_A \bar{G}(r)r^2 dA, \tilde{M}_z = \int\limits_A \bar{G}(r)\tilde{\theta}(r)r^2 dA, \bar{G} = G/G_m \approx \bar{E}. \tag{A.7b}$$

In this case thermal expansion does not cause shear strains and rotation angles of the cross sections ($\tilde{\theta} = \theta_T = 0$). But non-uniform heating of a rod of noncircular cross sections leads, under torsion, to their deplanation (warping) with point displacement along the axis z, which can lead to torsion deformation.

A.1.2 Pre-twisted Rods

For gas-dynamic reasons, gas and steam turbines blades are produced with an installation angle $\alpha(z)$ varying along the length of the blades profile (see above Fig. A.1b), i.e., are pre-twisted. Practical interest is represented by the rods of elongated profile, in which the "length" b of a cross section along one of the axis ξ essentially exceeds "thickness" δ along the other axis η (Fig. A.2a).

Let us designate initially an angle of rotation of the cross-section principal axes related to unit of the rod length (relative pre-twist) through $\theta_{in} = d\alpha(z)/dz$ and enter a parameter $\gamma = \theta_{in}b^2/\delta$ that can largely vary. At distance $R = (\xi^2 + \eta^2)^{0.5}$ from the rod axis, an inclination of the longitudinal "fiber" on a small angle $\beta_{in}(R) \approx \theta_{in}R \ll 1$ corresponds to the pre-twist θ_{in} (Fig. A.2b). If the twist increases in size θ due to rod deformation, the inclination increases in $\beta(R) \approx \theta R$ which causes the fiber lengthening in $\varepsilon \approx \beta_{in}\beta \approx \theta_{in}\theta R^2$ (as shown by bold site Δ) and appearence in Eq. (A.1) of a corresponding additional member.

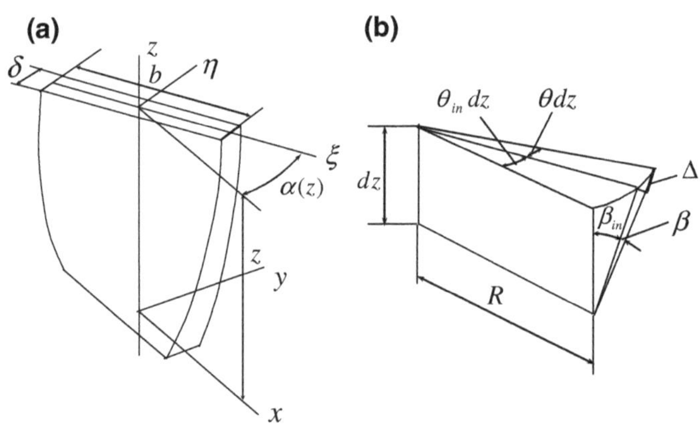

Fig. A.2 Scheme of a pre-twisted rod—(**a**), and elongation of an inclined fiber—(**b**)

For the elongated sections there are $R \approx \xi$, $R_{\max} \approx 0.5b$, and $\gamma = 2(\beta_{in})_{\max}(b/\delta)$, where the angle$(\beta_{in})_{\max} = 0.5\theta_{in}b$ refers to the inclined fiber, most removed from the rod axis. Analysis shows that the influence of pre-twist on the stress-strain state and frequency characteristics of rods of the elongated profile can be essential, despite angles β_{in} and β being small.

Equation (A.1), obtained by kinematic relations, is applicable at any temperature fields, additional deformations, and material properties. From Eq. (A.3) with an additional member $E(\xi,\eta)\theta\theta_{in}R_*^2$ it follows

$$\sigma(\xi,\eta) = E(\xi,\eta)\left[\varepsilon_0 + \chi_\xi\eta - \chi_\eta\xi + \theta\theta_{in}R^2 - \tilde{\varepsilon}_T(\xi,\eta)\right], \quad \text{(A.7c)}$$

and an additional member $E(\xi,\eta)\,\theta\theta_{in}R_*^2$ occurs in Eq. (A.4a, b) where

$$R_*^2 = R^2 - \frac{j_p^*}{A} - \frac{J_{p\xi}^*}{J_\xi^*}\eta - \frac{j_{p\eta}^*}{J_\eta^*}\xi, \quad \text{(A.8)}$$

$J_p^* = \int_A \bar{E}R^2 dA,\ J_{p\xi}^* = \int\limits_A \bar{E}R^2\eta\, dA,\ J_{p\eta}^* = \int\limits_A \bar{E}R^2\xi\, dA.$

The generalized coordinate R_*^2 can accept both positive and negative values. In the pre-twisted rod, a torque moment M_z is counterbalanced by the sum of the moment of tangential stresses M_τ, and the moment M_σ of projections of normal stresses in inclined fibers onto a cross-section plane

$$M_z = M_\tau + M_\sigma = G_m J_t^*\theta + \theta_{in}\int\limits_A \sigma R^2 dA, \quad \text{(A.9)}$$

where the torsion rigidity for a rod of elongated cross-section is

$$J_t^* \approx \frac{1}{3}\int\limits_b \bar{G}(\xi)\delta^3(\xi)d\xi, \quad \bar{G} \approx \bar{E}.$$

Having substituted the expression for normal stress σ from Eq. (A.7c) in Eq. (A.9), we find a relative angle of elastic twist

$$\theta = \frac{1}{G_m J_t^{**}}\left[M_z - \theta_{in}\left(\frac{F+\tilde{F}}{A}J_p^* + \frac{M_\xi + \tilde{M}_\xi}{J_\xi^*}J_{p\xi}^* - \frac{M_\eta + \tilde{M}_\eta}{J_\eta^*}J_{p\eta}^* - \tilde{B}^*\right)\right], \quad \text{(A.10)}$$

where $J_t^{**} = J_t^* + \theta_{in}^2\frac{E_m}{G_m}J_r^*$, at $J_r^* = J_{r0}^* - \frac{(J_p^*)^2}{A} - \frac{(J_{p\xi}^*)^2}{J_\xi^*} - \frac{(J_{p\eta}^*)^2}{J_\eta^*}$, $J_{r0}^* = \int\limits_A \bar{E}R^4 dA$, $\tilde{B}^* = \int\limits_A E\tilde{\varepsilon}R^2 dA$.

Tangential stresses of torsion at a profile contour are defined approximately as $\tau_{\xi z} = G(\xi)\delta(\xi)\theta$.

The basic features of the pre-twisted rods are the coupling of longitudinal, torsion and bending deformations, increase of torsion, and reduction of longitudinal rigidities with pre-twist increasing, and nonlinear distribution of normal stresses over the cross section.

Temperature expansion of a material causes additional effects.

At uniform heating of a rod, when displacements are restricted only in longitudinal direction, normal compressing stress at the rod axis will stay the same, as in prismatic one, while the stresses at the endpoints of the cross sections can become tensile.

Additional temperature loads, defined by Eqs. (A.4a, b), (A.9), and (A.10), arise at uneven heating.

A.2 Non-uniformly Heated Circular Plates

A.2.1 Axisymmetric Tension/Compression of Plates

A.2.1.1 Thin Plates

A thin plate is usually considered as being in generalized flat axis-symmetric stress state with lateral stress $\sigma_z \approx 0$. Then physical relations presented in polar coordinates r—in radial and φ—in circumference directions look like

$$\varepsilon_r = \frac{\sigma_r - \mu\sigma_\varphi}{E} + \tilde{\varepsilon}_r,\, \varepsilon_\varphi = \frac{\sigma_\varphi - \mu\sigma_r}{E} + \tilde{\varepsilon}_\varphi,\, \varepsilon_z = -\mu\frac{\sigma_r + \sigma_\varphi}{E} + \tilde{\varepsilon}_z, \tag{A.11a}$$

where σ_r, ε_r are radial, and σ_φ, ε_φ—circumferential stresses and strains.

Additional strains $\tilde{\varepsilon}_r, \tilde{\varepsilon}_\varphi, \tilde{\varepsilon}_z$ can include temperature expansion ε_T and inelastic strains.

Invert correlations are

$$\sigma_r = E\frac{\varepsilon_r + \mu\varepsilon_\varphi - (\tilde{\varepsilon}_r + \mu\tilde{\varepsilon}_\varphi)}{1 - \mu^2};\, \sigma_\varphi = E\frac{\varepsilon_\varphi + \mu\varepsilon_r - (\tilde{\varepsilon}_\varphi + \mu\tilde{\varepsilon}_r)}{1 - \mu^2}. \tag{A.11b}$$

As well as the above, for temperature expansion we accept $\tilde{\varepsilon}_r = \tilde{\varepsilon}_\varphi = \varepsilon_T$, for inelastic strains—$\tilde{\varepsilon}_r = \varepsilon_r^\circ$, $\tilde{\varepsilon}_\varphi = \varepsilon_\varphi^\circ$. Generally, $E = E(r)$ at $\mu \approx \text{const}$.

Total strains are connected with radial displacement $u(r)$ by geometrical relations

$$\varepsilon_r = \frac{du}{dr},\ \varepsilon_\varphi = \frac{u}{r}. \tag{A.12}$$

The equilibrium equation in a radial direction looks like

$$\frac{d(\sigma_r r\delta)}{dr} - \sigma_\varphi \delta + q_r r = 0, \tag{A.13}$$

where q_r is a distributed radial load referred to a plate thickness δ.

For an elastic plate of constant thickness with a constant elasticity modulus, solutions can be obtained in the closed kind. In this case, it follows from Eqs. (A.11a, b)–(A.13) a differential equation

$$\frac{d^2u}{dr^2} + \frac{1}{r}\frac{du}{dr} - \frac{u}{r^2} = \frac{d}{dr}\left[\frac{1}{r}\frac{d(ru)}{dr}\right] = -\frac{1-\mu^2}{E\delta} q_r + (1+\mu)\frac{d\varepsilon_T}{dr}, \tag{A.14a}$$

with general solution at $q_r = 0$

$$u(r) = \frac{1+\mu}{r}\int\limits_{r_0}^{r} r\varepsilon_T(r)dr + Ar + \frac{B}{r}. \tag{A.14b}$$

The constants A and B must be defined by the set boundary conditions at internal and external plate radii $\sigma_r(r_0) = \sigma_{r0}$ and $\sigma_r(R) = \sigma_R$.

A.2.1.2 Plate with Central Hole

For a plate with a central hole, temperature stresses $\sigma_r(r)$ and $\sigma_\varphi(r)$ are defined by equations

$$\begin{aligned} \sigma_r(r) &= \frac{E}{r^2}\left[\frac{r^2 - r_0^2}{R^2 - r_0^2}\int\limits_{r_0}^{R} r\varepsilon_T(r)dr - \int\limits_{r_0}^{r} r\varepsilon_T(r)dr\right], \\ \sigma_\varphi(r) &= \frac{E}{r^2}\left[\frac{r^2 + r_0^2}{R^2 - r_0^2}\int\limits_{r_0}^{R} r\varepsilon_T(r)dr + \int\limits_{r_0}^{r} r\varepsilon_T(r)dr - \varepsilon_T r^2\right] \end{aligned} \tag{A.15a}$$

and temperature displacements are

$$u(r) = \frac{1}{r}\left[(1+\mu)\int\limits_{r_0}^{r} r\varepsilon_T(r)dr + \frac{(1-\mu)r^2 + (1+\mu)r_0^2}{R^2 - r_0^2}\int\limits_{r_0}^{R} r\varepsilon_T(r)dr\right]. \tag{A.15b}$$

Formulas for a compact plate are resulted at $r_0 = 0$. In the plate center, the limiting transition $r_0 \to 0$ leads to $u(0) = 0$, and

$$\sigma_r(0) = \sigma_\varphi(0) = E\left[\frac{1}{R^2}\int\limits_{r_0}^{R} r\varepsilon_T(r)dr - 0.5\varepsilon_T(0)\right]. \tag{A.16}$$

A.2.1.3 Thermal Stresses in Plates

It follows from analysis of Eqs. (A.15a, b–A.16):

- At a *uniform temperature field* and $\varepsilon_T = \text{const}$ the displacements of a plate with free edges change linearly, i.e., $u(r) = \varepsilon_T r$, and thermal stresses do not arise. But if at least one of the plate edges is fixed the thermal stresses appear. So, as the outer edge of the plate is fixed, i.e., $u(R) = 0$, radial stresses at this radius are

$$\sigma_r(R) = -E\varepsilon_T \frac{R^2 - r_0^2}{(1-\mu)R^2 + (1+\mu)r_0^2}.$$

In this case the strains in a compact plate are zero at all radii and compressing thermal stresses

$$\sigma_r = \sigma_\varphi = -\frac{E\varepsilon_T}{1-\mu}$$

occur over the entire plate.

- *At a non-uniform temperature field* the stresses arise even if the edges of a plate are free.

Let the temperature field change step-wisely, so that the internal part of a plate I from $r = 0$ to $r = r_*$ is regularly heated upon $\Delta T > 0$ and peripheral part II—from $r = r_*$ to $r = b$—is regularly cooled upon the same size $\Delta T < 0$, so as $\varepsilon_T^I = \varepsilon_T > 0$ and $\varepsilon_T^{II} = -\varepsilon_T$.

Under free expansion of each of the parts there would be no stresses in them, and their radial displacements at the common radius $r = r_*$ would be identical in size and different in sign, $u^I(r_*) = \varepsilon_T r_* = -u^{II}(r_*)$, i.e., the plane parts would "crawl" against each other over a quantity $\Delta u = u^I(r_*) - u^{II}(r_*) = 2\varepsilon_T r_*$. At joint plane parts displacements, a radial stress $\sigma_r(r_*) = \sigma_{r*}$ appears between parts, which inhibits their crawling. It follows from previous equations that

$$\sigma_{r*} = -E\varepsilon_T\left[1 - (r_*/R)^2\right].$$

If a temperature of the internal plane part is higher than the peripheral one, compressing radial temperature stresses cover the entire plate (including colder peripheral part). At the border of the parts at temperature jump $2\Delta T$, circumferential stresses undergo discontinuity and have different signs within hot and cold parts. Under smooth radial temperature distribution there will be no circumference stress discontinuity.

A.2.2 Axisymmetric Bending of Plates

At axisymmetric bending of a thin circular plate stress state in all points is biaxial, normal stresses in the radial σ_r and circumferential σ_φ directions alter along the plate thickness δ but Eqs. (A.11a) and (A.11b) remain valid. Temperature may change along the plate radius and thickness, but thickness variation of the elasticity modulus can be neglected.

Radial displacements along a plate thickness $u(r, z)$ are defined on the basis of the condition of "a rigid normal" (Kirchhoff's hypothesis)

$$u(r, z) = u(r) - z\psi(r), \tag{A.17a}$$

where $u(r)$ is a displacement of a point at a median plane of the plate, z is a distance from it to a current point ($-0.5\delta \leq z \leq 0.5\delta$), $\psi(r)$ is a rotation angle of a normal to this plane at radius r.

For small displacements

$$\psi(r) = w'(r), \tag{A.17b}$$

where $w(r)$ is a plate deflection at a current radius.

At axisymmetric deformations Eq. (A.12), fair for any points, become

$$\varepsilon_r(r, z) = u'(r, z), \varepsilon_\varphi(r, z) = u(r, z)/r,$$

so

$$\varepsilon_r(r, z) = \varepsilon_r(r) - z\psi'(r), \varepsilon_\varphi(r, z) = \varepsilon_\varphi(r) - z\psi(r)/r, \tag{A.18}$$

at $\varepsilon_r(r) = u'(r)$ and $\varepsilon_\varphi(r) = u(r)/r$.

Using a linear statement, the problems of a plate bending and tension/compression can be considered irrespectively, accepting for the bend $u(r) = 0$. At uniaxial stress state, taking into account Eq. (A.11b) for $\tilde{\varepsilon} = \varepsilon_T$, the stresses will be defined as

$$\begin{aligned} \sigma_r &= -\frac{E}{1-\mu^2} z(\psi' + \mu\frac{\psi}{r}) - \frac{E}{1-\mu}\varepsilon_T z, \\ \sigma_\varphi &= -\frac{E}{1-\mu^2} z(\frac{\psi}{r} + \mu\psi') - \frac{E}{1-\mu}\varepsilon_T z, \end{aligned} \tag{A.19}$$

and "linear" bending moments at a radius r, will be

$$
\begin{aligned}
M_r &= -\int_{-0.5\delta}^{0.5\delta} \sigma_r z dz = D(\psi' + \mu\psi/r) + M_T, \\
M_\varphi &= -\int_{-0.5\delta}^{0.5\delta} \sigma_\varphi z dz = D(\psi/r + \mu\psi') + M_T
\end{aligned}
\tag{A.20a}
$$

where is $D = E\delta^3/12(1-\mu^2)$ and a "temperature" moment is

$$
M_T = \frac{E}{1-\mu}\int_{-0.5\delta}^{0.5\delta} \varepsilon_T z dz. \tag{A.20b}
$$

Thermal expansion of a material leads to a plate bend only when the temperature changes over its thickness. In this case the moment $M_T \neq 0$.

The equilibrium equation of a plate element under the action of the bending moments in absence of lateral forces is

$$
\frac{d(rM_r)}{dr} - M_\varphi = 0. \tag{A.21}
$$

Having input Eq. (A.20a) into Eq. (A.21), we obtain the differential equation of plate bending presented in angles of rotation ψ

$$
\frac{d^2\psi}{dr^2} + \frac{1}{r}\frac{d\psi}{dr} - \frac{\psi}{r^2} = \frac{1}{D}\frac{dM_T}{dr}, \tag{A.22}
$$

which formally coincides with Eq. (A.14a, b) for a plate tension/compression at $q_r = 0$.

Therefore the resolving equations of a plate deformations presented in angles of rotation and also formulas following from them, will have an identical appearance for both problems. Plate deflections $w(r)$ will be defined by integrating Eq. (A.17b).

The general calculation methods for bend of circular plates of variable thickness and various temperature distributions on radius and on thickness, including temperature-dependent material characteristics, were considered in a number of works, cited in Chaps. 2 and 3.

A.3 Non-uniformly Heated Truss

A problem of analysis of thermal stresses in a non-uniformly heated truss is presented in matrix term, generally used in FE codes.

A.3.1 Thermoelastic Deformation

A.3.1.1 Basic Set of Matrix Equations

Let us consider a truss consisting of two rods of a constant cross section A with a single common node 1 (Fig. A.3).

A force F is applied to this node on an angle β to a vertical axis, and a spring with rigidity k_{str} limits the node displacement along y direction. Nodes 2 and 3 at the other ends of the rods are fixed. Deformation of a cold rod I is assumed be elastic ($\varepsilon_I^p \equiv 0$), its length is equal to$l/\cos\alpha_I$, the elasticity modulus is $E_I = \text{const}$. The rod II of the length $l/\cos\alpha_{II}$ is heated up to temperature $T_{II} > T_I$, to which there corresponds the elasticity modulus $E_{II}(T) < E_I$ and temperature expansion $\varepsilon_T(T)$. Displacements are assumed be small, not affecting the system configuration.

Element strains ε_I and ε_{II} are expressed through the components u_x and u_y of the node 1 displacement by relations

$$\begin{aligned} \varepsilon_I &= \frac{\cos\alpha_I}{l}(u_x\cos\alpha_I + u_y\sin\alpha_I), \\ \varepsilon_{II} &= \frac{\cos\alpha_{II}}{l}(u_x\cos\alpha_{II} - u_y\sin\alpha_{II}). \end{aligned} \tag{A.23}$$

A constitution equation for a non-isothermal problem of a rod of a material with any characteristics is

$$\varepsilon(\sigma, T) = \frac{\sigma}{E(T)} + \varepsilon_T(T) + \varepsilon^{\circ}(\sigma, T), \tag{A.24a}$$

and vice versa

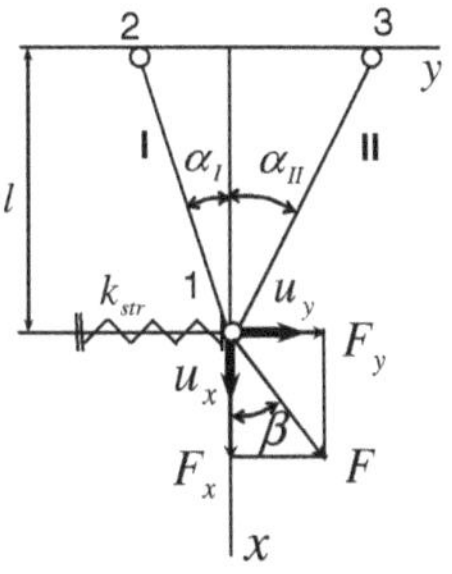

Fig. A.3 Scheme of a non-uniformly heated two-rod truss

$$\sigma = E(T)\,[\varepsilon(\sigma, T) - \varepsilon_T(T) - \varepsilon^{\circ}(\sigma, T)]. \tag{A.24b}$$

Here ε° is a nonelastic part of the strain.
Equations (A.23) and (A.24a, b) lead to

$$\begin{aligned}\sigma_I &= E_I \frac{\cos\alpha_I}{l}(u_x \cos\alpha_I + u_y \sin\alpha_I),\\ \sigma_{II} &= E_{II}\left[\frac{\cos\alpha_{II}}{l}(u_x \cos\alpha_{II} - u_y \sin\alpha_{II}) - \varepsilon_T - \varepsilon^{\circ}\right],\end{aligned} \tag{A.25a}$$

and external loads affecting the rods are

$$F_I = A\sigma_I,\;\; F_{II} = A\sigma_{II}. \tag{A.25b}$$

The equilibrium conditions of node 1

$$\begin{aligned}&F_x - F_I \cos\alpha_I - F_{II}\cos\alpha_{II} = 0,\\ &F_y - F_I \sin\alpha_I + F_{II}\sin\alpha_{II} - k_{str}u_y = 0\end{aligned} \tag{A.26}$$

yield, accounting Eq. (A.25a, b),

$$\begin{aligned}K_{11}u_x + K_{12}u_y &= F_x + F_x^T + F_x^{\circ},\\ K_{21}u_x + K_{22}u_y &= F_y + F_y^T + F_y^{\circ}\end{aligned} \tag{A.27}$$

where coefficients of the rigidity matrix $[K]$ are

$$\begin{aligned}K_{11} &= \frac{A}{l}(E_I \cos^3\alpha_I + E_{II}\cos^3\alpha_{II}),\\ K_{22} &= \frac{A}{l}(E_I \cos\alpha_I \sin^2\alpha_I + E_{II}\cos\alpha_{II}\sin^2\alpha_{II}) + k_{str},\\ K_{12} = K_{21} &= \frac{A}{l}(E_I \cos^2\alpha_I \sin\alpha_I - E_{II}\cos^2\alpha_{II}\sin\alpha_{II}),\end{aligned} \tag{A.28a}$$

and thermal and nonelastic components of the right parts are

$$\begin{aligned}F_x^T &= E_{II}A\varepsilon_T \cos\alpha_{II}, & F_y^T &= -E_{II}A\varepsilon_T \sin\alpha_{II},\\ F_x^{\circ} &= E_{II}A\varepsilon^{\circ}\cos\alpha_{II}, & F_y^{\circ} &= -E_{II}A\varepsilon^{\circ}\sin\alpha_{II}.\end{aligned} \tag{A.28b}$$

A matrix form of set of Eq. (A.27) is

$$[K]\{u\} = \{F\} + \{F^T\} + \{F^{\circ}\}. \tag{A.29}$$

After the displacements u_x and u_y are found from the solution of the equation set (A.27), all other variables of a problem in the elastic statement can be calculated.

For variable loading and heating regimes a rigidity matrix $[K(T)]$ changes against temperature according to change of the elasticity modulus $E(T)$.

A.3.1.2 Conditions of Thermal Stresses Arising

It follows from Eq. (A.27) that balance of node 1 in absence of the external loads ($F_x = F_y = 0$) and the spring ($k_{str} = 0$) is possible only for $\sigma_I = \sigma_{II} = 0$. In this case rod II may freely elongate and rotate on heating. Rod I turns together with rod II, so that their displacements at node 1 are identical but the length of rod I remains invariable. Thermal stresses in the system do not arise. In the presence of the spring ($k_{str} \neq 0$) the loads at node 1 differ from zero, and thermal stresses arise in both rods also at $F_x = F_y = 0$.

A.3.1.3 Variational Equations

In the FEM theory, matrix equations of type (A.29) are usually deduced from variational equation

$$\partial(W_{inn} - W_{out}) = 0,$$

where work variations of internal and external forces are

$$\partial W_{inn} = \int_V \{\sigma^T\}\{\partial\varepsilon\}dV,$$

$$\delta W_{out} = \int_V \{p^T\}\{\partial u\}dV + \int_S \{F^T\}\{\partial u\}dS.$$

Having expressed the strains ε_I, ε_{II} and the stresses σ_I, σ_{II} in elements of the considered system through the displacements u_x,u_y, and their variations, we come at $\partial u_x \neq 0$ and $\partial u_y \neq 0$ to the set of Eqs. (A.27)–(A.29).

A.3.2 Inelastic Deformations

A procedure of finite element modeling of any problem begins with elastic calculation. For the given truss system, this corresponds to $\varepsilon^\circ = 0$, $F_x^\circ = F_y^\circ = 0$. The "elastic" values of stresses $\sigma = \sigma^e$ and total strains $\varepsilon = \varepsilon^e$ are used for calculation of the first approach of nonelastic strains according to the corresponding theory of plasticity or creep. Values of additional deformations ε° are then refined by successive approximations.

Temperature field of a body, external forces, and loads from thermal expansion thus do not change.

For step-by-step procedure, Eq. (A.24b) is presented in increment terms

$$\Delta\sigma = E(T)[\Delta\varepsilon - \Delta\varepsilon_T(T) - \Delta\varepsilon^{\circ}(\sigma, T, \varepsilon, \ldots)], \tag{A.30}$$

where increment of the additional deformations $\Delta\varepsilon^{\circ}$ may depend on current values of stress, strain, temperature, and also parameters reflecting influence of the previous steps (see Chaps. 5 and 8).

MIX
Papier aus verantwortungsvollen Quellen
Paper from responsible sources
FSC® C105338

If you have any concerns about our products,
you can contact us on
ProductSafety@springernature.com

In case Publisher is established outside the EU,
the EU authorized representative is:
Springer Nature Customer Service Center GmbH
Europaplatz 3, 69115 Heidelberg, Germany

Printed by Libri Plureos GmbH
in Hamburg, Germany